Data analysis techniques for high-energy physics experiments

Data analysis techniques for high-energy physics experiments

R. K. Bock
European Organization for Nuclear Research (CERN)

H. Grote
European Organization for Nuclear Research (CERN)

D. Notz
Deutsches Elektronen-Synchrotron, DESY

and

M. Regler
Institute of High-Energy Physics of the Austrian Academy of Sciences

Edited by M. Regler

CAMBRIDGE UNIVERSITY PRESS

Cambridge

New York Port Chester Melbourne Sydney

Published by the Press Syndicate of the University of Cambridge
The Pitt Building, Trumpington Street, Cambridge CB2 1RP
40 West 20th Street, New York NY 10011, USA
10 Stamford Road, Oakleigh, Melbourne 3166, Australia

© Cambridge University Press 1990

First published 1990

Printed in Great Britain at the University Press, Cambridge

British Library cataloguing in publication data

Data analysis techniques for high-energy physics
experiments.
1. Physics. Data. Analysis
I. Bock, R. K.
530'.015

Library of Congress cataloguing in publication data

Data analysis techniques for high-energy physics experiments / R.K. Bock ··· [et al.] : edited by M. Regler.
 p.   cm.
Includes bibliographical references.
ISBN 0-521-34195-7
1. Particles (Nuclear physics)—Experiments—Data processing.
I. Bock. R. K. II. Regler, M.
QC793.412.D37 1990
539.7'6—dc20   17367 CIP

ISBN 0 521 34195 7

TM

Contents

	Preface	x
	Abbreviations	xii
	Introduction	1
1	**Real-time data triggering and filtering**	**7**
1.1	Definitions and goals of triggers and filters	7
1.1.1	General properties of particle accelerators	7
1.1.2	Secondary beams	8
1.1.3	Energy balance in scattering experiments	9
1.1.4	Luminosity	10
1.1.5	Time structure of accelerators	11
1.1.6	Event rates at different accelerators	13
1.1.7	Background rates	15
1.2	Trigger schemes	17
1.2.1	On-line data reduction	17
1.2.2	Dead time of electronic components	19
1.2.3	True and wrong coincidences, accidentals	22
1.2.4	Multilevel triggers	22
1.3	Queuing theory, queuing simulation and reliability	26
1.3.1	Queuing theory	26
1.3.2	Queuing simulation	36
1.3.3	Reliability theory	39
1.4	Classifications of triggers	45
1.4.1	Trigger on event topology	47
1.4.2	Trigger on type of particle	48
1.4.3	Trigger on deposited energy	50
1.4.4	Trigger on missing energy	51
1.4.5	Trigger on invariant mass	52
1.4.6	Trigger on interaction point (vertex)	52
1.4.7	Fixed and variable flow triggers	53
1.4.8	Logical and arithmetic triggers	55
1.4.9	Data and program driven trigger processors	56

1.5	Examples of triggers	56
1.5.1	Trackfinding with a lumped delay line	56
1.5.2	Trackfinding with memory look-up tables	56
1.5.3	Trigger on tracks with field programmable arrays	61
1.5.4	Trackfinders in the trigger with variable flow data driven processors	65
1.5.5	A hardwired processor with CAM and look-up tables	69
1.5.6	A microprogrammed track processor with CAM and look-up tables	71
1.5.7	Examples of triggers on energy	74
1.5.8	A data driven trigger on invariant mass	79
1.5.9	Examples of triggers on interaction point	82
1.5.10	A trigger on interaction point for short-lived particles with a microstrip detector	84
1.6	Implementation of triggers	85
1.6.1	Electronic components	85
1.6.2	Pipelines	95
1.7	Programmable devices	99
1.7.1	Bit-slice processors as event builders	99
1.7.2	Digital-signal processors (DSP)	102
1.7.3	Transputers	102
1.7.4	Emulators	103
1.7.5	Parallel processing	105
1.7.6	The Fermilab Advanced Computer Program multimicroprocessor project (ACP)	114
1.8	Communication lines, bus systems	115
1.8.1	Synchronous and asynchronous buses	117
1.8.2	Addressing	118
1.8.3	Data transfers	120
1.8.4	Control lines	121
1.8.5	Responses	121
1.8.6	Interrupts	122
1.8.7	Multiple masters, bus arbitration	124
1.8.8	Characteristics of CAMAC	125
1.8.9	Characteristics of FASTBUS	126
1.8.10	Characteristics of VME	126
1.8.11	Standardization of data buses	128
2	**Pattern recognition**	**129**
2.1	Principles and methods of pattern recognition	129
2.1.1	Detector systems	129
2.1.2	Proportional chambers	131
2.1.3	Drift chambers	132
2.1.4	Time projection chambers (TPCs)	135
2.1.5	Pattern space	136

2.1.6	Training sample and covariance matrix	137
2.1.7	Object classification	138
2.1.8	Feature space	141
2.1.9	Classes, prototypes, and metric	143
2.1.10	Template matching	145
2.1.11	Linear feature extraction	146
2.1.12	Minimum Spanning Tree (MST)	150
2.1.13	Compatibility graph	153
2.2	Numerical techniques for trackfinding in high-energy physics	155
2.2.1	Circles, polynomials, and splines for curve approximation	155
2.2.2	Interpolation and extrapolation	161
2.2.3	Parametrization	162
2.3	The task of trackfinding in high-energy physics	164
2.3.1	Point removal	166
2.3.2	Track quality	167
2.3.3	Working in projections or in space	168
2.3.4	Treating track overlaps	171
2.3.5	Efficiency	172
2.4	Methods of trackfinding in high-energy physics	174
2.4.1	A classification	174
2.4.2	Local methods	175
2.4.3	Global methods	179
2.5	Finding of particle showers	185
2.5.1	Some definitions	185
2.5.2	Physical processes in calorimeters	190
2.5.2.1	Electromagnetic showers	190
2.5.2.2	Hadronic showers	192
2.5.3	Calorimeter parameters	194
2.5.4	Shower parameters	200
2.5.4.1	Longitudinal shower shape	200
2.5.4.2	Lateral shower shape	203
2.5.5	Shower simulation	205
2.5.6	Calorimeter algorithms: case studies	208
2.5.6.1	Global energy flow in UA2	210
2.5.6.2	Missing energy in UA1	211
2.5.6.3	π^0 selection in TASSO	213
2.5.6.4	e/π separation in the Mark III electromagnetic shower counter	217
2.5.6.5	e/π separation in CHARM	219
2.5.6.6	Identification of e^- in UA1	222
2.5.6.7	Future DELPHI High-density Projection Chamber (HPC) electromagnetic shower recognition algorithms	223
2.5.6.8	Jet finding in calorimeters	226
2.5.6.9	Small correction and patchup algorithms	229

2.6	Identifying particles in ring imaging Cherenkov counters	231
2.6.1	The RICH technique	231
2.6.2	RICH applications	234
2.6.2.1	Electron identification in a test set-up	235
2.6.2.2	Particle identification in OMEGA and in E605	235
2.6.2.3	Projected RICH in SLD, with two radiators	236
2.6.2.4	Projects for fast RICH devices	239
3	**Track and vertex fitting**	**240**
3.1	The task of track fitting	240
3.2	Estimation of track parameters	243
3.2.1	Basic concepts	243
3.2.2	Global track fitting by the Least Squares Method (LSM)	246
3.2.3	A few remarks on estimation theory	248
3.2.3.1	Generalities	248
3.2.3.2	The LSM in practice	254
3.2.3.3	The χ^2 distribution	257
3.2.4	Test for goodness of fit	258
3.2.5	Recursive track fitting by the LSM (the Kalman filter)	262
3.3	Fitting the tracks of charged particles	269
3.3.1	The track model	269
3.3.1.1	The equations of motion	269
3.3.1.2	The choice of track parameters	272
3.3.1.3	Several types of track models	275
3.3.1.4	The field representation	288
3.3.1.5	The effects of matter on the trajectory	291
3.3.2	The weight matrix	298
3.3.2.1	The measurement error of a detector	298
3.3.2.2	Weight matrix formalism for multiple scattering	304
3.3.2.3	Resolution of magnet spectrometer	308
3.3.3	Track element merging	313
3.3.4	Numerical minimization technique	316
3.4	Association of tracks to vertices	319
3.4.1	Basic concepts	319
3.4.2	A fast global method for parameter estimation	321
3.4.3	χ^2 evaluation and track association strategy	323
3.4.4	Kinematical constraints	327
3.5	Final observations on track fitting	330
4	**Tools and concepts for statistical data analysis**	**334**
4.1	Data abstraction	335
4.2	Data selection	339
4.3	Data accumulation, projection, and presentation	347
4.4	Multidimensional analysis	352

4.5	Technical aids for high-energy physics data analysis	354
4.5.1	Basic graphical communication with computers	354
4.5.1.1	General problem areas in graphics programming	356
4.5.1.2	Graphics in high-energy physics	360
4.5.1.3	Graphics notions: a glossary	362
4.5.2	Data access methods and databases	374
4.5.3	Ready-made tools for statistical analysis	378
4.5.3.1	Home-made statistical packages	379
4.5.3.2	Commercial tools	382
5	**Program development and software management**	**384**
5.1	Characteristics of programs in large experiments	384
5.2	Main steps in event off-line analysis	386
5.3	Characteristics of real-time programs	387
5.4	Program development management	388
5.4.1	Program productivity and group size	389
5.4.2	Phases of software life-cycle	389
5.4.3	Some principles for developing complex systems	391
5.5	Structured Analysis, Structured Design (SASD)	392
5.5.1	Data-flow Diagrams (DFD)	392
5.5.2	Structure charts	395
5.5.3	Entity Relationship Diagrams (ERD)	397
5.5.4	State-Transition Diagrams (STD)	398
5.5.5	Data Dictionary (DD)	399
5.5.6	Process-Description Language (PDL)	400
5.6	Quality measures for good design	401
5.7	Support tools for program development	402
5.7.1	Final remark on software management	403
6	**Some final remarks**	**404**
	References	406
	Index	424

Preface

This book brings together for the first time all important data-handling aspects of today's particle physics experiments. For us this was the major reason for writing it. We hope that it will serve our intentions: to present the information which is currently scattered through journal articles, conference proceedings, official and informal laboratory reports, and collaboration notes in a single volume; to facilitate a global view of the problems involved; to show the close connection that exists between the different data-handling fields such as data acquisition, topological and kinematical event reconstruction, and how they are embedded in the framework of hardware and software. Our aim was thus to provide a useful introduction to the field for graduate students, a reference for physicists and engineers working in the field, and a guide to the efficient and successful planning of all data-handling aspects in a particle physics experiment. We hope that the book will, at the same time, prove to be useful for experimenters in fields other than particle physics who encounter similar problems in data acquisition and information extraction.

The first three chapters follow the chronological treatment of the data: real-time data triggering, filtering, and acquisition in the first chapter, recognition of the event topology by assembling tracks and showers from fine-grain raw data in the second chapter, and geometrical and kinematical event reconstruction ('fitting') in the third chapter. Then follow two chapters complementing these methods enumerated: a description of physics analysis methods, such as the principles of data abstraction and selection, and the use of graphics for data interpretation and event display in Chapter 4, and a treatment of the 'managerial' aspects of the software development work, such as the program life cycle and the problems of distributed code writing, in Chapter 5.

Wherever possible, the chapters contain both a presentation and a discussion of the basic principles, and of their application to existing or future experiments. In this way the book should turn out to be more useful than a 'recipe book' containing only detailed descriptions of applications which somehow never quite match the reader's requirements, and which tend to be outdated rather quickly because of the speed of progress in the fields of detectors, electronics, and large computers. We have nevertheless included many real-life examples and a large number of references to original papers for the active researcher in the fields described, since we are aware of the fact that a general description needs to be illustrated by examples, and that in many cases our condensed explanation does not give enough details of the actual implementation of a given method.

It is almost impossible to list all those to whom we are indebted for help, in one form or another, in the long process of the preparation of this book. A number of colleagues were kind enough to read and comment on some portions of the drafts. R. K. Bock expresses his thanks for valuable contributions to A. Clark, T. Hansl-Kozanecka, A. Putzer, B. Schorr, T. P. Shaw, and C. V. Vandoni. D. Notz would like to thank K. Rehlich for many fruitful discussions. He would like to record his special thanks to his wife Elke and his children Katrin, Annika, Dirk, and Wiebke for their patience, forbearance, and encouragement. M. Regler wishes to thank W. Mitaroff for his invaluable help during the preparation of this manuscript; thanks are also due to R. Frühwirth and M. Metcalf for many years of pleasant and stimulating collaboration in the field of track fitting. Several useful suggestions have been made by P. Billoir. H. Grote would like to thank F. Carena, J. C. Lassalle, and M. Metcalf for fruitful discussions. D. Notz and M. Regler would like to thank M. Dieckvoß, E. Ess, U. Kwapil and S. Karsky for their technical assistance in the preparation of the drawings and the text editing. Last but not least we would like to thank the people at Cambridge University Press for their encouragement and cooperation.

Abbreviations

ACP:	Advanced Computer Project (at FNAL)
ADC:	Analog-to-Digital Converter
AFS:	Axial Field Spectrometer (ISR/CERN)
ALU:	Arithmetic and Logic Unit
AMD:	Advanced Micro Devices
AMU:	Analog Memory Unit
BBQ:	Benzimidazo–BenzisoQuinoline–7–one, wavelength shifting material used in light collection
BGO:	$Bi_4Ge_3O_{12}$, a crystalline calorimeter material
CAD:	Computer-Aided Design
CAE:	Compuder-Aided Electronics
CAM:	Content Addressable Memory
CCD:	Charge Coupled Device
CERN:	European Organization for Nuclear Research (Switzerland)
CESR:	Electron Storage Ring at Cornell University
CLIC:	CERN Linear Collider (project)
CMS:	Code Management System
CMOS:	Complementary Metal Oxide Semiconductor
CPU:	Central Processing Unit
CRID:	Cherenkov Ring Imaging Detector
DAC:	Digital-to-Analog Converter
DASP:	Double Arm Spectrometer
DBS:	Data Base System
DBMS:	Data Base Management System
DC:	Direct Current
DD:	Data Dictionary
DESY:	Deutsches Elektronen-Synchrotron
DFD:	Data Flow Diagram
DMA:	Direct Memory Access
DORIS:	Storage ring at DESY
DSP:	Digital Signal Processor
DST:	Data Summary Tape
DVST:	Direct View Storage Tube

ECL:	Emitter Coupled Logic
EGS:	Electromagnetic Shower Simulation Program
ERD:	Entity Relationship Diagram
FADC:	Fast Analog-to-Digital Converter
FET:	Field Effect Transistor
FNAL:	Fermi National Accelerator Laboratory
FPLA:	Field Programmable Logic Array
FWHM:	Full Width at Half Maximum
GeV:	Giga electron volts (10^9)
GKS:	Graphics Kernel System
HERA:	Hadron Elektron Ring Anlage (DESY)
HPC:	High-density Projection Chamber
IC:	Integrated Circuits
IEC:	International Electrotechnical Commission
IEEE:	Institute of Electrical and Electronics Engineers
IHEP:	Institute for High-Energy Physics (Protvino)
ISR:	Intersecting Storage Rings (CERN, now dismantled)
KEK:	High-energy physics laboratory in Japan
keV:	kilo electron volts (10^3)
LAM:	Look-At-Me
LEAR:	Low-energy Antiproton Ring (CERN)
LED:	Light Emitting Diode
LEP:	Large Electron Positron Ring (CERN)
LHC:	Large Hadron Collider (project at CERN)
LSM:	Least Squares Method
MC:	Monte Carlo
MeV:	Mega electron volt (10^6)
MIMD:	Multiple Instruction Multiple Data
MLM:	Maximum Likelihood Method
MLU:	Memory Logic Unit
MS:	Multiple Scattering
MST:	Minimum Spanning Tree
MTBF:	Mean Time Between Failures
MWPC:	Multi Wire Proportional Chamber
NAn:	North Area Experiments (CERN)
NEP:	Number of Equivalent Particles
NIM:	Nuclear Instrumental Module
NMOS:	N-type Metal-Oxide Semiconductor
PAL:	Programmed Array Logic
PAW:	Physics Analysis Workstation
PC:	Personal Computer
PDL:	Process Description Language
PEP:	Storage ring at SLAC
PETRA:	Storage ring at DESY
PHIGS:	Programmer's Hierarchical Interactive Graphics Standard

PROM:	Programmable Read Only Memory
PS:	Proton Synchrotron (CERN)
PWC:	Proportional Wire Chamber
QCD:	Quantum ChromoDynamics
QED:	Quantum ElectroDynamics
RAM:	Random Access Memory
RICH:	Ring Imaging Cherenkov Counter
RISC:	Reduced Instruction Set Computers
RMS:	Root Mean Square
SASD:	Structured Analysis Structured Design
SIMD:	Single Instruction Multiple Data
SLAC:	Stanford Linear Accelerator Center
SLC:	Linear e^+e^- Collider at SLAC
SPEAR:	Storage ring at SLAC
SPS:	Super Proton Synchrotron (CERN)
SSC:	Superconducting Super Collider (project, USA)
STD:	State Transition Diagram
TDC:	Time-to-Digital Converter
TEA:	TriEthylAmine
TMAE:	Tetra(diMethylAmine)Ethylene
TeV:	Tera electron volt (10^{12})
TOF:	Time-Of-Flight
TPC:	Time Projection Chamber
TRISTAN:	Storage ring at KEK
TTL:	Transistor–Transistor coupled Logic
UAn:	Underground Area Experiments (SPS collider)
UNK:	Accelerator at IHEP (project)
VME:	Electronic bus
VSB:	Extension for VME bus
WAn:	West Area Experiments (CERN)
ZEUS:	Experiment at HERA

Introduction

In this book we are concerned with the data-handling aspects of experimental particle physics as it is performed today in laboratories around the world. Particle physics is the science of the fundamental structure of matter: it is a study of the properties of subatomic particles and the way in which they interact. Its ultimate aim is to find a complete description of the elementary constituents of matter and of the forces acting between them, a description which should be as simple as possible. As in all branches of the natural sciences, the field is approached both from the theoretical and the experimental points of view. Theory predicts phenomena which can be verified by experiments, and experiments very often provide new insight through unexpected results which, in turn, lead to an improved theoretical description.

Observing phenomena at the subatomic and subnuclear level, on a scale smaller than any other, requires extraordinary instruments. With visible light, objects of a size comparable to the wavelength of this light can be seen in an optical microscope. Smaller distances require quanta of shorter wavelength or, according to de Broglie's principle, quanta of higher energy. Thus electron microscopes operate with accelerated electrons with energies of several thousand electron volts (keV). Particle accelerators today achieve energies of up to 1 TeV (10^{12} eV) and in this way allow the investigation of quarks and gluon quanta deep inside nucleons, at distances down to 10^{-16} cm. Accelerators can thus be seen as the 'light source' for observing the ultimately small. The corresponding 'eye' is then the detector, where the incident particles lead to observable effects. The analogy goes even further: in a human being, the final image is the result of distributed brain functions operating on the incoming data. In a particle physics experiment, the final knowledge about a recorded event is the result of an analysis performed by the hardware and software components in the analysis chain. The purpose of our book is to describe precisely these 'brain functions'.

High energies are achieved in particle accelerators where long-lived or stable particles such as protons and electrons acquire the necessary energy in strong electric fields. Since electric fields cannot be made arbitrarily strong, the length of a linear accelerator limits the energy to which a particle can be accelerated. For this reason, the circular accelerator has become the more common type, because here a particle crosses the same accelerating field many times. Such machines are called synchrotrons. The energies achieved today range from several tens of thousands of millions of electron volts (1 GeV = 10^9 eV) to a million million electron volts (1 TeV = 10^{12} eV). These energies are considerably higher than the energy equivalent of the proton mass which is about 1 GeV. High-energy physics is therefore clearly dominated by relativistic effects.

Particle physics experiments can be divided into two groups: fixed-target experiments, and collider experiments. In the first case, a beam of highly energetic particles is directed at a solid or liquid target, and the resulting secondary particles are observed. The beam particles either come directly out of the accelerator (primary beam), or are created in an intermediate target by the primary beam particles (secondary beam). In this way it is possible to provide beams with a great variety of particles, and in particular neutral and short-lived ones, both of which types are unsuitable for direct acceleration. When particles interact at relativistic speeds, only the centre-of-mass energy E_{cm} is available for the interaction proper – to create new particles or to penetrate deeply into the target particles. The fixed-target technique has in this respect a disadvantage: in this type of interaction, E_{cm} increases only with the square root of the energy of the incident particle. The remaining energy is used to boost the outgoing particles into the forward direction. Doubling the interaction energy, therefore, requires a quadrupling of the beam energy, and so forth. This limitation has led to the second mode of experimentation: two counter-circulating beams of highly energetic particles are kept in orbit inside a storage ring and are made to collide at certain 'intersection points' along its circumference. When the two beams consist of packets of particles and their antiparticles, only one set of bending and focussing magnets and only one beam tube are needed. Furthermore, since in this case the centre of mass of the two colliding particles remains at rest in the laboratory system, the energy E_{cm} available for the interaction itself is now simply equal to the total energy (twice the beam energy). Note, however, that the centre of mass of the interacting particles does not remain at rest in the laboratory frame in all colliders. At the CERN 'Intersecting Storage Rings' (ISR), the crossing angle between

Introduction

the two proton beams was (before its shutdown) 15°, and at the DESY 'Hadron-Elektron Ring-Anlage' (HERA) the kinematics of proton–electron collisions is very asymmetric, so that its observational aspects more closely resemble those of a fixed target experiment.

The two modes of collision call for two types of experimental set-up: fixed-target experiments are confronted with energetic particles in the forward direction and, therefore, require long magnetic spectrometers with high-precision tracking detectors and good two-particle resolution in order to measure the momenta of the outgoing charged particles with sufficient precision from their deflection in a magnetic field. In collider experiments with equal beam momenta and zero crossing angles, no particular direction is privileged by the kinematics of point-like particle collisions (although in hadron colliders many secondary particles accumulate around the beam direction; they are the fragmentation products of those quarks that do not take part in the collision, the so-called 'spectator quarks'). Accordingly, a compact detector is needed which covers the full solid angle and records a maximum of information about all outgoing particles. Consequently, the lever arm for measuring the deflection of charged particles by a magnetic field is relatively short (the typical scale of a collider detector is 10 m, whereas for fixed-target detectors it may reach 100 m). As a result of the higher interaction energy, the particle multiplicity is on average much higher than in fixed-target experiments. Therefore, to compensate these two effects, high-precision tracking detectors with good two-particle resolution are required here as well.

Apart from the measurement of the momenta of charged particles, the experimental set-up has also to provide clues for their identification. These are given by electromagnetic effects such as Cherenkov radiation, transition radiation, and the relativistic rise of energy loss per unit length, all of which occur when charged particles pass through matter. The third task of the experimental device is to detect neutral particles such as photons, neutrons, or neutral kaons. As these particles do not give rise to significant ionization along their path through matter, they must be measured in a destructive way by firing them into a block of matter where they are absorbed and, thereby, create detectable showers of secondary particles. This type of detector is called a 'calorimeter', although the minute amount of heat produced is, of course, not measured directly. Instead, one looked for another detectable signal which is in some way proportional to the energy deposited, for instance, ultra-violet light emitted by scintillators suitably interspersed with absorbers. Calorimeters serve not only for the detection of neutral particles, but are also used to measure the energy of single charged particles and of highly collimated bundles of particles (the so-called 'jets'). They also allow the identification of certain particles such as muons, and enable us to distinguish electrons from hadrons. For highly energetic particles such as electron–positron pairs from Z^0 decays, the energy resolution of calorimeters is superior to that achieved by curvature measurements in magnetic spectrometers.

In order to provide as much information on secondary particles as possible, huge detector systems are the rule. They frequently consist of a dozen or more subdetectors, each one concerned with a specific task. The yoke of the spectrometer magnet in a typical collider experiment has a diameter of 5–10 m and weighs around 1000 tons. The interaction region is surrounded by an arrangement of sophisticated particle detectors with a depth of several metres. Hundreds of thousands of electronic channels perform the analog-to-digital conversion or the time digitization of the many signals coming out of the detector. Complicated electronics is needed for the fast-decision logic that 'triggers' on good events in the presence of a huge background of unwanted ones. High event rates and a high total number of interactions produce a huge amount of data to be processed (this is more dramatic for hadronic than for electron–positron collisions). Furthermore, rare events (which are, of course, normally very interesting) must be selected from complex raw data which may contain a million times more events.

So we finally come to the subject of this book, the handling and analysis of these data. This calls in a first stage, during the real-time operation of a detector, upon the most recent technical achievements in the field of electronics and computers, since the analysis begins while an event is still being recorded. Among the hardware components we find: highly integrated electronic components which allow 'distributed intelligence', i.e. small data processing and storage units at many places in the detector electronics; the latest semiconductor memory chips which can store 4 Mbit; fast hardwired processors that execute selection algorithms within a few microseconds; microprocessors which allow complex decisions to be taken within a few milliseconds; readout systems and standardized bus interfaces that have been speeded up by at least one order of magnitude during the last decade; distributed parallel processing units that allow sophisticated multilevel triggering, and a considerable reduction in the amount of raw data at an early stage, certainly before they are recorded. Real-time data triggering and filtering are covered by Chapter 1.

Introduction

Historically, one used to distinguish between the real-time 'on-line' analysis, and the subsequent steps of the analysis which were performed 'off-line'. The border used to be drawn at the point where the data were recorded on a mass storage device and could thus be 'replayed'. This sharp distinction tends to disappear in modern large experiments where enough real-time computing capacity exists to perform an increasing number of analysis tasks. The first step of this analysis (once an event has been accepted by the real-time filtering) consists normally of 'pattern recognition', where typically the signals belonging to each track are associated, vertices are found, and showers in calorimeters are reconstructed. This task, which is inherently of a combinatorial nature, has in the past mainly been performed on mainframes with large storage systems and high processing speeds. The situation is changing: if enough experience has been gained from the off-line analysis of real data, part of the mass storage for raw data recording can be saved by implementing more and more pattern recognition steps in the on-line filtering stage. This will, however, normally require from the outset the firm intention to achieve this goal, and a careful planning of the detector in parallel with pattern recognition studies. It goes without saying that data rejected on-line are lost forever, whereas recorded data can be reprocessed should the need arise. If the pattern recognition software has not reached a sufficient state of maturity, this risk should be avoided given today's fast and powerful mass storage devices. Pattern recognition without real-time constraints is discussed in Chapter 2.

The next important task of the data analysis is to extract the ultimate information on charged tracks as provided by the detector's high-precision tracking devices. This then permits a final test of the decisions taken in the course of the pattern recognition stage: an association of tracks to primary and secondary vertices, and the calculation of appropriate input quantities for further physics analysis. This requires the fast solution of the equation of motion for charged particles in a magnetic field, including the efficient storage of the field map, and flexible handling of the matrix operations involved. Track and vertex fitting is covered by Chapter 3.

An important technique in modern data analysis is the application of statistical methods to data abstraction. The development of such methods can be greatly helped by graphical presentation of the data, and by interaction. Graphics also allow the analysis program to be tuned interactively, and may, in addition, help in the recognition of unforeseen relationships between different components of the data. A large number of possible approaches to this problem exist, and the different solutions attempted so far have not yet converged. Broad methods of analysis, including the use of graphics, are treated in Chapter 4.

Large-scale detectors are typically planned, built, and exploited by groups of several hundred physicists and engineers. The fact that many of them take part in the software development imposes certain constraints on the way in which this software is designed, written, and managed. A further complication arises when the analysis programs have to run on different computers (and still have to give the same results), a requirement whose solution is far from obvious. The management of the program life cycle and the problems of distributed software development are discussed in Chapter 5.

1

Real-time data triggering and filtering

1.1 Definitions and goals of triggers and filters

The task of a trigger system is to select rare events and to suppress background events as efficiently as possible. To illustrate the trigger problem let's suppose that one has to find a friend among the 13 000 000 inhabitants of Mexico City. To find this friend requires a trigger sensitivity of $1:10^7$. With some further knowledge the selection problem can be reduced. He or she is in the city only for a short time and probably lives in a hotel. From 13 000 000 choices one is now down to the 10 000 hotel guests. Assuming that he or she lives in a hotel near the centre of the city reduces the search to a group of 400 people. This example demonstrates the various trigger levels which can be used to reduce the number of choices.

Data taking is the limiting factor in many experiments. The high rates and the volume of data which must be readout by the data acquisition system require there to be some time during which no data can be taken. This time is called '*dead time*'. In order to reduce dead time one has to improve the quality of the trigger by sophisticated processors which increase the number of good events per time unit.

1.1.1 *General properties of particle accelerators*

In scattering experiments a high-energy particle beam produced in an accelerator is directed either onto a target where scattering takes place or towards a highly focussed beam coming from the opposite direction. Electrons are liberated from a high-voltage triode tube, while protons originate from hydrogen which dissociates into positive ions with the help of oscillating electrons. These charged particles are then accelerated by linear accelerators which have a set of cavities, one behind the other, to accelerate the beam and quadrupoles to focus the beam. They are used at low energies (400 MeV) to inject particles into *circular accelerators* or at high energies to accelerate heavy ions or electrons. At the Stanford Linear Accelerator Centre (SLAC) electrons reach an energy of up to 60 GeV. *Linear accelerators* are proposed for the future which will accelerate electrons up to very high energies (1000 GeV for the CERN Linear Collider (CLIC)) (Johnsen 1987a) to avoid the difficulties encountered when extracting the beam and to avoid energy loss from synchrotron radiation. The energy loss per revolution for electrons in a circular accelerator is

$$\Delta E[\text{keV}] = 88.5 E^4 [\text{GeV}]/r[m] \qquad (1.1)$$

where E is the energy of the electrons and r is the radius of the accelerator in metres. The energy loss for a machine with $r = 200$ m and $E = 18$ GeV is therefore $\Delta E = 46$ MeV. The loss increases to 124 MeV if the beam energy is increased to 23 GeV.

The beams of a linear accelerator are of high quality. They have a good energy resolution, are well focussed and have very little halo around them. Linear accelerators are long (3.2 km at SLAC), they need many accelerating elements and deliver short pulses.

One can overcome these disadvantages by using circular machines in which particles are forced by *dipole magnets* to stay in a closed orbit. The cavities are used to accelerate the same particles in each cycle. One needs less space but the beam quality is not very good: the beam spot is larger with more halo, the energy resolution is not so good and one has to compensate for the loss of synchrotron radiation. In electron synchrotrons the magnetic field, and therefore the energy is ramped like a sine wave. The beam is extracted at the maximum of this wave. For long extraction times the energy of the extracted beams varies within certain limits. The typical time taken to accelerate electrons to 7 GeV is 10 ms and to accelerate protons to 500 GeV is 50 s. The intensity of the extracted beam is of the order of 10^{13} particle/pulse.

1.1.2 Secondary beams

If one wants to investigate the scattering of *neutral* particles or *short-lived* particles by matter one has first to produce these particles and then to direct them as secondary beams towards the target in which the scattering takes place. The extracted electron or proton beam hits a primary target.

To produce a γ *beam*, electrons are first directed onto a target of high atomic number a and then deflected by a magnetic field. The energy spectrum of the photons is inversely proportional to the momentum of the γs, thus many low-energy photons are produced. By having lithium hydride in the beam the number of low-energy photons can be reduced.

When the photons are directed towards another target they produce e^+e^- pairs. The *positrons* can then be selected by a magnetic field and a collimator. A *beam transport* system containing a set of *dipole magnets* and *quadrupoles* can be used to study positron scattering by selecting positrons of a narrow energy band and sending them onto a target. *Charged hadrons* ($\pi, K, \bar{p}$) are produced by interactions of high-energy protons in a target. Particles of a certain angle, charge, and momentum interval are transported by the beam transport system.

1.1.3 Energy balance in scattering experiments

In fixed-target experiments part of the energy of the incoming particles is not available for the interaction but is wasted in boosting the particles in the forward direction. Storage rings or colliders are used to increase the energy available in an interaction. In these, two beams from opposite directions interact with each other. The energy W available in an interaction can be expressed in a *Lorentz invariant* form:

$$W^2 = s = (p_1 + p_2)^2 \tag{1.2}$$

where s is the square of the invariant energy, $p_1 = (m_1, \mathbf{p}_1)$ is the four vector of the incoming particle, and $p_2 = (m_2, \mathbf{p}_2)$ is the four vector of the target particle ($p_2 = (m_2, 0)$ for a fixed target). When a proton of energy $E = 400$ GeV hits a stationary hydrogen target (protons) the square of the invariant mass is

$$\begin{aligned} s &= m_p^2 + m_p^2 + 2Em_p = 754 \text{ GeV} \\ (m_p &= 0.938 \text{ GeV}) \\ W &= s^{\frac{1}{2}} = 27.5 \text{ GeV} \end{aligned} \tag{1.3}$$

Subtracting twice the proton mass for the surviving protons (baryon number conservation) gives the energy available in the reaction

$$M_x = W - 2m_p = 25.6 \text{ GeV} = 2 \times 12.8 \text{ GeV}$$

which means that two colliding protons each of 12.8 GeV release as much energy as a proton of 400 GeV hitting a proton at rest. The equivalent beam energy for a fixed-target machine for a given storage ring beam energy E

Table 1.1. Colliders

		Particles	Beam energy (GeV)	Luminosity (cm^{-2}s^{-1})	E_b equivalent (GeV)	Crossing (µs)
SPS	CERN	$\bar{p}p$	315	5×10^{29}	211 000	4
TEVATRON	FERMILAB	$\bar{p}p$	1000	10^{30}	2 132 000	3.5
SPEAR	SLAC	e^+e^-	4.2	10^{32}	69 000	0.78
DORIS	DESY	e^+e^-	5.5	10^{32}	118 000	0.5
PETRA	DESY	e^+e^-	23	10^{32}	2 070 000	4
CESR	CORNELL	e^+e^-	8	10^{32}	250 000	0.84
PEP	SLAC	e^+e^-	18	10^{32}	1 268 000	2.3
LEP	CERN	e^+e^-	55	1.5×10^{32}	11 839 000	23
HERA	DESY	e^-p	30e + 820p	10^{31}	52 000	0.096
SLC	SLAC	e^+e^-	50	6×10^{30}	9 784 000	5500
TRISTAN	KEK	e^+e^-	30	2×10^{31}	3 522 505	5
UNK(II)	SERPUKHOV	pp	3000	10^{32}	19 189 765	0.005

and a particle mass m is then:

$$E_{\text{fixed target}} = 2E^2_{\text{storage ring}}/m \tag{1.4}$$

Three different types of colliders can be distinguished: hadron–hadron machines, lepton–lepton machines and hadron–lepton machines. Several of these machines are listed in Table 1.1. The column headed 'E_b equivalent' gives the equivalent energy which a beam particle must have when hitting a particle at rest to produce the same *center-of-mass* energy.

1.1.4 Luminosity

The number of interactions in a fixed-target experiment is proportional to the *cross section* σ for the type of interaction, the *particle flux* and the *number of atoms* per cubic centimetre in the target multiplied by the length l. The inverse of the last quantity has the dimension of an area and is called the *target constant*. The number of particles per cubic centimetre is given by the *Avogadro* constant $N_A \times$ density ρ/atomic weight A.

$$F = A/(N_A \rho l) \tag{1.5}$$

For a liquid hydrogen target ($\rho = 0.071$ g cm^{-3}) with a length of 11 cm the target constant is

$$F = 1 \text{ g}/(6.022 \times 10^{23} \times 0.071 \text{ g/cm}^3 \times 11 \text{ cm})$$
$$= 2.1 \times 10^{-24} \text{ cm}^2 = 2.1 \text{ b}$$

Cross sections are measured in barns [an '*area*' of $1 \text{ b} = 10^{-24} \text{ cm}^2$].

1.1 Definitions and goals of triggers and filters

Assuming a given cross section one can estimate the number of reactions per time interval by

$$\frac{N_{\text{events}}}{\text{second}} = \sigma \frac{N_{\text{flux}}/\text{second}}{F} = \sigma \times \text{Luminosity} \quad (1.6)$$

The *luminosity* is a measure of sensitivity and gives directly the number of events per second for a cross section of $1\,\text{cm}^2$. For the target described above and a flux of 10^7 particles per second the luminosity is

$$L = 4.8\,10^{30}\,\text{cm}^{-2}\,\text{s}^{-1} = 4.8\,\mu\text{b}^{-1}\,\text{s}^{-1}$$

A cross section of $1\,\mu\text{b}$ would result in 4.8 events per second.

In a storage ring the luminosity depends on several parameters such as the number of particles per bunch N_b, the number of bunches in each beam k_b, the distance between bunches or the revolution frequency f and the beam radii of the bunches σ at the crossing point.

$$L = N_b^2 f k_b / 4\pi\sigma^2 \quad (1.7)$$

All these quantities depend on the energy and on the beam dynamics. A bunch with a high density may increase in diameter when interacting with a bunch coming from the opposite direction, with the mirror charge on the beam pipe walls or due to head–tail interactions. It is therefore hard to calculate the luminosity for a given storage ring.

The luminosities for some storage rings are given in Table 1.1. They range from 10^{30}–$10^{32}\,\text{cm}^{-2}\,\text{s}^{-1}$ which is of the same order of magnitude as the example given above. One would therefore expect that the event rates would be of the same order of magnitude.

1.1.5 Time structure of accelerators

Time structure at fixed-target accelerators

The particles in an accelerator are accelerated by electric fields which are generated by a *radiofrequency system*. This defines a timing structure and requires particles that are packed into *bunches*. Because of this timing structure an experiment receives particles for only a fraction of the overall time: this is called the *duty cycle* of the machine. The duty cycle is a measure of the efficiency of an accelerator.

duty cycle = available beam time/total time

= duration of a bunch × number of bunches per second/second (1.8)

The duty cycle at the linear accelerator SLAC is:

$$\text{duty cycle (SLAC)} = 1.2\,\mu s \times 360/s = 0.04\%$$

At the electron synchrotron DESY there are 50 acceleration cycles per second or 50 spills with a spill length of 1 ms. The duty cycle is then

$$\text{duty cycle (DESY)} = 1\,\text{ms} \times 50/1\,\text{s} = 5\%$$

The magnetic field and therefore the energy are ramped like a sine wave. With the help of interference with higher modes of the accelerating field the shape of the acceleration can be modified in such a way so as to produce a *flat top* which prolongs the spill and can improve the duty cycle by some fraction.

In proton accelerators the cycle time to accelerate particles is of the order of 2–50 s and the burst time is of the order of 300 ms–15 s giving values between

$$\text{duty cycle} = 300\,\text{ms}/2\,\text{s} = 15\%$$

and

$$\text{duty cycle} = 15\,\text{s}/50\,\text{s} = 30\%.$$

A data acquisition system at fixed-target accelerators must be organized in such a way that data are collected rapidly during the spill time. Filtering, monitoring, and recording of data preferably take place in the time between the spills.

Time structure at colliders

In storage rings both rings are filled with particles which are then accelerated to the nominal energy. The time taken to fill a storage ring is of the order of 5–15 min for e^+e^- or pp colliders. A $\bar{p}p$ collider can only be filled a few times per day because it takes several hours to collect enough antiprotons.

The acceleration of the stored particles ranges from some minutes to about half an hour depending on the mass of the particles and the nominal energy. After tuning the collider for high luminosity with the help of focussing quadrupoles near the interaction regions the experiments take data for some hours until the intensity of the beams is so low that a new filling procedure is justified.

The time between bunches depends on the circumference of the machine and the number of bunches. The Low-Energy Antiproton Ring (LEAR) at CERN (2 GeV $\bar{p}p$) is a direct current (DC) machine with a Poisson

1.1 Definitions and goals of triggers and filters

distributed bunch structure. In the proposed colliders UNK (II) and LHC the time between bunch crossings will be 5 and 25 ns, respectively. At HERA the crossing time will be 96 ns. It is not possible to reach a trigger decision within such a short time. All incoming data must be time delayed for several microseconds while the trigger processors are active (see Subsection 1.6.2). In colliders such as PEP (2.3 μs), LEP (23 μs) and SLC (5500 μs) there is enough time to define a trigger between bunch crossings with hardwired processors.

1.1.6 Event rates at different accelerators

Event rates at fixed target accelerators

The event rates increase proportionally with the cross section, the flux, the target length, and density or, if one combines the last three items, with the luminosity. At energies above 10 GeV the *pp* cross section is about 40 mb and increases at higher energies. It is easy to produce a high-intensity proton beam because one can extract the proton beam from the accelerator. When 10^{13} protons per pulse with a pulse length of 10 s hit an 11 cm long hydrogen target ($F = 2.1$ b) one gets a peak luminosity of

$$L_{peak} = 10^{13}/(10\,\text{s} \times 2.1\,\text{b}) = 0.48 \times 10^{12}\,\text{b}\,\text{s}^{-1}$$
$$= 0.48 \times 10^{36}\,\text{cm}^{-2}\,\text{s}^{-1}$$

The number of events per second which must be recorded during the pulse is therefore 19.2×10^9. This rate is remarkably high. In a typical experiment one measures the differential cross section in which particles are scattered through a certain angle. In these cases the measured cross section goes down by up to 10 orders of magnitude giving 0.2 events per second. Taking the acceleration time of 20–50 s into account the rate decreases to 0.004 events per second and the luminosity decreases to $L = 10^{35}\,\text{cm}^{-2}\,\text{s}^{-1}$.

The total photoproduction cross section γp for energies above 4 GeV is of the order of 120 μb. Photons are produced in secondary beams with an intensity of 10^3 'energy-tagged' photons per 1 ms pulse. The peak luminosity for an 11 cm long hydrogen target is then

$$L_{peak} = 10^3/(0.001\,\text{s} \times 2.1\,\text{b}) = 0.48 \times 10^{30}\,\text{cm}^{-2}\,\text{s}^{-1}$$

giving an event rate of

$$\text{rate} = 120 \times 0.48 = 58\,\text{events per second}$$

Event rates in $\bar{p}p$ or pp colliders

In $\bar{p}p$ or pp colliders the total cross section is 60 mb and above. In $\bar{p}p$ colliders the luminosity is limited by the number of antiprotons and these cannot be produced in large quantities. A typical value for the luminosity is of the order of 10^{30} cm^{-2}s^{-1} which results in an event rate of 6×10^4 events per second. For selected reactions like the production of the *intermediate boson* Z^0 the cross section is only 2 nb which results in a production rate of one Z^0 per 5×10^3 s or $19\,Z^0$ per day. The planned hadron–hadron colliders will consist of two proton rings with magnetic fields in opposite directions. The pp cross section is expected to go up to 135 mb at $\sqrt{s} = 17$ TeV. It consists of three parts: the *elastic, diffractive* and *inelastic* cross sections. Only the inelastic part contributes to the event rate of a general purpose detector. The elastic and diffractive events send outgoing particles almost exclusively along the beam line. With an inelastic cross section of 60% of the total cross section and an expected luminosity of $L = 10^{33}$ cm^{-2}s^{-1} the event rate becomes $(6-8) \times 10^7$ Hz. With 25 ns between bunch crossings the average number of observed interactions in each crossing is 1.5! If one wants to trigger on the missing energy, which is sensitive to reactions with an escaping neutrino, one has to limit the luminosity in such a way that one gets only one event per bunch crossing:

$$L_{\text{limit}} = \langle n \rangle / (\text{Time between bunches} \times \sigma_{pp}). \tag{1.9}$$

Even for $\langle n \rangle = 1$ there is more than one event per crossing in 26% of the interactions. (Poisson distribution $P_1(>1) = 1 - P_1(0) - P_1(1) = 1 - e^{-1} = 0.26$.)

Event rates in e^+e^- colliders

In e^+e^- machines on the other hand the cross section is very small. The *one photon* exchange cross section is determined by the cross section for $e^+e^- \to \mu^+\mu^-$ which is

$$\sigma_\sigma = \frac{4\xi\alpha^2}{3s} = \frac{21.9 \text{ nb GeV}^2}{E_{\text{beam}}^2} \tag{1.10}$$

where α is the fine structure constant. The ratio of hadron production to $\mu^+\mu^-$ production is given by the sum of the squared charges of possible quarks multiplied by 3 for the three colours:

$$R = 3 \sum_{i=1}^{5(6)} Q_i^2 \approx 4. \tag{1.11}$$

1.1 Definitions and goals of triggers and filters

At LEP energies ($E_{beam} = 55\,\text{GeV}$, $L = 1.5\,10^{31}\,\text{cm}^{-2}\,\text{s}^{-1}$) this small cross section gives an event rate of

$$N/s = L \times \sigma = 1.5 \times 10^{31}\,\text{cm}^{-2}\,\text{s}^{-1} \times 4 \times 21.9 \times 10^{-33}\,\text{cm}^2/55^2$$
$$= 0.0005/s$$

or 26 events per day. These rare events must be selected from a background which has mainly two sources: beam–gas interactions and showers in the beam pipe. The rates for these processes are of the order of 10^3–10^4 per second. If one operates LEP at the energies necessary to produce Z^0s the production cross section increases to 50 nb which yields an event rate of 1.8×10^4 events per hour.

Event rates at electron–proton colliders

In electron–proton colliders the cross section has two parts: a neutral current (γ, Z^0) and a charged current ($W^\pm$) contribution. These contributions can be calculated and depend mostly on the momentum transfer Q^2. Above $Q^2 = 1000\,\text{GeV}^2$ the cross section for 30 GeV electrons and 820 GeV protons is of the order of 150 pb. For a luminosity $L = 10^{31}\,\text{cm}^{-2}\,\text{s}^{-1}$ this results in 4 events per hour.

1.1.7 Background rates

A trigger system should select good events and suppress those reactions which are not interesting. Fig. 1.1 summarizes the requirements for a trigger in e^+e^- colliders (Waloschek, 1984). At a beam energy of 20 GeV at medium luminosities one expects an event rate of 1 event per 5 min. Interactions of the beam with the gas which is left inside the beam tube even under good vacuum conditions or interactions of beam particles with the walls of the beam tube cause high background rates of the order of 10^3–10^4 per second. Particles lose energy due to synchrotron radiation or other instabilities of the accelerator. These particles are no longer kept in a closed orbit at the nominal beam position and hit the collimators or walls of the vacuum system. Background processes of this type do not create tracks coming from the interaction region: the tracks are boosted in the forward direction. A good trigger system must therefore reject tracks which do not have their origin at the interaction point (*vertex detectors*). Cosmic ray events may appear at any time; they are neither correlated with the timing structure of the beam nor with the interaction point.

On the other hand some background events can be useful for the

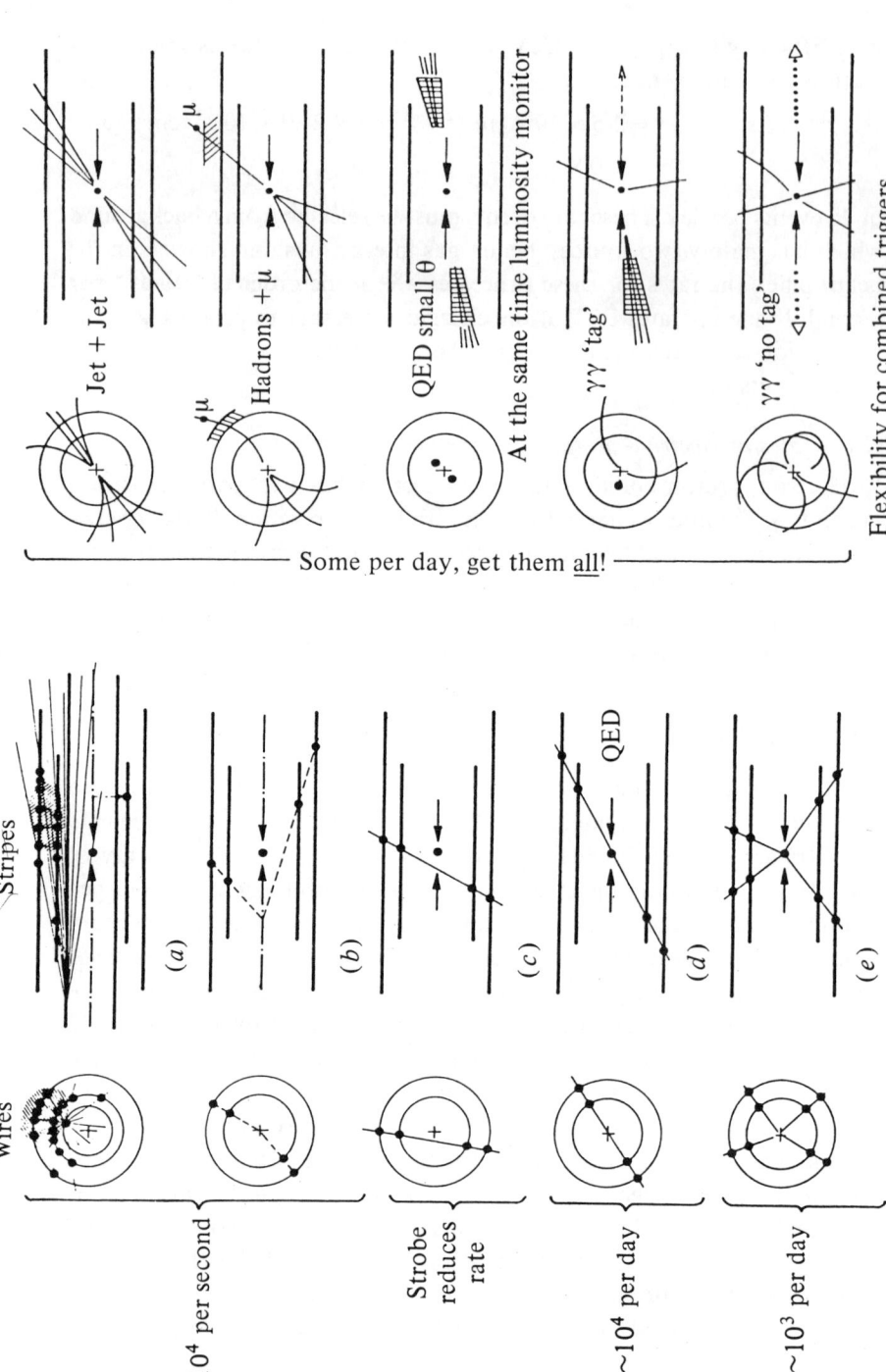

experiment. QED events such as $e^+e^- \to e^+e^-$ (*Bhabha scattering*) are used for normalization to measure the luminosity of the collider. They produce two coplanar and colinear tracks. Another source of background comes from electrons emitting photons by bremsstrahlung. These photons can react with the photons in the opposite beam: photon–photon or $\gamma\gamma$ reactions. In high-energy electron beams (1 TeV) the *two photon cross section* becomes very large for small two photon invariant mass.

The main source of background in electron-proton colliders comes from interactions between the proton beam and the gas while the background rates in proton–proton colliders are small compared with the high event rate.

1.2 Trigger schemes

1.2.1 On-line data reduction

A trigger system should select all good event candidates and reject most of the background events. Let us consider a detector with 100 000 detector elements which operates in a storage ring with 0.1 µs time between beam crossings and which generates a raw data rate of approximately 2×10^{12} bytes per second (1 byte contains 8 bits of information). The time to write data onto a storage medium is of the order of 1 Mbyte per second. The on-line data acquisition system must therefore reduce the incoming raw data by at least 6 orders of magnitude. The border line between on-line and off-line data handling is defined here as follows: on-line data acquisition operates with the data from their first appearance in digital form up to writing them onto a permanent storage medium; off-line analysis is any analysis which reads the information from that medium.

In old experiments a simple trigger requiring a coincidence of some electronic signals was used and data were written directly to tape. Nowadays with the advent of microprocessors more sophisticated event filtering can be performed. The speed of the electronic components in an experiment ranges from nanoseconds to seconds and will be briefly discussed here.

Fig. 1.1 Rates and topologies of e^+e^- background and events. Some selected event topologies are shown on the right. A trigger must be sensitive to as many event candidates as possible and should discriminate against background as much as possible. (*a*) Shower originating in the vacuum tube; (*b*) beam–gas interaction; (*c*) cosmic ray background; (*d*) *coplanar* and *colinear* beam–beam interaction; (*e*) *multibody* beam–beam event.

The incoming and outgoing particles of a reaction are measured by a detector. One of the most commonly used particle detectors is a *scintillation counter* in which a fraction of the energy lost by a charged particle is used to excite atoms in a scintillating medium. Part of the energy in the deexcitation can produce light. The rise time of the light output is of the order of 1 ns and the decay time varies for some plastic scintillators between 1.3 and 4.0 ns.

The scintillator light is converted into an electrical signal and amplified by a *photomultiplier*. Here electrons are emitted in the *photocathode* by the *photoelectric effect* and are accelerated onto the dynodes of the tubes. The amplification for a 12 stage tube is of the order of 10^7.

Another fast detector is the *Cherenkov counter*. Particles travelling faster than the speed of light in the radiator of a Cherenkov counter produce light at a certain angle. The light is focussed by mirrors onto the photocathode of a photomultiplier.

The spatial resolution of scintillation counters is limited to approximately 1 cm. If better resolution is required one can use *multiwire proportional chambers*, *drift chambers* or *semiconductor detectors* (see Subsection 2.1.2). Years ago *bubble chambers* or *emulsion detectors* were used if high resolution was required. These detectors were slow and so are no longer used.

The electric signals of a photomultiplier vary in both length and amplitude. In order to perform logical operations such as AND or OR among several detectors one has to use standard pulses with fixed amplitudes, short rise times, and small variations in length. *Pulseformers* or *discriminators* are used to generate a standard output signal if the input pulse exceeds a given *threshold*. Typical output pulse lengths and amplitudes are of the order of 20 ns and 0.8 V, respectively (see Subsection 1.6.1). Fast electronic circuits perform logical operations such as AND or OR and operate in the same speed range as discriminators.

When a trigger condition is fulfilled the entire information in the detector is readout into a memory. Simple yes/no information which just indicates whether there was a particle or not is stored in a *flip-flop* or a *latch*. The latches are connected to a data acquisition bus and are readout within approximately 100 ns (see Section 1.8).

Analog information from detectors which measure, for example, particle energy is converted to digital information by Analog-to-Digital Converters (ADC). An ADC has a reset time of approximately 1 μs. A second event can not be recorded during that time. If the trigger condition is fulfilled the analog information in the ADC must be converted; this takes from 50 μs to

1.2 Trigger schemes

a 1 ms. Many ADCs are then readout by a *voltage ramp* and a *comparator circuit*. The trigger system has to reduce the raw input rate to 1 kHz if ADCs of this speed range are used. If higher rates are required one can use *Flash ADCs* (FADC) which convert data within 10–100 ns. FADCs currently have the disadvantage that their resolution is only of the order of 8 bits.

The time information needed to measure the speed of particles between two detectors such as Time-Of-Flight (TOF) counters or to measure the drift time in a drift chamber is digitized by *Time-to-Digital Converters* (TDC). TDCs need the same amount of time for conversion as ADCs.

If one assumes an event length of 100 kbytes and a trigger rate of 1 kHz the transfer rate to a computer is of the order of 100 Mbytes per second which exceeds the bandwidth of a 'normal' computer bus or data acquisition bus system. Further intermediate stages are needed.

The event is stored *in parallel* by different detector components called '*subdetectors*'. With the help of special processors, e.g. *bit-slice processors* and *digital-signal processors* the input rate may be reduced to 10–100 Hz (see Subsections 1.7.1, 1.7.2). At this rate a complete event of 100 k bytes can be transferred to a computer but the data rate is still too high to be recorded onto tape.

Further reduction with filter algorithms is needed. These algorithms can now operate on the entire event and can search for tracks, clusters, or interaction points and look for example for correlations between track elements found in different subdetectors. The computation may take place in microprocessors which can reduce the event rate to 5–10 Hz leading to a data rate of 0.5–1 Mbytes per second which matches the tape speed (see Subsection 1.7.6). A good trigger scheme must be designed in such a way that dead time is minimized and the number of good events processed per time unit is increased.

1.2.2 Dead time of electronic components

Each particle detector has a specific resolution time. Particles following each other within a short time interval are not detected as two separate particles. The nonsensitive period of the detector or the electronics is called *dead time*. The loss caused by dead time must be corrected for and must be kept small for reasons of efficiency. In detectors the dead time ranges from several nanoseconds for scintillators up to micro and milliseconds for devices which must be recharged with high voltage such as wire chambers.

The dead time due to *electronics and the data acquisition system* covers a wide range. Fast electronics has a dead time of nanoseconds while the dead time of a computer can be of the order of minutes if a tape has to be rewound and a new tape mounted.

Pulseformers or *discriminators* have a dead time due to the discharging of capacities or to the reestablishment of the initial conditions in circuits with back coupling (see Subsection 1.6.1). For discriminators instead of speaking about dead time one uses the double pulse resolution as a characteristic quantity. The double pulse resolution is of the order of 10 ns. If the second pulse appears more than 10 ns after the first pulse and if the first pulse was short (5 ns) then the second pulse will be registered by the discriminator (Fig. 1.2). To minimize dead time one tries to make the detector signals short with the help of a *clipping cable* or by pulse differentiation. For pulse clipping the signal from the scintillator is sent to the discriminator and to a short coaxial cable which is short-circuited at the end (Fig. 1.3). Reflections at the end of the cable force the voltage to zero and the pulse is shorter.

The input signals appear statistically. One must find a way to correct for losses due to dead time. The entire dead time of a system can be measured directly if each component has a dead time of more than 20 ns. Either this dead time is fixed or the components generate a busy signal of variable length. Variable dead time occurs in computers when for example writing data to tape, mounting tapes or waiting for response of an operator. The

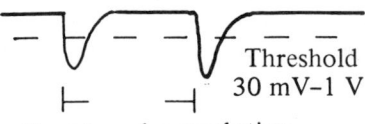

Double pulse resolution

Fig. 1.2 Double pulse resolution for a discriminator. Knowledge of the double pulse resolution is needed to estimate the maximum rate which can be taken by a trigger system.

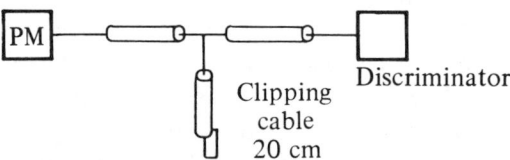

Fig.1.3 Clipping cable to make pulses short. For some applications such as long calorimeter signals it might be necessary to shorten pulses with the help of a clipping cable (PM = photomultiplier).

1.2 Trigger schemes

dead time is measured by comparing the counts of an ungated clock to a clock with is gated with the OR of all busy signals (Fig. 1.4).

To measure the dead time of fast electronic circuits one uses two radioactive sources and compares the single rate of each to the combined rate of both sources. We assume that pulses which appear within the dead time do not increase the dead time of the whole system (Stuckenberg 1968). The procedure goes like this: If a scaler has registered n pulses per second it cannot count $n\tau$ pulses for a system with a dead time τ. The number of real pulses is then $N = n/(1 - n\tau)$. This equation can be approximated for small dead times by $N = n(1 + n\tau)$.

Let A and B be the true rates from each source and Z be the zero rate if no source is present. The zero effect, source A alone, source B alone and A plus B together are then measured and from the equations

$$A + Z = n_a(1 + n_a\tau)$$
$$B + Z = n_b(1 + n_b\tau)$$
$$A + B + Z = n_s(1 + n_s\tau)$$

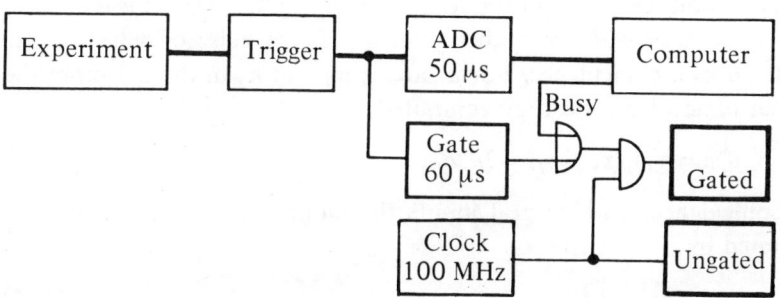

Fig. 1.4 Measurement of dead time in a data taking system. The dead time is measured by comparing the gated and ungated clocks. The busy signal of the computer has a variable length (e.g. the wait for tape mounting). A gate is used to take the ADC dead time and the computer respond time into account.

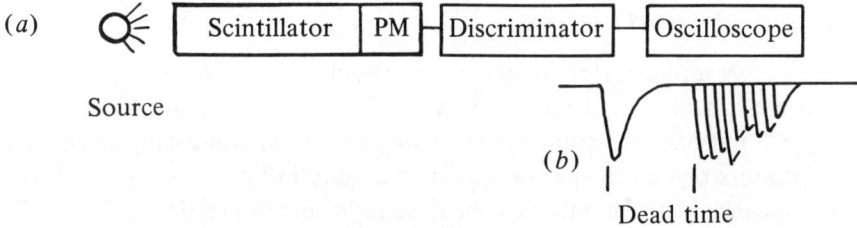

Fig. 1.5 Measuring the dead time with an oscilloscope (PM = photomultiplier).

one can compute the dead time

$$\tau = (n_a + n_b - n_s - Z)/(n_s^2 - n_a^2 - n_b^2) \tag{1.12}$$

All the other quantities are known and one can therefore compute τ. Care must be taken that both sources have the same geometrical acceptance. With a fast oscilloscope one can measure the dead time directly if the input rate is high. Using the set-up of Fig. 1.5(a) one observes at the oscilloscope the pattern shown in Fig. 1.5(b) which gives the dead time directly.

1.2.3 True and wrong coincidences, accidentals

In a reaction which produces two particles both particles can be measured with two detectors using a logical AND. If both particles are produced at the same time one has a *true coincidence*. On the other hand coincidences can be simulated due to the fixed *resolution time* of the detector components. Two independent particles which appear within the resolution time of the detector and the electronics are registered as a coincidence. The rate of these *accidentals* depends on the single rate of each detector and the resolution time. Detector A produces pulses of length τ_1, detector B those of τ_2. A pulse in A which is τ_1 before or τ_2 after a pulse of detector B is registered as a coincidence. For a rate of n_A and n_B in the detectors the number of accidentals can be computed by

$$acc = n_A n_B (\tau_1 + \tau_2) = 2 n_A n_B \tau. \tag{1.13}$$

For coincidences with several inputs the number of accidentals can be computed by

$$acc = k\tau^{k-1} \prod n_k. \tag{1.14}$$

The time resolution of a coincidence can be measured with a variable delay line and two scalers (Fig. 1.6).

1.2.4 Multilevel triggers

A sophisticated trigger system should be able to reduce the input rate from background processes in an efficient way without losing good events. In order to improve the quality of an experiment by using fast processors one has a specific aspect of quality in mind, i.e. the statistical significance or the number of *good* events recorded per time unit.

To reduce dead time and to include complex decisions in the trigger,

1.2 Trigger schemes

several trigger levels are required. At each trigger level more information is available to perform better filtering. Assuming the cross section is so high that the experiment is limited by the tape speed, one can only record events with a rate of 7 Hz. In this case one can use three trigger levels.

(1) Trigger 1 acts on the prompt information which is available in scintillation counters, proportional chambers or drift chambers with small gaps. Hardwired processors and fast electronics select a rough trigger which should not run above 10 kHz.

(2) The input rate at level 2 is 100 μs per event. Within this time data can be digitized by ADCs. Special processors e.g. *bit-slice processors* and *digital-signal processors* (see Subsections 1.7.1, 1.7.2) or special processors with *Content Addressable Memories* (CAMs) can do a better track or energy cluster search or vertex fitting. The conventional microprocessors (von Neumann computers) are not adequate. Depending on the readout time of the entire event the output rate should be below 20–100 Hz.

(3) At level 3 the complete event information and digitizations are available. The track filtering can be improved and correlations between various detector elements can be utilized to reduce the data rate to 7 Hz which matches the tape speed. At this level microprocessors or minicomputers can be used to finish their task in 100 ms. If 100 ms are

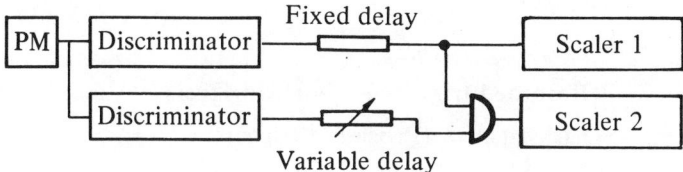

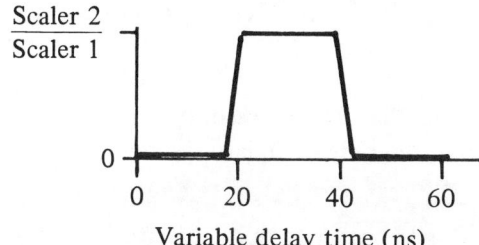

Fig. 1.6 Measurement of the resolution time of a coincidence. Knowledge of the resolution time for a coincidence is needed to estimate the maximum rates.

not enough the events can be handed over to several parallel processors. But one must take care when designing a high speed bus. Between each trigger level the events must be buffered. This decouples the different processor speeds from the statistical arrival of the data. The optimal buffer length will be given in Section 1.3.

With fast processors one can *decrease* the number of recorded background events and at the same time *decrease* the dead time and *increase* the number of processed events per time unit. In order to study the effect of multilevel triggers in a quantitative way (Lütjens 1981) let us assume a given event rate of n_e per second. The mean waiting time for an event is then $t_e = 1/n_e$. With a recording time of t_R the number of triggers recorded per second is:

$$1\,\text{s} = n_R(t_R + t_e) \tag{1.15}$$

The fraction of recorded events is then

$$E = n_R/n_e = (1 + t_R/t_e)^{-1}$$

The dead time caused by a long recording time t_R can be reduced by a second stage veto. A fast processor with a good algorithm should be able to detect and reset a trigger caused by a background event within a *'processing time'* $t_p < t_R$. With a fast processor the number of triggers aborted per second is Kn_t and the number of triggers recorded per second is n_t. Equation (1.15) then becomes:

$$\begin{aligned} 1\,\text{s} &= n_t(t_R + t_e) + Kn_t(t_p + t_e) \\ &= \text{Recorded triggers} + \text{Aborted triggers} \\ &= n'_t(t'_R + t_e) \quad \text{(Processed triggers)} \end{aligned} \tag{1.16}$$

The fraction of processed events is therefore

$$E' = n'_t/n_e = (1 + t_R/t_e)^{-1} \frac{K+1}{1 + K(1 + t_p/t_e)/(1 + t_R/t_e)} \tag{1.17}$$

$$E' = EG$$

G is limited by the fact that the processor itself produces dead time. The limit at large rates t_R/t_e (high dead time caused by long recording time) is

$$G < G_{\max} = \frac{1+K}{1 + k(t_p/t_R)}$$

A second stage veto reduces the amount of recorded data by a factor of G/K.

$$n_t = (G/K)n_r \tag{1.18}$$

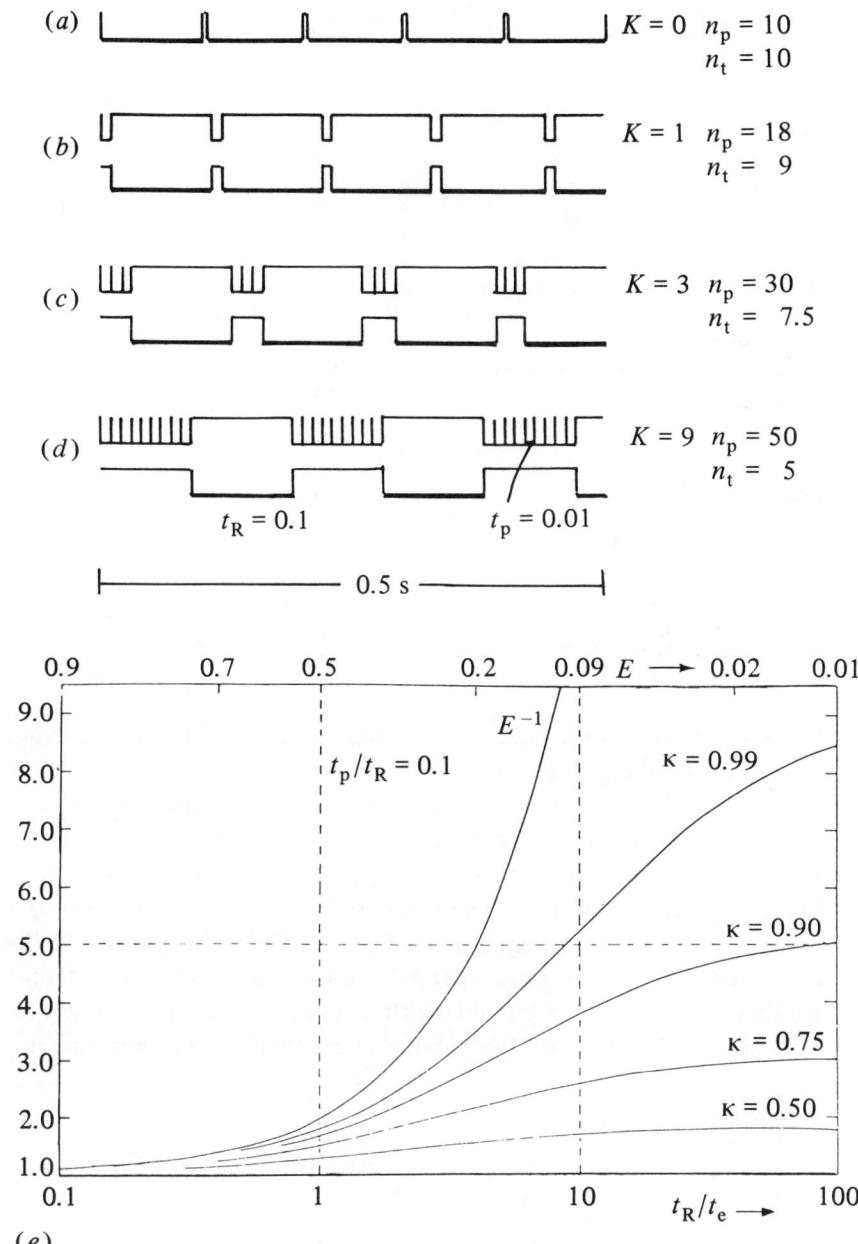

Fig. 1.7 Gain factor obtained by using a second level trigger: t_e is the average waiting time for an event, t_R is the recording time (≈ 0.1 s), t_p is the processing time for the second stage trigger (here $t_p/t_R = 0.1$, $t_p \approx 0.01$ s). Parts (a)–(d) indicate in which way the number of processed events increases for rejection factors $K = 1$ to $K = 9$. In (e) κ is the rejection rate of the second stage trigger. If events occur at a rate of $t_e = 0.001$ s ($t_R/t_e = 100$) and if 90% of the triggers are rejected by the second level processor the gain factor is 5.

Table 1.2. *Influence of fast second stage trigger processors on trigger efficiency, dead time and maximum gain.*

Rejection factor K	0	1	3	9	99
Rejection rate $\kappa = K/(K+1)$	0	0.5	0.75	0.90	0.99
Trigger Efficiency $E[\%]$	0.99	1.8	3.0	5.0	8.4
Number of recorded triggers n_t per sec	9.9	8.9	7.5	5.0	0.84
Number of aborted triggers n_a per sec	0	8.9	22.4	45.0	83.2
Number of processed triggers n_p per sec	9.9	17.8	29.9	50	84.0
Gain factor G	1	1.80	3.02	50.1	84.9
Reduced dead time $t_{R'}$ [s] per trigger	0.1	0.055	0.033	0.019	0.011
Maximum gain G_{max}	1	1.82	3.08	5.26	9.17

The number of processed events (recorded plus aborted) is

$$n'_{t'} = (1 + K)n_t$$

The 'reduced' dead time caused by a second stage veto is

$$t'_R = \frac{t_R + K t_p}{1 + K} \qquad (1.19)$$

Fig. 1.7 summarizes the effect of a second stage veto for different trigger rates t_R/t_e and the gain in sensitivity for various rejection rates $\kappa = K/(K+1)$. In this figure we assume that a second stage processor is ten times faster than the recording time: $t_p/t_R = 0.1$, e.g. $t_r = 100$ ms, $t_p = 10$ ms. If the average waiting time of an event is $t_e = 1$ ms one gets the results shown in Table 1.2. The recording time is several orders of magnitudes slower than the time between input triggers at the first stage. With fast processors (fast compared to the recording time), which select good events and reject background events, it is possible with a sufficient algorithm to decrease dead time and to increase the number of events processed per time unit.

1.3 Queuing theory, queuing simulation, and reliability

1.3.1 Queuing theory

The results of queuing theory can be used to answer the following questions which can be illustrated by Fig. 1.8 (Morse 1958; Allen 1978; Margenau and Murphy 1964). The events occur independently of each other and enter the system at a rate $\lambda = 5$ events per second. These events are handled by computers with different rates and different buffer lengths.

1.3 Queuing theory and reliability

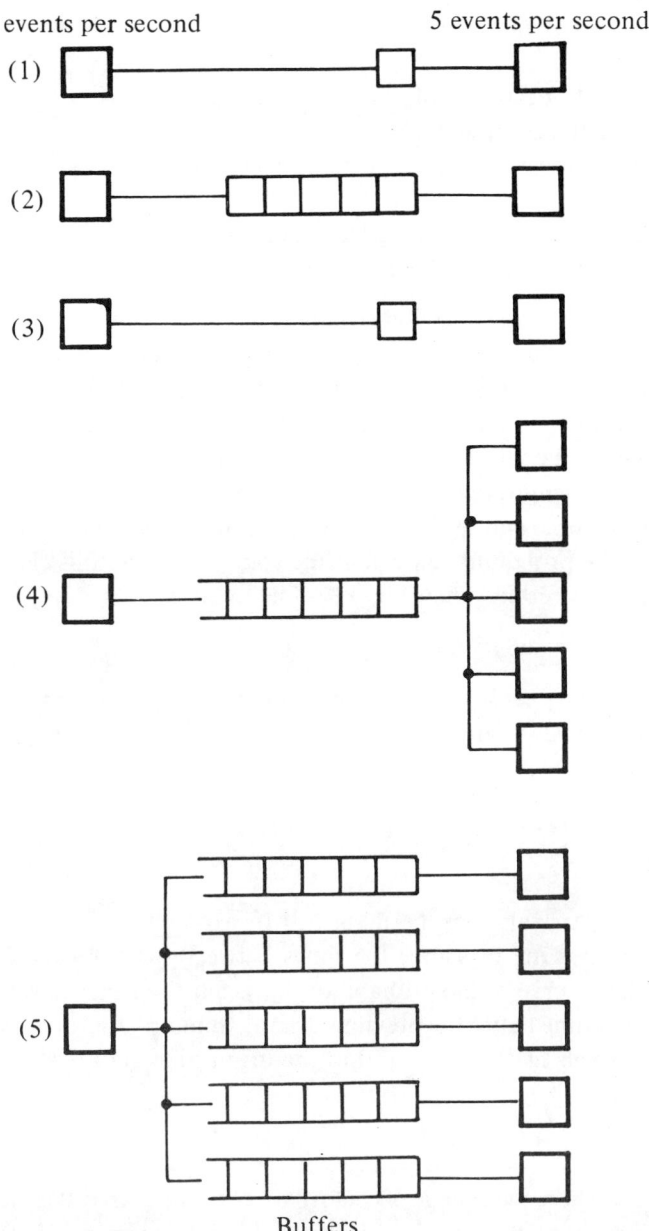

Fig. 1.8 Examples of queuing theory. Which system has little dead time and short waiting queues? Data enter from left at a rate of 5 events per second and are processed by one or several processors.

(1) What is the dead time of a system with an input rate λ and a processing rate $\mu = \lambda$?
(2) What is the dead time of a system as before but with five event buffers in front of the computer?
(3) How fast must a computer with a single buffer be to get the same dead time as in (2)?
(4) What is the average queue length of system (4)?
(5) What is the average queue length of system (5)?

The answers to these questions will be given at the end of this subsection. The time a computer or a variable flow trigger processor needs to handle an event depends on the event's complexity i.e. the number of tracks, the magnitude of the energy deposited in clusters, the background hits, etc. If one measures the time each event needs and arranges this sequence in order of decreasing length one can plot the number of events which need longer than a given time and by dividing by the total number of cases one gets the curve for the probability $S_0(t)$ that the computation in this class will take longer than a certain time (Fig. 1.9). The quantity

$$s(t) = - dS_0(t)/dt \tag{1.20}$$

is the *probability density* that an operation is completed in time t. It is a rate since its dimensions are probability divided by time. The average computing time is

$$T = \int_0^\infty t s_0(t)\, dt \tag{1.21}$$

Irregular arrivals may be described in a manner quite analogous to service times. One measures the times between arrivals, and from these constructs a curve of the probability $A_0(t)$ that the next arrival will come later than a time t after the previous arrival. Similarly one defines the mean rate of arrivals as the reciprocal of the mean time between arrivals

$$T_a = 1/\lambda = \int_0^\infty A_0(t)\, dt \tag{1.22}$$

In the case where the change of occurrence of the next arrival is independent of the time since the last arrival, the probability corresponds to

$$A_0(t) = e^{-\lambda t}$$

The probability density $a(t)$ which defines the chance that the next arrival

1.3 Queuing theory and reliability

comes between t and dt after the previous one is $a(t) dt$:

$$a(t) = -dA_0(t)/dt \tag{1.23}$$

We sometimes wish to know the probability that n arrivals will occur within an interval of duration t. This probability is

$$A_n(t) = \int_0^t a(x) A_{n-1}(t-x) dx \tag{1.24}$$

For exponential arrivals we have

$$a(t) = \lambda e^{-\lambda t}, \quad A_0(t) = e^{-\lambda t}$$

$$A_1(t) = \int_0^t \lambda e^{-\lambda x} e^{-\lambda(t-x)} dx = \lambda e^{-\lambda t} \int_0^t dx = \lambda t e^{-\lambda t} \tag{1.25}$$

$$A_2(t) = \int_0^t \lambda e^{-\lambda x} \lambda x e^{-\lambda(t-x)} dx = \lambda^2 e^{-\lambda t} \int_0^t x\, dx = \lambda^2 t^2 e^{-\lambda t}/2 \tag{1.26}$$

$$a_n(t) = (\lambda t)^n e^{-\lambda t}/n! \quad \text{(Poisson distribution)} \tag{1.27}$$

We will now discuss a simple system with a single exponential service channel with a queue of maximum allowed length $N-1$ with Poisson

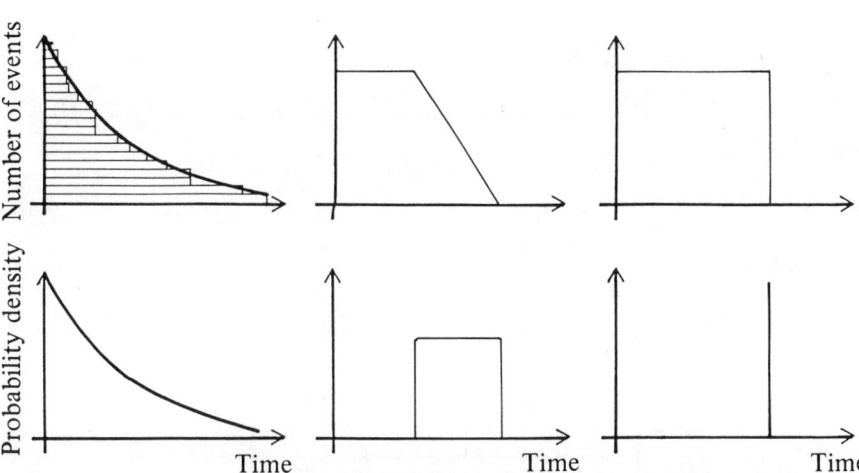

Fig. 1.9 Distributions of service times. The right hand distribution represents a system with constant service times the left hand one represents a system with an exponential service time while the middle one needs fixed minimum and maximum service times. The lower diagrams show the probability densities that an operation is completed in time t.

distributed arrivals. The mean arrival rate should be $\lambda = 1/\tau_a$ and the mean service rate $\mu = 1/\tau_s$. The various states of this system can be characterized by the total number of units in the system, the number in service plus the number in the queue. We can expect that the system will settle down to a statistical steady state so that for example the numbers of units in the queue is independent of time. To show the principle of how one can solve this problem we write the conditions which are required to find one event in the queue in the time interval $(t, t + dt)$ (Fig. 1.10):

(1) At time t there was one event in the system. No new event came in, no event left the system in the time interval dt. The probability is

$$(1 - \lambda\, dt)(1 - \mu\, dt) P_1(t)$$

(2) At time t there was one event in the system. One event entered and one event left the system. The probability is

$$\lambda\, dt\, \mu\, dt\, P_1(t).$$

(3) At time t there was no event in the system. One event entered, no event left the system. The probability is

$$\lambda\, dt(1 - \mu\, dt) P_0(t)$$

(4) At time t there were two events in the system. One event left. The probability is

$$(1 - \lambda\, dt)\mu\, dt\, P_2(t).$$

The probability $P_1(t + dt)$ is the sum of all the probabilities given above:

$$P_1(t + dt) = P_1(t) + dP_1 = P_1(t) - (\lambda + \mu)\, dt\, P_1(t)$$
$$+ \lambda\, dt\, P_0(t) + \mu\, dt\, P_2(t) + \text{high order terms} \qquad (1.28)$$

This leads to

$$dP_n = [\lambda P_{n-1} + \mu P_{n+1} - (\lambda + \mu) P_n]\, dt \qquad (1.29)$$

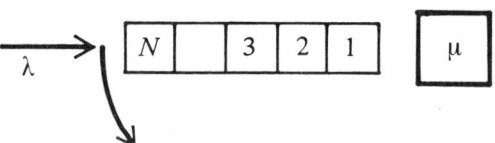

Fig. 1.10 Queue for a system with a finite queue length. If the queue is full the system will cause dead time.

1.3 Queuing theory and reliability

If this is equal to zero then P_n will be independent of time. This gives

$$\mu P_{n+1} + \lambda P_{n-1} - (\lambda + \mu)P_n = 0, \quad \text{for } n > 1 \tag{1.30}$$

For $n = 0$ this has a special form because there can be no P_{-1}. Condition (1) needs to be modified and condition (3) is not possible. This leads to

$$\lambda P_0 - \mu P_1 = 0 \tag{1.31}$$

This set of equations can easily be solved. If we express all Ps in terms of P_0 we have

$$P_n = (\lambda/\mu)^n P_0 = \rho^n P_0 \tag{1.32}$$

$$\rho = \lambda/\mu = \tau_s/\tau_a \tag{1.33}$$

We now want to compute the dead time of a system with N buffers or an upper limit of the queue. The equations hold for n from 0 to $N - 1$. For $n = N$ one gets

$$\lambda P_{N-1} - \mu P_N = 0 \tag{1.34}$$

and the solution $P_n = \rho^n P_0$ holds for $0 \leq n \leq N$. We can derive P_0 by adding all P_n and requiring that the sum of all Ps is unity:

$$1 = \sum_{n=0}^{N} P_n = P_0(1 + \rho + \cdots + \rho^N) \tag{1.35}$$

Using

$$(1 + \rho + \cdots + \rho^N)(1 - \rho) = 1 - \rho^{N+1}$$

gives

$$P_0 = \frac{(1-\rho)}{(1-\rho^{N+1})}$$

$$P_n = \frac{(1-\rho)\rho^n}{(1-\rho^{N+1})} \tag{1.36}$$

The system cannot accept more events and will produce dead time if the queue is full. The dead time is therefore

$$\tau = \frac{(1-\rho)\rho^N}{(1-\rho^{N+1})}, \quad \rho = \lambda/\mu \neq 1, \quad N = \text{number of buffers} \tag{1.37}$$

$$\tau = \frac{(1)}{(N+1)}, \quad \rho = 1 \tag{1.38}$$

The dead time for different buffer lengths and processor speeds is given in Fig. 1.11. If one waits long enough there will be a steady state situation with an average *queue length* in the system of

$$L = \sum_{n=0}^{N} nP_n = \rho \frac{1 - (N+1)\rho^N + N\rho^{N+1}}{(1-\rho)(1-\rho^{N+1})} \qquad (1.39)$$

which reduces to

$$L = \begin{cases} \rho + \rho^2 & \rho \ll 1 \\ N/2 + N(n+2)(\rho-1)/12 & \rho \to 1 \\ N - (1/\rho) & \rho \gg 1 \end{cases} \qquad (1.40)$$

For $\rho \geq 1$ the solutions are not stable. The queue is increasing to infinity. From the queue length one can compute the average waiting time which is $W = L/\lambda$.

We will now discuss systems with an infinite number of buffers (infinite queue length). Each incoming event will enter the queue, there is no dead

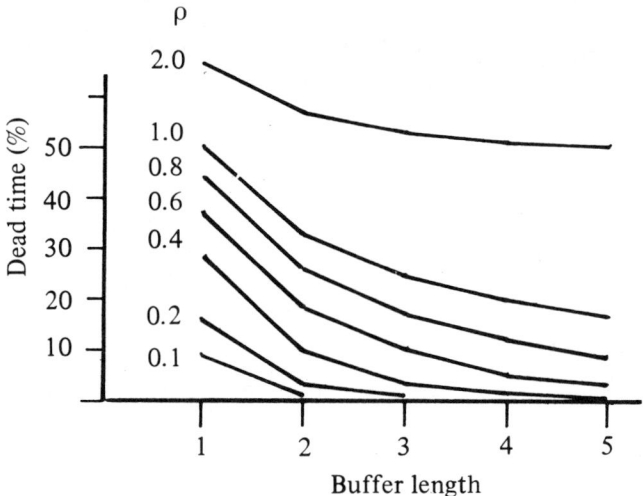

Fig. 1.11 Dead time as a function of buffer length and processor speed. ρ is the ratio of service rate to arrival rate; $\rho = 1$ means arrival time equals service time. In this case the dead time drops from 50% to 16.6% if five buffers are used.

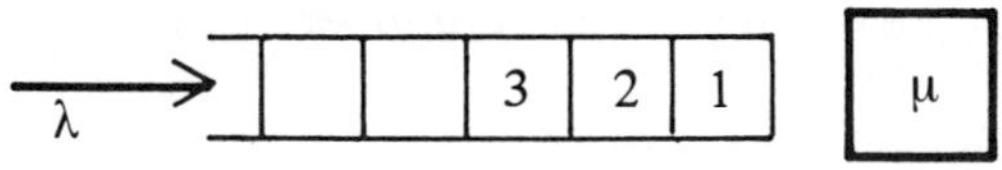

Fig. 1.12 Infinite queues. Each event enters the queue.

1.3 Queuing theory and reliability

time (Fig. 1.12). For an infinite queue the steady state solutions are

$$P_n = (1-\rho)\rho^n \tag{1.41}$$
$$L = \rho/(1-\rho)$$

A system with $\rho = 1$, i.e. arrival rate = service rate will not give a stable solution. The queue length diverges to infinity. We will now discuss systems with one queue and several computers, let us say M computers (Fig. 1.13). Condition (4) now changes. If there are two events in the system and one event leaves we get the probability

$$(1-\lambda)\,dt\,2\mu\,dt\,P_2(t)$$

because two events are handled by two computers (if $M \geqslant 2$). The equations of detailed balance for steady state operation are therefore

$$\left. \begin{array}{l} \mu P_1 - \lambda P_0 = 0 \\ (n+1)\mu P_{n+1} + \lambda P_{n-1} - (n\mu + \lambda)P_n = 0, \quad 0 < n < M \\ M\mu P_{n+1} + \lambda P_{n-1} - (M\mu + \lambda)P_n = 0, \quad M \leqslant n \end{array} \right\} \tag{1.42}$$

For a system with a maximum queue length N the equation for $n = N$ is

$$\lambda P_{N-1} - (M\mu + \lambda)P_N = 0 \tag{1.43}$$

The solution for this system is

$$\left. \begin{array}{l} P_n = (M\rho)^n P_0/n!, \quad 0 \leqslant n < M \\ P_n = M^M \rho P_0/M!, \quad M \leqslant n \leqslant N \\ \rho = \lambda/M\mu \end{array} \right\} \tag{1.44}$$

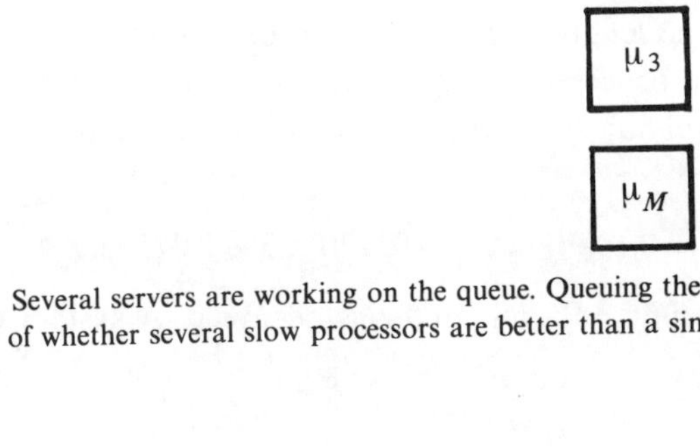

Fig. 1.13 Several servers are working on the queue. Queuing theory should answer the question of whether several slow processors are better than a single fast processor.

An experiment with five processors of processing rate μ and no extra buffers will then produce the following dead time:

$$M = N = 5$$
$$P_1 = M\rho P_0/1, \quad P_2 = (M\rho)^2 P_0/2!, \quad P_3 = (M\rho)^3 P_0/3!$$
$$P_4 = (M\rho)^4 P_0/4!, \quad P_5 = (M\rho)^5 P_0/5!$$

Normalizing

$$1 = \sum P_n = P_0[120 + 120M\rho + 60(M\rho)^2 + 20(M\rho)^3 + 5(M\rho)^4 + (M\rho)^5]/120$$

leads to dead time $P_5 = 5^5 \rho^5 P_0/5!$. For $\rho = 0.6$ this gives 8.05%. For a system with infinite queue length P_0 is given by

$$P_0 = \frac{1}{\sum_{n=0}^{M-1}[(M\rho)^n/n! + (M\rho)^M/M!(1-\rho)]} \tag{1.45}$$

The queue lengths for different systems are shown in Fig. 1.14.

We can now answer the questions from the beginning of this subsection.

(1) For a system with one buffer and an arrival time which equals processing time i.e. $N = 1$ and $\rho = \lambda$ the dead time is

$$\tau = 1/(N+1) = 50\%$$

(2) With five buffers the dead time drops to

$$\tau = 1/(N+1) = 1/6 = 16.6\%$$

By using just derandomization the dead time drops by a factor of 3!

(3) A fast processor with a single buffer should produce 16.6% dead time.

$$\tau = 16.6\% = (1-\rho)\rho/(1-\rho^2) = \rho/(1+\rho) \qquad \rho = 0.2$$

The processor must be five times faster than a processor in case (2). But buffers are much cheaper than fast processors.

(4) The average queue length is $L = \sum n P_n$. In our example we have five processors with a processing power which is as high as the incoming rate, $\rho = 0.2$. Applying the formula mentioned above for P_0 and P_n with $M = 5$ gives

$$L = 0.36782 + 0.18391 \times 2 + 0.06130 \times 3 + 0.01533 \times 4 = 0.98084$$

For most of the time only one buffer is occupied. A system with five

1.3 Queuing theory and reliability

buffers will generate a dead time of ($P_0 = 120/(120 + 120 \times 1 + 60 \times 1^2 + 20 \times 1^3 + 5 \times 1^4 + 1^5) = 0.368$):

$$\tau = 1^5 \times P_0/5! = 0.31\%$$

(5) The average queue length of a single queue is $L = \rho/(1 - \rho)$. The input rate is divided equally among five queues. Each processor is then five

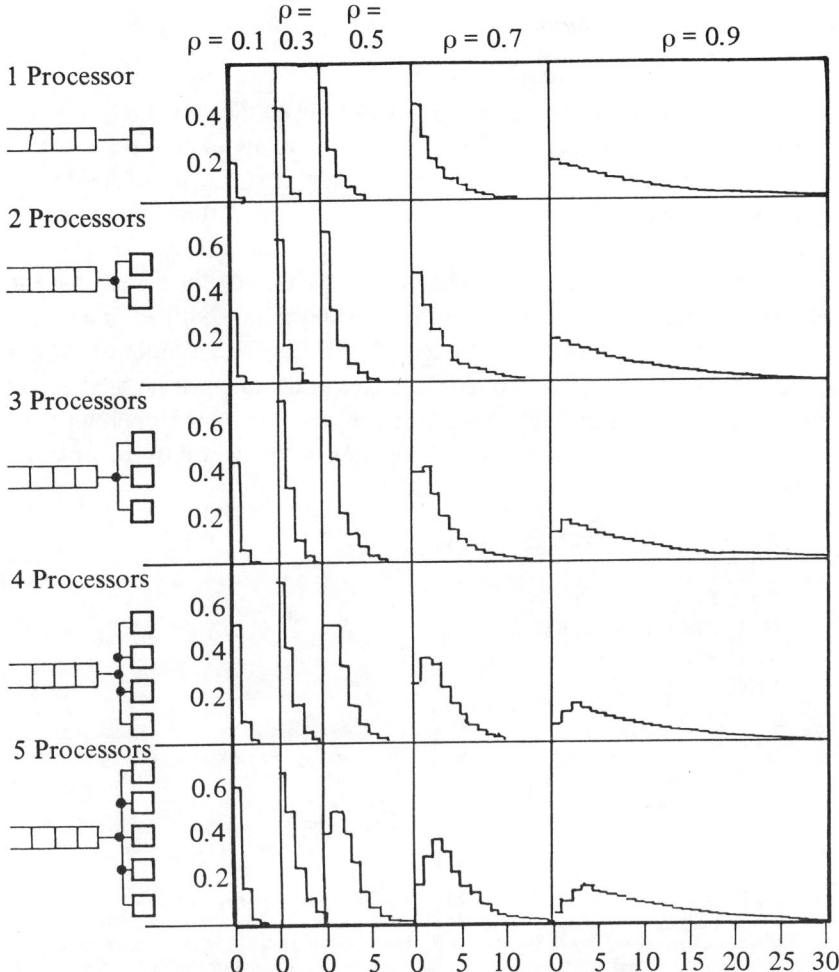

Fig. 1.14 Queue lengths for different systems. One to five processors are used to process data. From left to right the processors are slower or the input rate gets higher. One can imagine that the queue length goes to infinity if service rate is equal to the arrival rate.

times faster than the incoming rate, $\rho = 0.2$. The queue length in each queue is $L = 0.2/0.8 = 0.25$. Multiplying this by 5 gives $L = 1.25$. The queue length is a bit longer than in case (4) but the dead time is smaller because we have more buffers in the system. Dead time occurs if one queue is full.

$$\tau = (1 - \rho)\rho^5/(1 - \rho^6) = 0.8 \times 0.2^5/(1 - 0.2^6) = 0.026\%$$

1.3.2 Queuing simulation

It is not easy to derive a simple formula to describe the behaviour of queues in complicated readout systems with several levels of triggers and processors with a wide range of speed. This systems are often simulated in a computer to optimize speed, buffer length, and cost (Dewdney 1985).

We have discussed in the previous subsection a system with average arrival (λ) and service (μ) rates. We will now describe a simple program to generate exponential arrivals. The method can be illustrated by the following picture: Suppose we have a wall of 100 m long with an opening of 5 m. In fixed time interval, say 1 s, somebody reaches a random position along the wall. If he or she happens to arrive at the opening he or she may pass through. The time difference between arrivals at the opening then follows an exponential function and depends on the width of the opening (Fig. 1.15).

```
    INTEGER FUNCTION ITDIST(W)
    ITDIST = 0
  2 ITDIST = ITDIST + 1
    IF(RNDM(ITDIST).GT.W)GOTO 2
    RETURN
    END
```

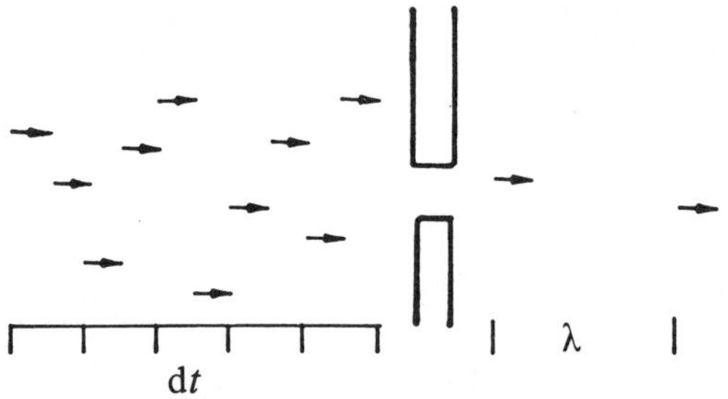

Fig. 1.15 Simulation of exponential arrivals. People arrive in fixed time intervals at a wall with an opening w. The time difference between people passing through the opening follows an exponential function.

1.3 Queuing theory and reliability

Another way of producing noninteger arrivals is the *importance sampling method*. We do not want to weight equally distributed random numbers with an exponential weight. This would give some events with a high weight because the exponential function varies greatly. We want to produce arrivals which should behave like an exponential function without needing to weight them

$$da(t)/dt = A e^{-\lambda t}$$

The time values should appear in the interval $t = 0$ to $t = T$. The random number generator of a computer produces a uniform distribution. We are therefore looking for a uniform variable z:

$$da(t) = A e^{-\lambda t}(dt/dz)dz$$

Then

$$e^{-\lambda t}dt(z)/dz = \text{const} = b \qquad (1.46)$$

z is a random number

$$dz/dt = (1/b)e^{-\lambda t}$$
$$z = C - (1/b\lambda)e^{-\lambda t}$$

from which we can deduce t:

$$t = -(1/\lambda)\ln[(z-C)(-\lambda b)] \qquad (1.47)$$

The integration constant C and the constant b are adjusted in such a way that $z = 0 \rightarrow t = 0$ and $z = 1 \rightarrow t = T$.

$$\left.\begin{array}{ll} -\lambda b(0-C) = 1 & \rightarrow C = 1/(\lambda b) \\ -(1/\lambda)\ln[-\lambda b(1-C)] = T & \rightarrow b = (e^{-\lambda T}-1)/(-\lambda) \end{array}\right\} \qquad (1.48)$$

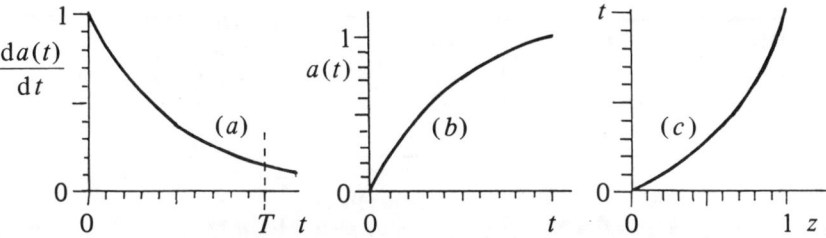

Fig. 1.16 The method of importance sampling. The left hand function (a), should be generated. The function must be integrated (b), and than inverted (c). If the constants are normalized correctly a uniformly distributed random number between 0 and 1 will generate events which follow the original left hand function.

The principle of the method is shown in Fig. 1.16.

```
FUNCTION TDIST(ALAM, AMAX)
B = (EXP(-ALAM*TMAX) - 1)/(-ALAM)
C = 1/(ALAM*TMAX)
TDIST = -ALOG((RNDM(TDIST) - C)*(-ALAM*B))/ALAM
RETURN
END
```

With these routines we can now construct and simulate a queue. TIME is the running time starting at TIME = 0. TARR is the arrival time and TSERV is the service time of the running process. LENG is the length of the queue. The flow chart of a program with $\lambda = 5$ and $\mu = 10$ which simulates a system with one input, one output, and an infinite number of buffers is shown in Fig. 1.17. With this simulation program one can simulate readout systems with several buffers, processors with different speeds, and trigger systems with several levels.

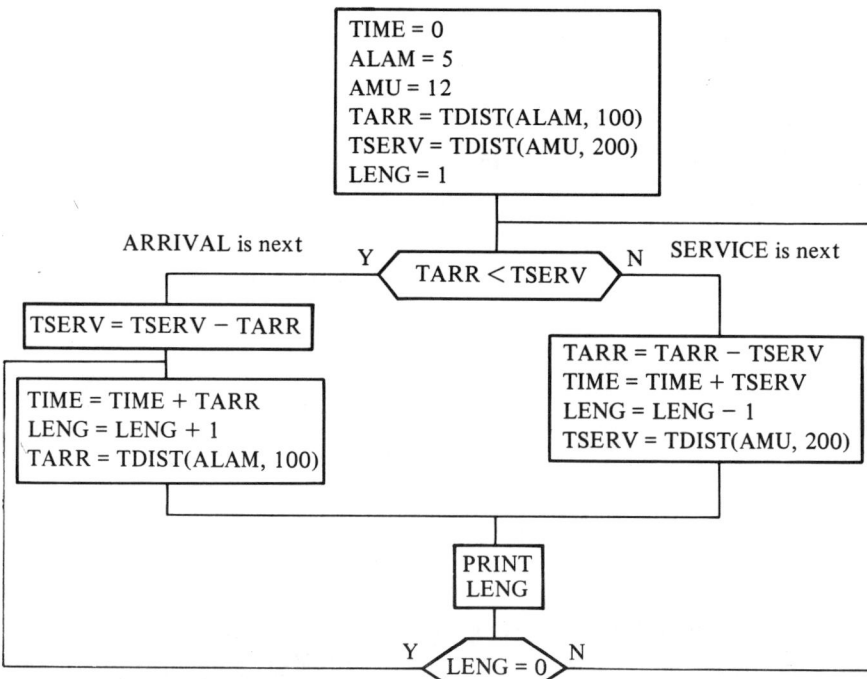

Fig. 1.17 Program to simulate a queue.

1.3.3 Reliability theory

The results from *reliability theory* can be used to find an answer to the following questions (Schorr 1974; Lala 1985).

(1) We have computers with given *Mean Times Between Failure* (MTBF). For a given task it is necessary to operate two computers in series which will reduce the MTBF or decrease the reliability. How many computers must be placed in parallel so that the entire system is as reliable as a single system (Fig. 1.18)?
(2) How many computers must be used under the conditions shown in Fig. 1.19? Here the system can operate if (AB) or (AD) or (CB) or (CD) are operational.
(3) How many computers must be used under the conditions shown in Fig. 1.20? Here the stand-by computers are only switched on with an ideal switch in zero time if one of the operating computers fail. We assume that stand-by computers do not age.

The answers to these questions will be given at the end of this subsection.

But what is reliability? The probability $R(t)$ that an unrepairable system performs a specific function without failure under certain conditions for a specific time of length t is called the reliability or reliability function of the

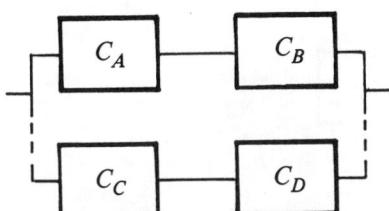

Fig. 1.18 Example of realiable theory. How many computers must run in a *series-to-parallel interconnection* to get the same reliability as that of a single computer?

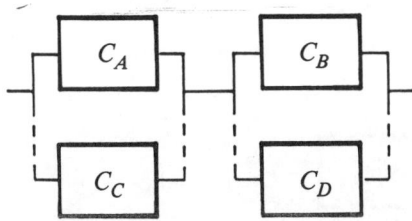

Fig. 1.19 Example of reliability theory. How many computers must run in a *parallel-to-series* interconnection to get the same reliability as that of a single computer?

system. Unrepairable means that failures of the system during operation lead in practice, to unrepairable consequences. The probability that a system operates correctly shortly after switching on is nearly 1 and that it still works after an infinite amount of time is 0. This leads to $R(0) = 1$ and $R(\infty) = 0$. An example for $R(t)$ could be $R(t) = e^{-\lambda t}$.

The failure distribution $F(t)$ describes the probability of failure of a system and is defined by

$$F(t) = 1 - R(t) \tag{1.49}$$

For an exponential reliability function $F(t)$ is then $F(t) = 1 - e^{-\lambda t}$. The failure density function $f(t) = F'(t)$ describes the lifetime of a system. The failure rate function

$$\lambda(t) = f(t)/R(t) \tag{1.50}$$

describes the probability $\lambda(t)dt$ that the system, having reached the age t, will fail during the interval $(t, t + dt)$. For an exponential reliability function $\lambda(t)$ is a constant. Very often one does not know the reliability function but one can say something about the average lifetime of the system. One determines the MTBF which is defined as the expected lifetime of the system.

$$\text{MTBF} = \int_0^\infty t\,dF(t) = \int_0^\infty tf(t)dt = \int_0^\infty R(t)dt \tag{1.51}$$

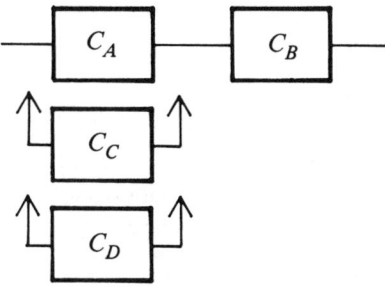

Fig. 1.20 Example of reliability theory. How many computers must stand by to get the same reliability as that of a single computer?

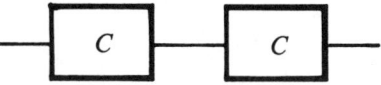

Fig. 1.21 Computers operating in series. If one computer goes down the whole system is down. How does the reliability change?

1.3 Queuing theory and reliability

We will now discuss how the reliability changes if two computers operate logically in series (Fig. 1.21). That means, if one computer fails the whole system goes down. T_i is the lifetime of the computer C_i. We can then say that $R(t)$ is the probability that $T > t$ or

$$R(t) = P(T > t)$$

For a coupled system we require that

$$R(t) = P(T_1 > t, T_2 > t)$$

As the individual lifetimes are independent it follows that

$$R(t) = \prod_{i=1}^{2} P(T_i > t) = \prod_{i=1}^{2} R_i(t) \tag{1.52}$$

The failure rate function of the series connection is then

$$\lambda(t) = \sum_{i=1}^{2} \lambda_i(t)$$

For constant component failure rates the failure rate of the system sums as

$$\lambda = \sum \lambda_i$$

If both computers have the same reliability we get

$$R(t_i, n\lambda_0) = e^{-n\lambda_0 t}$$
$$\text{MTBF} = 1/\sum \lambda_i = (1/n)\text{MTBF}_{\text{component}} \tag{1.53}$$

Two computers in series with the same MTBF will have only half the MTBF of a single computer.

In systems with parallel computers one may distinguish between *hot parallel systems*, in which all components are in the operating state, and *cold parallel connections* or stand-by components, which are available to replace an operating component (Fig. 1.22). Let T be the lifetime of the

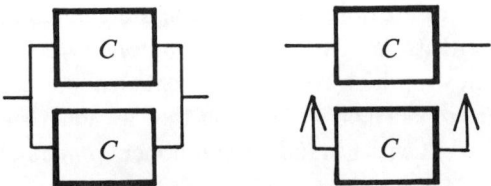

Fig. 1.22 Computers operating in parallel. One can distinguish cold and hot redundancy. In cold redundancy a stand-by computer takes over if the operating computer goes down.

system. Then we have

$$R(t) = 1 - P(T < t) = 1 - P(T_1 < t, \ldots, T_n < t)$$

The system fails if all the components have failed.

$$R(t) = 1 - \prod_{i=1}^{n} P(T_i < t) = 1 - \prod_{i=1}^{n} [1 - P(T_i > t)] \qquad (1.54)$$

A system of computers $C_1, \ldots, C_n$ in hot parallel connection has the reliability

$$R(t) = 1 - \prod_{i=1}^{n} [1 - R_i(t)]$$

For a system with computers of the same reliability the entire reliability improves to

$$R(t) = R(T; n\lambda_0) = 1 - (1 - e^{-\lambda t})^n$$

The failure rate $\lambda(t)$ is no longer a constant but

$$\lambda(t, n\lambda_0) = \frac{n\lambda_0}{\sum_{j=1}^{n-1} (1 - e^{-\lambda_0 t})^{-j}} \qquad (1.55)$$

This function increases with time and converges to 1 for $t \to \infty$. The MTBF changes to

$$\text{MTBF} = (\sum 1/k) \text{MTBF}_{\text{component}}$$

Two parallel computers increase the MTBF to

$$\text{MTBF}_2 = 1.5 \times \text{MTBF}_{\text{component}}$$

or only by 50%. But the time interval at 90% reliability increases by a factor of 4 (Fig. 1.23).

We will now discuss the situation of stand-by computers (Fig. 1.22) with cold connection. The components $C_1, \ldots, C_n$ are called cold redundant or stand-by components if they are not in the operating state but are available to replace failing components. We assume that components do not age as long as they are on stand-by which is true for mechanical devices but not necessarily for electronics. Further we assume that an ideal switch can replace the failed component with a stand-by component immediately after failure.

The system should consist of just one operating computer C_0 for which

1.3 Queuing theory and reliability

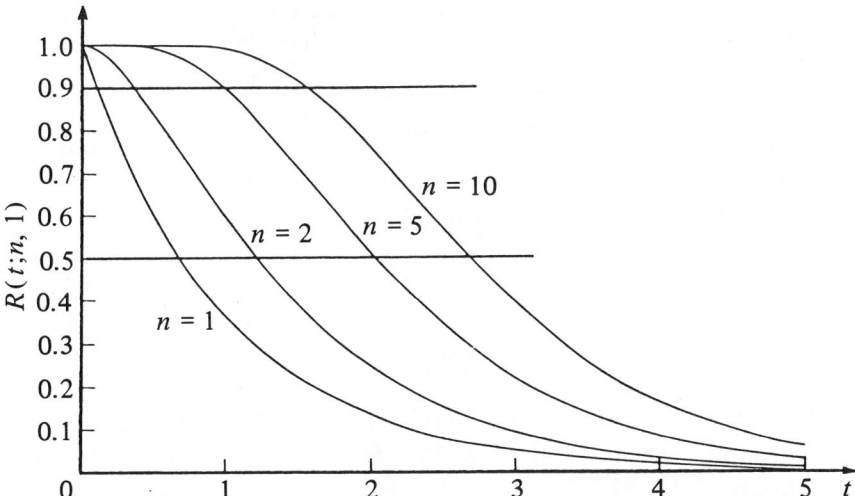

Fig. 1.23 Reliability of coupled systems with hot connection. The components have equal failure rates. On average one does not gain very much for long periods of time, but the time with a reliability of 90% increases by a big fraction.

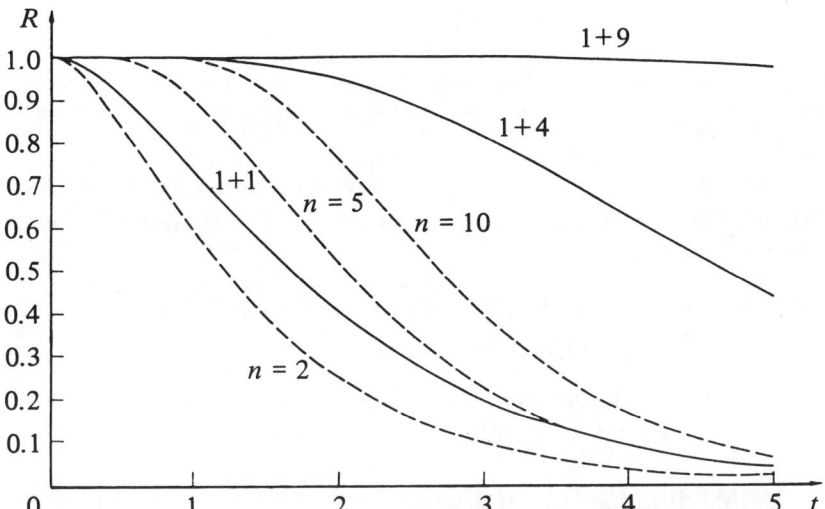

Fig. 1.24 Reliability for systems with cold connections. The reliability of systems with cold connections or stand-by computers is higher than for systems with hot connections. Stand-by computers do not age. The dashed lines show the corresponding systems with hot connections.

all the stand-by computers are spare parts. If C_0 fails C_1 will replace it and so on. The system fails if all the components have failed. This leads to

$$\text{MTBF} = \sum \text{MTBF}_{\text{component}} \tag{1.56}$$

In a system of m operating components in a series connection and n computers as stand-bys we get a failure after $n + 1$ components have failed. For equal components with constant failure rates one gets

$$\text{MTBF} = [(n + 1)/m]\text{MTBF}_{\text{component}} \tag{1.57}$$

Generally speaking a system with cold redundancies is much more reliable than those with hot redundant connections (Fig. 1.24). This comes from the fact that stand-by components do not age.

We can now answer the questions at the beginning of this subsection. In the first example the MTBF for two serial computers is $\text{MTBF}_{\text{serial}} = 0.5 \, \text{MTBF}_{\text{component}}$. The MTBF of the whole system of several computers connected in series by a hot connection is

$$\text{MTBF}_{\text{system}} = \left(\frac{1}{2} \sum_{k=1}^{n} 1/k \right) \text{MTBF}_{\text{component}} \tag{1.58}$$

For which n is $\text{MTBF}_{\text{system}} \geqslant \text{MTBF}_{\text{component}}$?

$$\frac{1}{2} \sum_{k}^{n} 1/k \geqslant 1; \; (1 + \tfrac{1}{2} + \tfrac{1}{3} + \tfrac{1}{4})/2 \geqslant 1$$

We need eight (four parallel branches with two computers in each branch) computers to get an MTBF for the whole system which is at least as high as that for a single computer.

In the second example a four-computer system can operate if (AB) or (AD) or (CB) or (CD) work. The reliability of the parallel-to-series interconnection is

$$R_{\text{system}} = [1 - (1 - R_A)(1 - R_C)][1 - (1 - R_B)(1 - R_D)]$$
$$= R^4 - 4R^3 + 4R^2$$

for computers with the same reliability $R = R_A = R_B = R_C = R_D$. The MTBF for exponential reliability $R(t) = e^{-\lambda t}$ then becomes

$$\text{MTBF}_{\text{system}} = \int_0^\infty R(t)_{\text{system}} dt \tag{1.59}$$

$$= \frac{4}{2\lambda} - \frac{4}{3\lambda} + \frac{1}{4\lambda} = \frac{11}{12\lambda}$$

which is nearly the MTBF of a single system. Using six computers (= two series of three computers), MTBF = (73/60) MTBF$_{component}$. With six computers the MTBF of the whole system is at least as high as the MTBF of a single computer.

In the third example, for a system with two computers in series we need only one spare computer to get the same reliability as that of a single computer, $n = 1$, $m = 2$:

$$\text{MTBF}_{system} = (2/2)\text{MTBF}_{component} = \text{MTBF}_{component} \qquad (1.60)$$

We need only three computers for cold redundant connection compared to eight computers for hot redundant connection and six computers for parallel-to-series connection.

1.4 Classifications of triggers

Classification of triggers based on physical processes

When designing a trigger for an experiment one has to search for physical criteria to select event candidates from background processes. After having defined the trigger conditions one has to check seriously with Monte Carlo studies the way in which a trigger can influence the results of an experiment.

In accelerators or storage rings the beams are very small in diameter and have a fixed timing structure. In an accelerator beam particles are extracted from the accelerator at the end of the acceleration cycle. The trigger electronics should therefore only accept events if they appear within a short time window ('*spill gate*'). At colliders the beam signal is generated by a pick-up coil at each bunch crossing.

The various final states of an interaction define which class of trigger may be sufficient. Some of these classes are:

(1) Track multiplicity of event.
(2) Direction of particles.
(3) Deflection or curvature of particles to measure momentum.
(4) Coplanarity of event.
(5) Type of particle.
(6) Deposited energy in all or part of the detector to measure total or transverse energy.
(7) Missing energy.
(8) Invariant mass.

(9) Interaction point of event ('vertex') or secondary interaction ('kinks', 'V^0').

Classification of triggers based on realization in electronics

In the classical spectrometer experiment using fixed-target machines simple coincidences are used to count the number of scattered particles and to define the particle flux in the beam. Their ratio multiplied by the target constants and correction factors determines the cross section. The trigger decision is available within 100 ns. Data are then either recorded by a computer, or scalers are incremented to measure total rates. Thanks to technological advances in the field of microelectronics new and powerful tools are available to upgrade the triggering system. The advent of microprocessors and the increasing availability of special integrated circuits with a better cost/performance ratio has allowed the implementation of sophisticated multilevel trigger schemes in the event selection. This leads to wide ranges in time from 100 ns to 10 ms and in complexity from 2 coincidences to programs with 1000 statements and 15 000 cycles. Also the prices range from 1000 Swiss franks for a coincidence or a microprocessor to 100 000 Swiss franks for special hardware devices filling an entire rack of electronics.

Because of this variety one tries to group trigger systems into various classes (Conetti 1984):

(1) Fixed flow triggers.
(2) Variable flow triggers.
(3) Logical triggers.
(4) Arithmetical triggers.
(5) Program driven processors.
(6) Data driven processors.

As a first step one can distinguish *fixed flow* and *variable flow triggers*. Fixed flow triggers produce their trigger signal at a given time, independent of the complexity of the event. Variable flow triggers contain counters, loops, and programs. Only a minimum and maximum time are known. Fixed flow processors can be grouped into *logical triggers* dealing with logic operations like AND and OR and *arithmetic triggers* asking for a certain pulse height above threshold or a certain number of counters. Variable flow processors are either *data driven* or *program driven*. In a *data driven processor* an operation is executed when all data needed for the execution

1.4 Classifications of triggers

are available. This is implemented by sending tokens or data ready signals. A *program* or *demand driven processor* demands an execution which requires evaluation of its arguments.

1.4.1 Trigger on event topology

Trigger on track multiplicity

The *track multiplicity* of an event depends on the available energy, the physical process, and the average energy of the emerging particles. In an experiment investigating hadronic final states, in general, the production of hadrons can be described by a *multistring model* leading to multiplicities of 30 charged particles at an energy $s^{\frac{1}{2}} = 500$ GeV and about 70 charged particles at $s^{\frac{1}{2}} = 20$ TeV (Kunszt 1987). To trigger on these events one can surround the interaction region by several scintillation counters and then trigger on the number of scintillation counters which give a hit.

New phenomena or new particles can result in events with low multiplicities. Examples of new particles are the production of Higgs H or *supersymmetric particles* like *sleptons* $\tilde{l}$ or *winos* $\tilde{W}$. These events show only a few tracks and cannot be detected by a multiplicity trigger because background events often also show high rates of low multiplicities.

Trigger on direction of particles

The *directions* of particles together with their momenta can be used to trigger on events with high momentum transfer. In fixed-target experiments or in experiments at storage rings most events are produced at low momentum transfer with particles going in the forward direction. When probing matter at small distances one is interested in events with high transverse momentum p_T. To trigger on these events one has to detect particles emerging from the interaction region perpendicular to the incoming beam particle and having high energy. This is preferably done in calorimeters.

Trigger on momentum of particles

To investigate in fixed-target experiments, the inelastic scattering of electrons, for example, the scattered electron is detected at a given angle and a given energy. This then defines a fixed momentum transfer. The detectors used are called spectrometers. They can be rotated perpendicular to the beam, around the interaction region. The momentum of the electron is measured by its *deflection* in a magnetic field. In front of and behind the magnet are multiwire chambers to reconstruct the tracks and behind the last track are scintillation counters to generate a fast trigger.

Trigger on coplanarity

Coplanar events with well-defined simple kinematics and cross sections are used for measuring the incoming flux in a separate reaction. In e^+e^- storage rings one uses the coplanar and colinear reaction $e^+e^- \rightarrow e^+e^-$ (*Bhabha scattering*) of which the cross section is known, to measure the luminosity.

Three jet events in e^+e^- reactions are coplanar if one jet is the fragment of a gluon which is radiated by quark–gluon bremsstrahlung. The fact that in this case the jets are coplanar does not mean that there are no particles outside the 'plane'.

1.4.2 Trigger on type of particle

For many experiments a topological trigger is not sufficient. Often one wants to trigger on a certain type of particle. In a search for a Higgs particle one could use the decay mode $H \rightarrow ZZ \rightarrow llll$; l is a lepton, $l = e, \mu$ or τ. Another decay of a Higgs could be $H \rightarrow WW \rightarrow l\nu l\nu$ resulting in two leptons and two escaping neutrinos. In charge exchange reactions such as $K^-p \rightarrow \bar{K}^0 n$ one would trigger on $\bar{K}^0$. A good overview of triggering on different types of particles can be found in Fernow (1986) and is summarized in Table 1.3. In addition one can use TOF measurements or record dE/dx ionization losses.

Photons and neutral pions, γ and π^0

When high energy photons interact with dense matter they create e^+e^- pairs. The electrons then radiate photons again by bremsstrahlung. A single photon thus creates many electromagnetic particles. A collection of all these particles is called an electromagnetic shower. Thin lead sheets can be used to convert the photons. The electrons created are then detected in scintillators or *liquid argon*. Another way to detect photons is the production of light in heavy transparent materials as in *lead-glass* Cherenkov counters or in NaI scintillation counters. More detailed information can be found in Section 2.5. Neutral pions are reconstructed in detectors with fine granularity by measuring the two photons and reconstructing the invariant mass.

1.4 Classifications of triggers

Table 1.3. *Triggers for type of particle*

Particle	Identification method
γ, π^0	Photons and neutral pions, γ and π^0
	Electromagnetic showers in lead-scintillator, NaI, Bismuth–Germanium-Oxide (BGO), lead glass, lead–liquid argon
e	Electromagnetic showers as for γ
	Cherenkov identification
	Transition radiator
μ	Penetration through magnetized iron
$\pi^\pm$	Cherenkov identification
$K^\pm$	Cherenkov identification
K^0	Multiplicity increase $n \to n+2$
	Tracks not from interaction point
	Charged particle veto
$p, \bar{p}$ (fast)	Cherenkov identification
p (recoil)	Range
	Solid state detector
$n, \bar{n}$	Plastic scintillator
	Liquid scintillator
	^{3}He-filled Proportional Wire Chambers (PWC)
	n–p elastic scattering
	Charged particle veto
$\Lambda \bar{\Lambda}$	Multiplicity increase $n \to n+2$
	Fast p

Electrons, e

Electrons are detected by their electromagnetic showers as photons. To distinguish them from photons a scintillation counter in front of the shower counter indicates the appearance of a charged particle. In certain momentum ranges one can use Cherenkov counters.

Another method of detecting electrons uses the phenomenon of transition radiation. A moving particle emits radiation when crossing different dielectrics. The transition radiation is emitted in a small cone around the particle's direction with typical angle of $\Theta = 1/\lambda, \lambda = E/m$. The local intensity increases linearly with λ. Here λ denotes the *Lorentz factor*. To achieve sufficient intensity, particles must cross a few hundred radiator boundaries.

Muons, μ

Muons have the ability to penetrate a considerable depth of matter before being absorbed. A muon detector consists of a hadron absorber (for example 1 m iron) and a detector for charged particles such as a multiwire chamber or a hodoscope of scintillation counters.

If one wants to measure the momentum of the muons one uses segmented *magnetized* iron. The major background comes from hadron '*punch through*' or '*shower leakage*'. It is possible for a small fraction of hadrons to reach the hodoscopes.

Charged pions, π

In hadronic events with high multiplicity it is more likely for any track to be a $\pi^{\pm}$ than anything else. One therefore assumes that the observed tracks are pions. If one has to discriminate pions from kaons or protons one uses Cherenkov counters or one measures TOF or ionization loss.

Cherenkov counters have the disadvantage that they can only mainly be used for particles travelling on the optical axis. To overcome this *Ring Imaging Cherenkov Counters* (RICH) are used for particles diverging from an interaction point. A spherical mirror focusses the cone of the Cherenkov light onto a 'ring image' on a spherical detector (see Section 2.6).

Kaons, K

To discriminate fast charged kaons from pions one uses two threshold Cherenkov counters. One counter sets a veto for pions while the second gives a positive signal for kaons which are above a certain threshold. This second counter gives also a signal for pions. One needs the first counter to flag the pion. Neutral kaons K_s decay after a short distance (of the order of centimetres) into charged pions. The multiplicity then increases by two. The background produced by Λ production is reduced in the off-line analysis in which the invariant mass of the $\pi^+\pi^-$ or the $p\pi^-$ system is computed.

Protons, p

Fast protons are identified with Cherenkov counters. Slow protons can be measured by TOF or ionization loss. Fig. 1.25 shows a set-up of several Cherenkov counters in a hadron beam to select π, K or p.

1.4.3 Trigger on deposited energy

The resolution of wire chambers in terms of track energy decreases with energy while that of calorimeters increases. Therefore, with high-energy accelerators at energies above 100 GeV calorimetry takes a prominent place. A particle hitting a shower counter deposits energy there.

1.4 Classifications of triggers

The light output of the scintillator or, generally speaking, the pulse height is a measure of the energy of the incoming particle or jet.

In a calorimeter which surrounds the interaction region hermetically one can simply trigger on total energy or transverse energy above a certain threshold. This threshold can be rather high, of the order of half the total centre-of-mass energy. In experiments with high granularity one generates different types of triggers such as the global sum for the entire detector, local sums if part of the detector gets high energy or sums which depend on the angle or transverse momentum p_T.

1.4.4 Trigger on missing energy

A trigger on missing energy is used in experiments which search for reactions with escaping neutrinos. In *ep* colliders, for example, one distinguishes neutral current and charged current events. In neutral current events the electron is scattered and a photon or a Z^0 is exchanged. In charged current events the electron couples to a neutrino and a W^-. The neutrino is not detected resulting in a large amount of missing energy. To distinguish these two processes one can either detect the electron or for a *hermetic* detector one can build a trigger for charged current events which have a large fraction of missing momentum and a small amount of total energy.

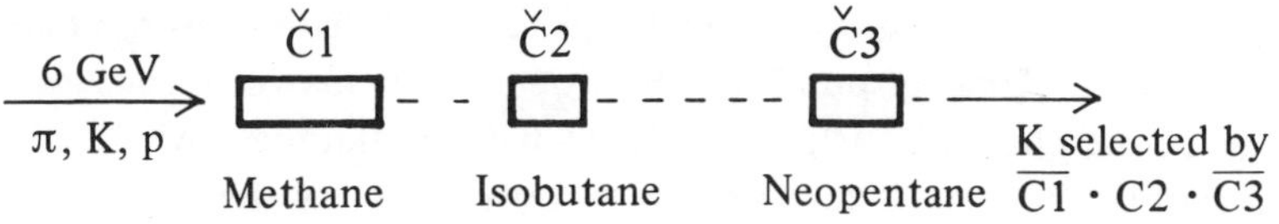

	Č1 Methane	Č2 Isobutane	Č3 Neopentane
l	1 m	0.3 m	0.5 m
p	2.1 atm	3.3 atm	1 atm
n	1.00090	1.0045	1.00175
$Thr(\pi)$	3.3 GeV	1.5 GeV	2.3 GeV
$Thr(K)$	11.7 GeV	5.2 GeV	8.4 GeV

Beam: 6 GeV π, K, p → Č1 -- Č2 ----- Č3 → K selected by $\overline{C1} \cdot C2 \cdot \overline{C3}$

Fig. 1.25 Threshold Cherenkov counter to select K. The beam at 6 GeV contains π, K and p. The experiment should study reactions with Ks which are selected by Cherenkov counters. The threshold is adjusted in such a way that the contamination of $\bar{p}$ for a K^- signal is below 1% but for a $\bar{p}$ signal the K^- contamination is 20% (l = length of the counter, p = pressure, n = refractive index, Thr = threshold).

1.4.5 Trigger on invariant mass

The gauge boson Z^0 can decay into an e^+e^- pair. e^+e^- pairs are also produced at high rates by photons in electromagnetic interactions. Triggers on invariant mass are important for spectrometer experiments looking for *di-muon events* with high mass. These events are produced at low rates. Therefore spectrometers with a large aperture are used (Greenhalgh 1984). From the trigger point of view one has to find tracks in the chambers behind the magnet and compute the invariant mass

$$M^2 = (P_1 + P_2)^2 = (E_1 + E_2, \mathbf{p}_1 + \mathbf{p}_2)^2 = E_1^2 + 2E_1 E_2 \\ + E_2^2 - \mathbf{p}_1^2 - \mathbf{p}_2^2 - 2\mathbf{p}_1 \mathbf{p}_2 (1 - \Theta^2/2) \quad (1.61)$$

$(\cos \Theta \approx 1 - \Theta^2/2)$

Neglecting the particle masses at high energies leads to $M^2 \approx \mathbf{p}_1 \mathbf{p}_2 \Theta^2$. The invariant mass of two particles at high energies is thus approximately given by $M = \Theta (\mathbf{p}_1 \mathbf{p}_2)^{\frac{1}{2}}$ where $\mathbf{p}_1$ and $\mathbf{p}_2$, are the momenta of the two particles and Θ is their opening angle at production.

1.4.6 Trigger on interaction point (vertex)

In storage rings the position of the interaction point ('vertex') is determined by the beam crossing. Transverse to the beam direction it is determined by the beam size, usually to an accuracy of better than one millimetre and in the longitudinal direction it is accurate within several centimetres. In e^+e^- colliders many background processes are due to beam–gas interactions. A good vertex trigger can be used to discriminate against this sort of background (see Section 3.4). In many storage ring experiments the charged particles are detected by chambers with wires aligned parallel to the beam line. The only way to determine the longitudinal position of outgoing tracks along these wires is with the help of *charge division* (see Subsection 1.5.9).

In fixed-target experiments a vertex trigger is necessary to study short lived particles like F or Λ_c. These particles contain the *charm quark* and are produced at currently accessible energies at a rate which is 1/1000 that of 'normal' hadrons with light quarks. The main difficulty consists in finding a very selective signature to separate the signal from the combinatorial background in exclusive decay channel mass plots. A distinct signature is the finite lifetime in the range 10^{-13}–10^{-11} s. This implies that tracks of decaying particles, when extrapolated to the primary vertex, have an average impact parameter of 30–3000 µm. The requirements for detectors to resolve secondary vertices from the decay of short lived particles are:

(1) Very good spatial resolution.
(2) Very good two-particle separation since they operate close to the interaction point.
(3) High rate capability.

1.4.7 Fixed and variable flow triggers

Fixed-flow triggers operate in a time range of 50 ns to 1 µs. Their results appear at a fixed time after the operation has been initiated. The synchronization of several trigger systems and the gating for analog signals is tedious but in principle easy. Delay lines can have a fixed length. They are used at the first level of a trigger system.

Figure 1.26 gives a typical example of such a trigger. The number of particles hitting the target is defined by the beam counters $A \cdot B$ and a veto of a counter C with a hole to discriminate against particles not coming along the beam. The scattered particles are detected by the counters D and E. The cross section can then be computed by

$$\sigma = \text{Number of events} \times F/\text{Flux}$$

where F is the target constant. In this example one would like to measure the flux of the incoming and scattered particles.

Particles produce a light pulse in the *scintillator* which is transformed to an electrical signal by a *photocathode* and amplified by a *photomultiplier*. Then the scintillator signals are sent to a discriminator which produces a

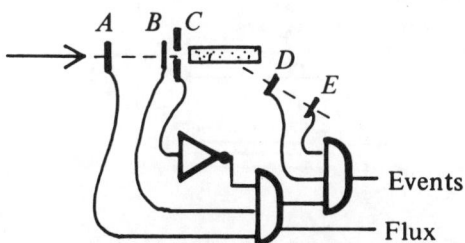

Fig. 1.26 Typical experiment to measure angular distributions. The scattered particles are counted in counters D and E. The incoming flux is measured by $A \cdot B \cdot \bar{C}$. The ratio of events and flux times the target constant and some correction factors determines the cross section.

54   *Real-time data triggering and filtering*

pulse of −0.8 V (NIM standard) and a variable length of between twenty and several hundred nanoseconds. These standard pulses enter a coincidence to establish an AND operation to count the beam particles and another coincidence to detect the scattered particles. Detected events can then be recorded by a computer. The pulse height must be digitized by an ADC.

Another example describes a processor which finds curved tracks in a cylindrical drift chamber operating at a storage ring. Neighbouring wires are grouped together. More than 5000 coincidences check in parallel for possible tracks. This processor needs 100 000 wrapped wires and occupies a rack of electronics. The trigger information is available 350 ns after the drift time of the chamber (Fig. 1.27).

To avoid having these many wires and gates, one can use variable flow processors. These processors will handle the trackfinding problem of Fig. 1.27 in the following way: The chamber is split into segments and the segments connected to a trackfinding device. If there is no track in a segment the next sector will be tested. If there are some hits, a track search will start for this segment. This sort of processor needs more time than hardwired devices but it is cheaper and more compact. Another example of a variable flow processor is the microprocessor which is used to search for

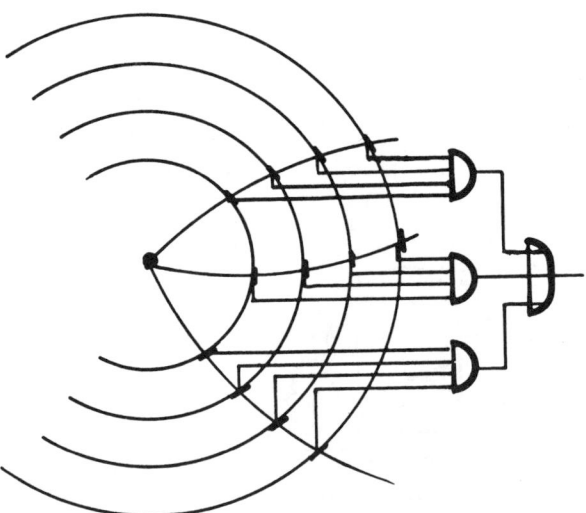

Fig. 1.27 Trigger processor for drift chambers. To find many curved tracks in a cylindrical wire chamber many coincidences (1000–5000) are required.

1.4 Classifications of triggers

tracks and to compute invariant masses. All variable flow processors need provisions for time out to avoid long waiting times.

1.4.8 Logical and arithmetic triggers

In a logical trigger one writes down the conditions under which a trigger should appear. An example of a logical trigger is given in Fig. 1.26. The flux is counted if the flux counters are set and if the particles travel along the beam line. An event occurs if a particle is scattered through a certain angle. The trigger condition is then

$$\text{Event} = A \cdot B \cdot \bar{C} \cdot D \cdot E$$
$$\text{Flux} = A \cdot B \cdot \bar{C}$$

All the rules from *Boolean algebra* must be obeyed in the design of a trigger. The main rules are:

$\cdot$ = logical AND
$+$ = logical OR

Commutative law	$U \cdot V = V \cdot U$	$U + V = V + U$
Associative law	$(U \cdot V) \cdot W = U \cdot (V \cdot W)$	$(U + V) + W = U + (V + W)$
Distributive law	$(U + V) \cdot W = U \cdot W + V \cdot W$	

(1.62)

An important rule is the *de Morgan's law*. If the signals are inverted one has to interchange an AND with an OR operation.

$$C = A + B \Rightarrow \bar{C} = \bar{A} \cdot \bar{B} \tag{1.63}$$

As has already been shown one also has to take the timing and the pulse lengths into account to get correct results.

Arithmetic triggers ask whether a value lies above a certain threshold or within two limits. A simple application is the use of a window discriminator to find within nanoseconds whether the number of counters is, for example,

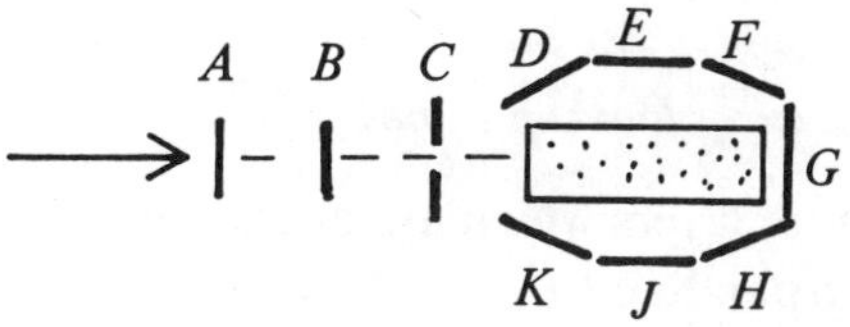

Fig. 1.28 Experiment to measure total cross section. Whenever four or more counters have fired the trigger condition should be fulfilled.

at least of the order of 4 (Fig. 1.28). The trigger condition for an event is then: $P = \text{TRUE}$, if four of the counters $D, E, \ldots, K$ got a hit

$$\text{Event} = A \cdot B \cdot \bar{C} \cdot P$$

This condition combines a logical and an arithmetic trigger.

1.4.9 Data and program driven trigger processors

A data driven system produces a result as soon as input data are available. The last example showed such a system. The electronics 'waits' for input data at the input lines and generates an output trigger after some internal propagation delay. The main characteristic of data driven systems is that they can operate without an internal clock. Program driven devices are clocked. The program needs data from a memory location which are loaded after decoding the instruction. All microprocessors or trigger computers are program driven.

The input data are strobed in some latches or registers and are then transferred to the Central Processing Unit (CPU), digested there and stored into another memory location. Program driven devices are not as fast and modular as data driven trigger systems.

1.5 Examples of triggers

1.5.1 Trackfinding with a lumped delay line

The MARK III detector at SLAC uses two layers of drift chambers with a 1 cm drift space to trigger on tracks in 200 ns (Lankford 1984a). The drift cells of the two layers are displaced by half a cell. The trigger decision makes use of the fact that the sum of the drift times in the two overlapping layers is a constant for high-energy (straight) tracks originating from the interaction point. It uses a *chronotron* composed of a *'lumped element delay'* line with ten *'taps'* to define a hit (Fig. 1.29 shows only six taps).

1.5.2 Trackfinding with memory look-up tables

In chambers with more layers which are operating in a magnetic field one defines masks corresponding to possible tracks. The number of masks can be very high because the position and the curvature of a track can vary. For chambers with a few layers which are separated by more than

1.5 Examples of triggers

a few centimetres high-energy tracks can be handled like straight tracks. This reduces the number of masks considerably. For a circular chamber with 500 wires per layer and 4 layers one has to test 500 positions and for each position three directions, for positive, negative and straight tracks. This gives 1500 masks. A trigger processor can be built by connecting 1500 AND circuits to all the wires in the correct combination (Fig. 1.27).

Such a processor would have the following three disadvantages:

(1) It would be very bulky; it would need many wire connections.
(2) It would not be flexible; if a wire did not work one could not reconfigure the processor.
(3) The processor would not allow for inefficiencies: at high rates a wire has some dead time and a single inefficient wire could inhibit the finding of tracks even if the other three layers show a clean hit.

To gain flexibility and to allow for inefficiencies one can use memories which are initialized in such a way that they output a valid trigger for each allowed input (including inefficiencies). In the TASSO experiment at DESY a special memory, the RAM C10115, is used for the central proportional chamber (Jaroslawsky 1977; Synertek 1976; Platner 1976). In this memory one does not specify a single address to get the contents of a cell. The address lines are organized like a 7×5 matrix and the output is the OR of all addressed cells. Two wire layers can then be connected to the columns

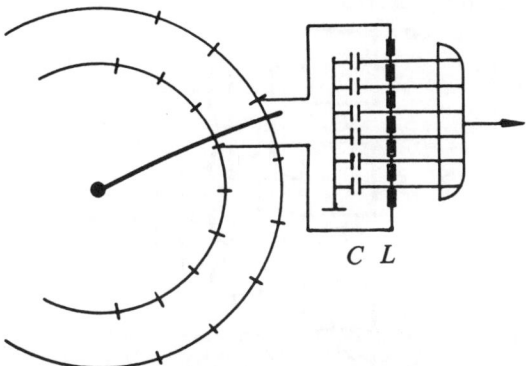

Fig. 1.29 Trackfinding with a lumped delay line in MARK III (Lankford 1984a). The principle of this trackfinder with two drift chambers displaced by half a cell is based on the fact that the sums of drift times in the two cells transversed by a track is a constant. The maximum drift time and a delay time of the same order should give a trigger at a fixed time after beam crossing.

and rows of the address inputs. Possible tracks are marked by a 1 in the diagonal. The output gives a track candidate for two layers. This output is then combined with possible track elements for two other layers. In this way one can take wire inefficiencies into account and fake a signal if the corresponding wire is broken (Fig. 1.30).

Processors described in this subsection can be used for drift chambers at storage rings with a time between beam crossings longer than 1.5 µs or at a second trigger level. The processors use 0.5 µs to reach a trigger decision and 1 µs is needed to reset the electronics (ADCs, TDCs) if no trigger has occured. The ADCs and TDCs are open at each beam crossing. Either they

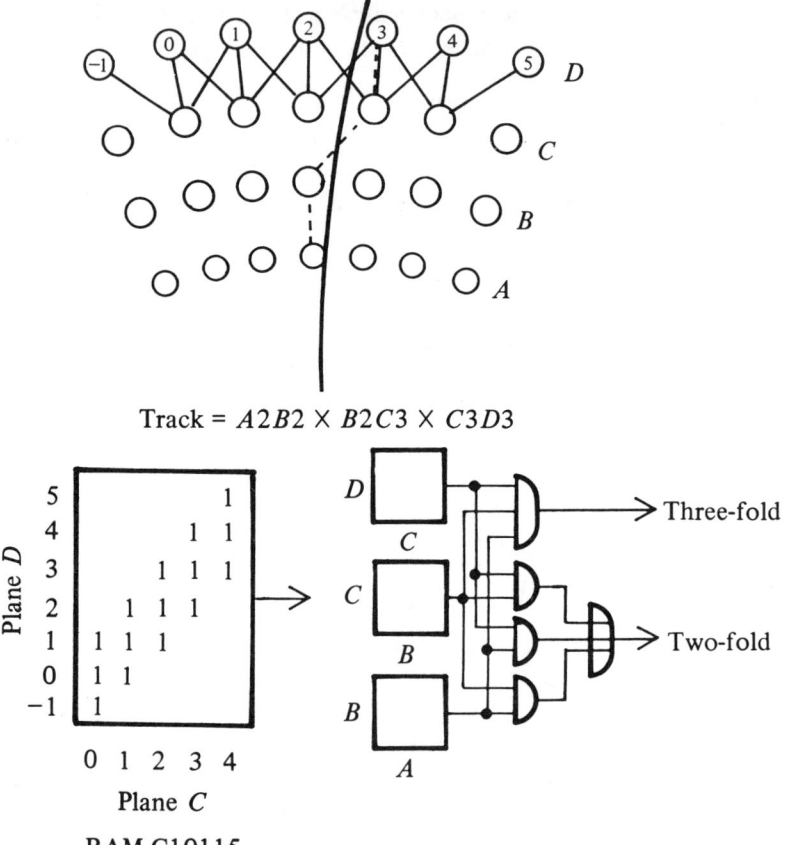

Fig. 1.30 Proportional chamber processor with RAM C10115. The processor should find tracks in four cylindrical layers of a wire chamber. Possible tracks and road widths can be generated by software.

1.5 Examples of triggers

are reset or they start conversion to digitize the event for the computer. The time limit of 500 ns is given not by the processors but by the drift time of the drift chamber. This requires short cells or a cell structure as shown in Fig. 1.31, which guarantees that each emerging particle from the interaction region passes close to at least one wire. The trigger processors rely on the fact that the particles must have their origin in a small vertex region and that high-energy particles can be approximated by straight lines for short detector devices.

The vertex detector of the TASSO experiment has eight layers which can be split into two parts (Rehlich 1980; Notz 1984). Track elements are then searched in each part separately by defining masks. The two parts are then split into sectors of ten wires. Ten wires can produce 1024 linear combinations, of which a few are real track candidates. An ideal track without inefficiency is found if the address lines 0, 2, 5, 7 are set (Fig. 1.32). This address cell must therefore contain a one. This technique is called *memory look-up* (see Subsection 1.6.1). To allow for inefficiencies a track should also be generated if lines 0, 2, 5 or 0, 2, 7 or 2, 5, 7 are set. In this way one can get in one memory cycle, track segments for four layers.

In a second cycle, track segments are combined to generate a track of all layers. It is a disadvantage of such a trigger that it does not work reliably if two layers (for example near the beam) do not work. The time needed to

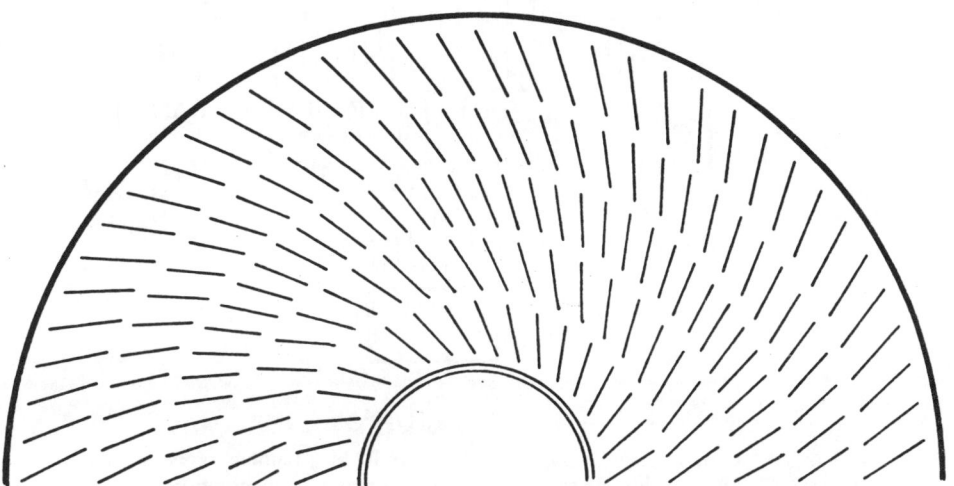

Fig. 1.31 Cell structure for short drift times. To avoid a trigger delay due to long drift times a special cell structure can guarantee that high-energy tracks from the interaction pass close to a wire.

generate a trigger is given by the drift time (100 ns), the cable length, the propagation time in the preamplifiers, and the setting time for the memories (50–100 ns).

In the previous processor, track candidates are oriented as mushrooms starting from the innermost layer. The following processor of the CELLO experiment at DESY uses circle segments which have common points at the interaction vertex and in the outer layer (Schroeder 1981). The detector consists of five cylindrical multiwire proportional chambers interspersed with seven drift chambers. To obtain the flexibility necessary to adapt to different experimental conditions like background rate, chamber inefficiencies or momentum cuts a parallel look-up table technique is needed. The trigger is separated into two systems, one for the plane normal to the beam axis containing the interaction point and one plane perpendicular to the

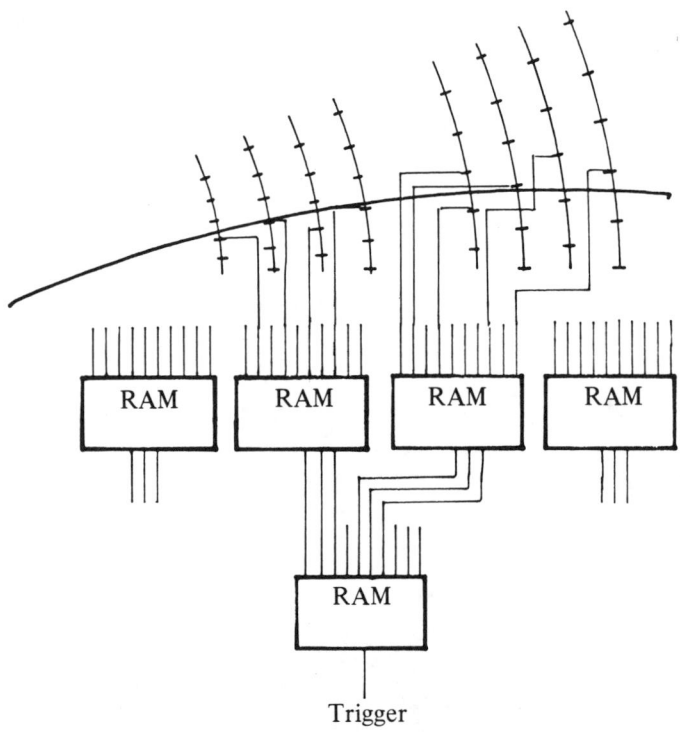

Fig. 1.32 TASSO track chamber processor with three RAMs (Random Access Memories) per sector. Chambers with more than six layers would require large memories to define all masks and coincidences. In these cases one can split the chamber into subchambers and look first for track segments which are then linked together (Rehlich 1980; Notz 1984).

1.5 Examples of triggers

storage ring and containing the beam line to search for the vertex along the beam direction. The various chamber signals to be used for the trigger have been grouped so that 64 equally sized angular sectors are obtained in $R\phi$. The acceptance region for the $R\phi$ trigger is ± 10 mm around the beam axis. More proportional chamber layers are used in the trigger instead of drift chamber signals due to the shorter sensitivity time.

The 64 sectors × 7 layers wire chamber signals are strobed into an input register 800 ns after bunch crossing. The chamber signals are fed via a fanout unit into the address inputs of the RAM (Random Access Memory) array containing the mask logic. The RAMs are loaded with the patterns of all physical meaningful curved tracks (Fig. 1.33). The RAMs have a capacity of 1024 four-bit words. The bits correspond to the curvature of a track. A straight track hits all layers in sector 0, a curved track hits sector 1 in layer 1, 2, 3, and sector 0 in layers 4, ..., 7. In this manner 24 masks with the same sign of curvature are stored in six RAMs of each sector, the oppositely signed masks are loaded into another six RAMs. By appropriate programming of the RAMs, chamber inefficiencies are taken into account. But if for example chamber 1 is missing, the last mask of RAM1 and the first mask of RAM2 are identical. To avoid these ambiguities, which will result in a higher trigger rate, possible inefficiencies are only loaded into one RAM. The output bits are first summed for each sector and then in all 64 sectors to produce a trigger.

1.5.3 Trigger on tracks with field programmable arrays

The TASSO processor can be used without pretriggers at storage rings with a bunch crossing time above 2 µs or as second level trigger (Notz 1981; Stuckenberg 1981): 1 µs is used for the trigger decision and 1 µs is required to reset the electronics. The processor gets its input from six drift chamber layers with 72–216 wires, from 48 TOF scintillation counters and from 48 bits of track information from a faster proportional chamber processor. The processor should produce a trigger if there are more than three, four or five tracks with a minimum transverse momentum of at least 200–300 MeV, but it should also give a trigger for QED events such as $e^+e^- \to e^+e^-$ or $\mu^+\mu^-$. These events are characterized by two high-energy particles emerging in opposite directions from the interaction point.

To perform at high speed many operations run in parallel. In a first step one has to find tracks. For each of the 72 wires of the innermost chamber 15 possible tracks are checked in each cycle. The 15 tracks have positive or

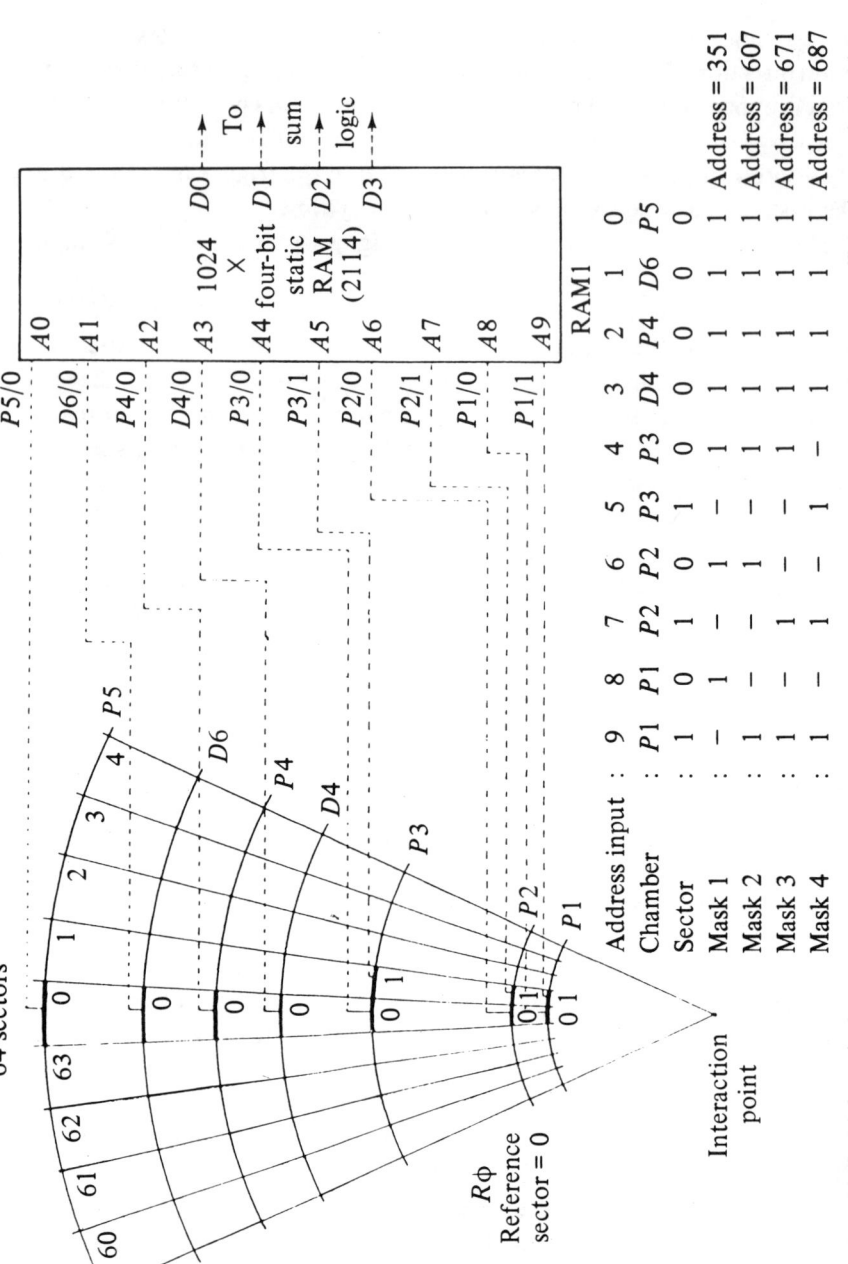

Fig. 1.33 CELLO track chamber processor with RAMs. Sixty-four sectors with their tracks and allowed inefficiencies are checked in parallel. The road width and trigger condition can be changed dynamically (Schroeder 1981).

1.5 Examples of triggers

negative signs and different momenta; 72 reference wires × 15 track candidates gives 1080 possible masks. These possible masks are hardwired. Within each mask one has to combine several neighbouring wires by a wired OR to form a track point for the trigger masks.

From the six possible track points one has to decide whether there is a valid track or not. This is very simply achieved by a six-fold coincidence. But if one also wants to take inefficiencies into account one has to use several five-fold coincidences and combine their results by an OR. This requires a lot of hardware. Another possibility is the use of *Programmable Read Only Memories* (*PROMs*) or *RAMs* as discussed in the previous subsection. For each valid address the generated output represents a trigger. If one wants to gain speed one has to use gates instead of memories (Fig. 1.34). One can benefit from the speed of gates and use nevertheless compact electronics with the help of *Field Programmable Logic Arrays* (*FPLAs*) or *Programmed Array Logic* (*PAL*) (see Subsection 1.6.1). The basic logic of a FPLA consists of a programmable AND array whose outputs feed a programmable OR array. The designer decides how to configure the AND and OR connections. With a special program device the FPLAs are programmed by the customer by burning the fuses. Another approach are PALs which operate in a similar way to FPLAs, but in which the OR combinations are fixed (see Subsection 1.6.1).

At the output of the possible tracks one gets 1080 bits which must be further reduced to a yes/no trigger (Fig. 1.35). The 1080 tracks are now

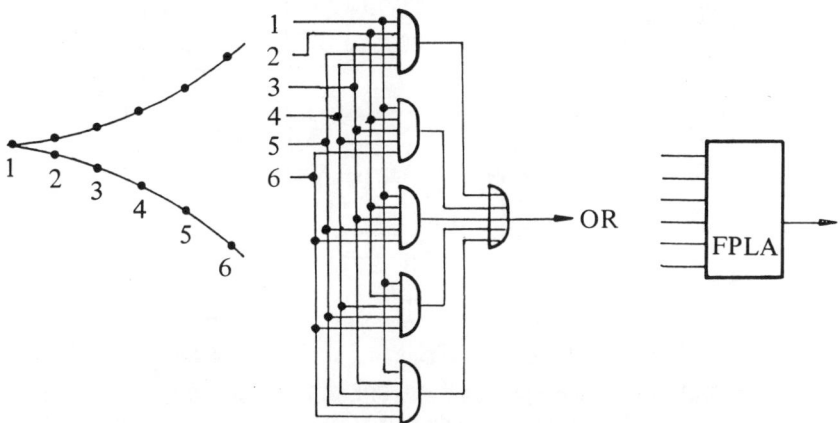

Fig. 1.34 Use of FPLA in a trackfinding processor. A single field programmable logic array can replace many ANDs and ORs which are needed if one wants to allow inefficiencies in wire chamber (Notz 1981; Stuckenberg 1981).

compared to the 48 tracks of the proportional chamber. Only those tracks in the processor which also have a hit in the proportional chamber are allowed. To reduce transverse momentum a mask selects only high momentum tracks. The final tracks are then combined with information from scintillation counters surrounding the drift chamber. Forty-eight possible bits remain which are then sent to a majority logic to discriminate against low multiplicity events. In a separate branch high-energy tracks

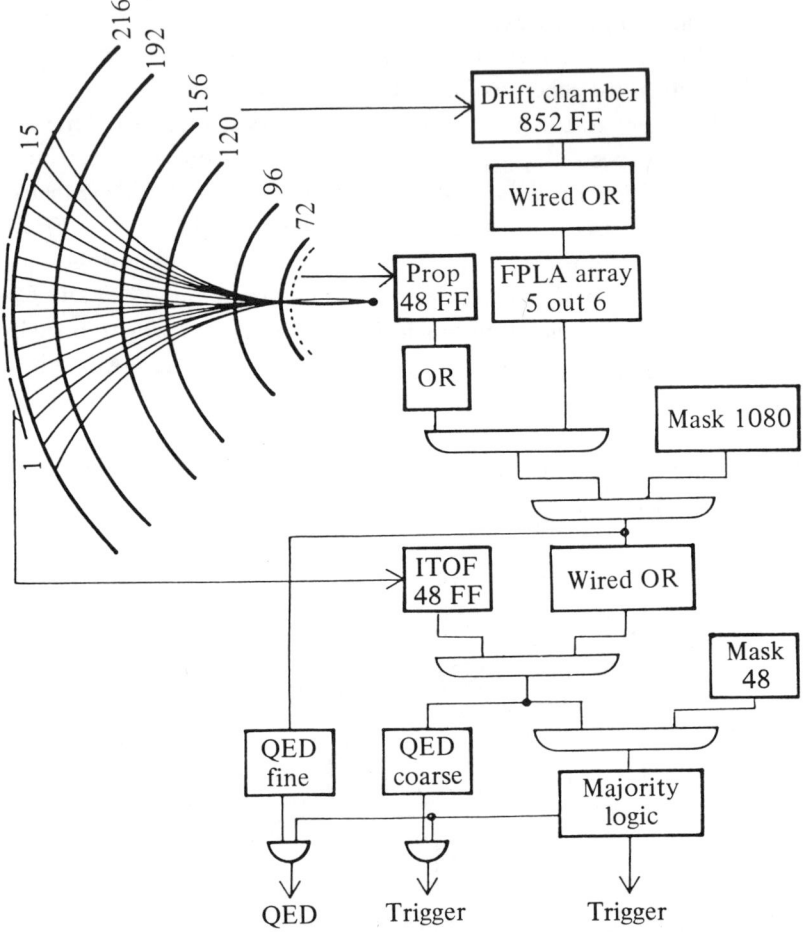

Fig. 1.35 TASSO track chamber processor with FPLAs. For each of the 72 inner wires 15 masks are generated and compared to the data. The 1080 masks are done by hardwired FPLAs to avoid these many ANDs and ORs. A trigger is generated for hadronic events (majority $\geqslant 3$ tracks) or QED events as e^+e^- or $\mu^+\mu^-$ pairs (Notz 1981; Stuckenberg 1981). (ITOF is an inner time-of-flight counter.)

1.5 Examples of triggers

emerging from the opposite direction from the interaction point are combined and masked as QED events. Typical rates at 15 GeV beam energy are of the order of some triggers per second:

Beam crossing rate	260 kHz
QED rate e^+e^-	1 Hz
Two-prong events	7.5 Hz
Three-prong events	1.3 Hz
Four-prong events	0.5 Hz

1.5.4 Trackfinders in the trigger with variable flow data driven processors

Hardwired parallel processors can compute the number of tracks in a few cycles by comparing the wire chamber hits with predefined masks. These processors are huge devices with several thousand wrapped wires and electronic circuits. Variable flow processors are small because they compute sequentially the possible tracks from all combinations. They need more time (5 μs–2 ms) to find tracks but the tracks are more precise because these processors act on single wires and not on wire clusters. Figure 1.36 shows one view of a set of wire chambers – drift chambers or proportional chambers. In the case of drift chambers we use only the information that a wire was hit, we do not use the drift time.

A typical trackfinding algorithm works as follows: Combine each hit from chamber 1 with those from chamber 3, fit a straight line and predict the position of a hit in chamber 2. If there is a hit within a certain road width a track is formed. For N hits per plane this algorithm needs N^3 operations. The algorithm to find a hit in chamber 2 must be very fast. A track is given by $y = Ax + B$. The coordinates of points in chamber 1 are (x_{11}, y_1), (x_{12}, y_1), (x_{13}, y_1),... and in chamber 3 (x_{31}, y_3),..., respectively. First we have to find the parameters A and B of the track and then one can predict a

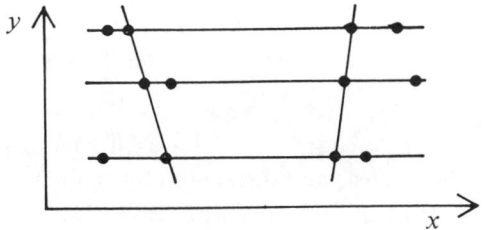

Fig. 1.36 Wire chambers in a spectrometer experiment. The beam is coming from below. The coordinate system is used in the algorithm.

point in chamber 2:

$$y_1 = Ax_{11} + B$$
$$y_3 = Ax_{31} + B$$

$$A = \frac{y_1 - y_3}{x_{11} - x_{31}}$$

$$B = y_1 - x_{11}\frac{y_1 - y_3}{x_{11} - x_{31}}$$

The predicted value x_{21} in chamber 2 is then

$$x_{21} = (y_2 - B)/A$$
$$= \frac{(y_1 y_1 - x_{11} y_3)/(x_{11} - x_{31}) + y_2 - y_1}{(y_1 - y_3)/(x_{11} - x_{31})} \quad (1.64)$$

$$x_{21} = x_{11}\left(\frac{y_2 - y_3}{y_1 - y_3}\right) + x_{31}\left(\frac{y_1 - y_2}{y_1 - y_3}\right) = x_{11}a + x_{31}b \quad (1.65)$$

The chambers are in fixed positions y_1, y_2, y_3. Therefore a and b are constants and can be computed at the beginning of the experiment. In a von Neumann computer one would write a program with three loops to find the tracks. In order to save computer time in a special purpose processor one can get rid of the innermost loop with the help of a Content Addressable or Associative Memory (CAM). Let's take the following example to explain the function of a CAM: A teacher would like to know whether somebody in the class is ten years old. He or she can ask each pupil if he or she is ten years old. This is a sequential process like the inner loop. Or he can ask that those who are 10 years old should raise their hand. This indicates a match in the operation of a CAM. In our case the information of chamber 2 is not stored in an ordinary memory but in a CAM. After prediction of a value x_2 one adds some bits for the road width and checks in one cycle whether the CAM contains that value. In this case a track is found (see Subsection 1.6.1.).

We can now decompose the algorithm into several steps and try to produce a modular system for trackfinding.

(1) Store all x coordinates in separate buffers for each chamber.
(2) For the loops, store data in lists and then count by an index generator.
(3) Take a hit from chamber 1 and multiply it by a. Take a hit from chamber 3 and multiply it by b.
(4) Add the two results and predict a hit in chamber 2.

1.5 Examples of triggers

(5) Match the predicted position with a possible hit.
(6) Count the number of tracks.

These steps can be realized with electronic modules which were developed by LeCroy (Levit and Vincelli 1985) and by Nevis Laboratories and which are in use at Fermilab (Kostarakis *et al.* 1981). The algorithm can be performed by the following system (Fig. 1.37).

The modules are data driven. As soon as input data are ready the operation is executed. Each module implements a simple operation such as adding two 16-bit numbers, comparing a quantity with upper and lower boundaries, computing a function with a look-up table. The design cycle time is 25 ns.

The protocol between modules is handled by four control lines to exchange control information and to synchronize the processes (Avilez 1984). These control lines are:

V *Valid.* Here is a data word.
H *Hold.* Couldn't accept this word.
C *Complete.* End of block.

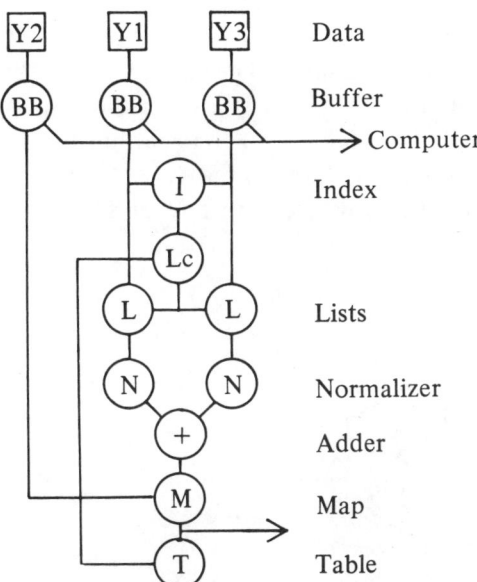

Fig. 1.37 Variable flow, data driven, trackfinding processor. The processor consists of separate building blocks which are connected by a data and control path. An operation is executed whenever data are ready (Kaplan 1984).

B *Block reset*. Abort this block.

The signals H and B travel against the flow of data from the receiving module to the sending module. With the help of internal buffers a new operation can begin every cycle. The modules are constructed on two-layer printed circuit boards. Chips of the ECL 10000 series are used to reach the high speeds. If more operations are needed more modules are added to the system to keep the speed high. Approximately 15 different modules have been produced. We will discuss only those modules which were used in the trackfinding processor of our example.

(1) *Block buffer*. The block buffer is a 256-word memory with three ports: One port to read data from the wire chambers, one port to write data to the next processor modules and a third bidirectional port to transfer data to the computer for data acquisition or to get data from the computer to test the trigger.

(2) *Index generator*. The index generator puts out an index pair (8 + 8 bits) which calls all possible hypotheses from the lists.

(3) *List counter*. The list counter generates multiple passes around the loop for each hypothesis.

(4) *Normalizer*. The normalizer works on the basis of a look-up table and can therefore compute several functions. It consists of two 256-word 16-bit tables whose outputs are summed. The high order eight bits are sent to one table the lower order eight bits to the other allowing any function of the form $f = g$(eight bits high order) $+ h$(eight bits low order) to be computed by preloading (Fig. 1.38).

(5) *Arithmetic operators*. These devices can add or subtract two 16-bit numbers.

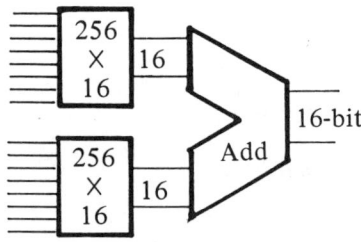

Fig. 1.38 Memory look-up table as a normalizer. To avoid time consuming multiplications the result of a function or multiplication is stored into a memory. The normalizer in this picture computes two functions and adds their results.

1.5 Examples of triggers

(6) *Map modules.* The map module checks whether a predicted hit matches a real hit. These modules use CAMs.

A processor of this type is being used at Fermilab to study phenomena of 10^{-11} of the total proton–nucleon cross section. With ten hits per wire plane it finds tracks in 5 µs.

1.5.5 A hardwired processor with CAM and look-up tables

The processors described in the previous subsection are useful at storage rings which have a bunch crossing time of more than 2 µs if they are used as first level triggers. At storage rings with shorter crossing rates one has to use a coarse pretrigger consisting of a set of scintillation counters surrounding the interaction region. These counters can be used for *TOF* measurements and produce a pretrigger rate of the order of 10–100 kHz. At an average rate of 1 kHz a trackfinding processor with 1 µs operation time would introduce 0.1% dead time. We have seen that these processors with their high degree of parallelism are very large and expensive. Toleration of a dead time of 5% leads to a processor which has 50 µs to find tracks. Today's commercial microprocessors cannot be used in this time range, they would perform only 200–300 cycles. As an example we will discuss the processor of the ARGUS experiment which uses 36 layers of drift cells (Schulz and Stuckenberg 1981; Schulz 1984). Eighteen of these have their signal wires spanned parallel to the beam axis and 18 are stereo layers. The number of cells in a cylinder varies from 60 on the inside to 264 on the outside. The drift cells are small ($18 \times 18 \,\text{mm}^2$). The processor does not use the drift time information. The trackfinder must identify circular tracks emerging from the interaction region, count the number of tracks needed to produce a trigger and store all information about tracks for later use and monitoring.

A layout of the processor is shown in Fig. 1.39. Eight layers of the drift chamber are used in the trackfinding algorithm. In a sequential operation with 170 ns cycle time, mask after mask of possible tracks are applied to the chamber. The chamber signals are taken from the LeCroy 4291B TDC hit register and stored in *flip-flops*. For each chamber the corresponding wires inside a track road must be ORed. The masks are stored in a mask memory of $2k \times 12$ bits. Three bits indicate chambers $1,\ldots,8$, five bits point to the first wire in a road and three bits contain the width $(1,\ldots,8)$ of the road. To select quickly from these $5 + 3$ bits the right hits per layer, a PROM is used which has one output for each wire. This output is enabled and strobes the

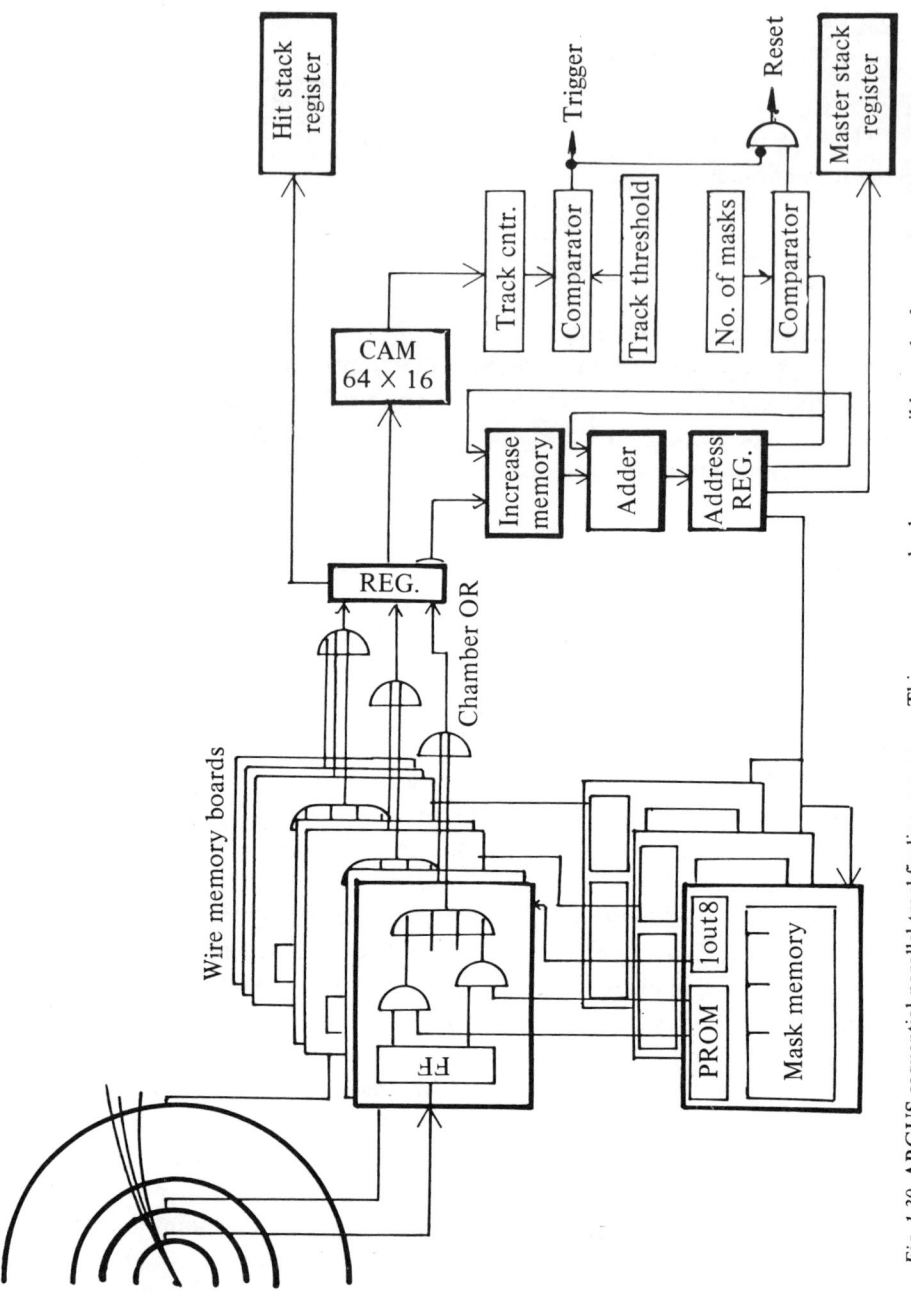

Fig. 1.39 ARGUS sequential parallel trackfinding processor. This processor checks one possible track after another. If a track mask is empty, all tracks belonging to part of this mask are skipped (Schulz and Stuckenberg 1981; Schulz 1984).

1.5 Examples of triggers

wire information for this mask into a CAM register. With a clock frequency of 50 MHz and a cycle time of 20 ns the masks select the drift chamber information in eight cycles. An overflow bit in the mask memory is used in those cases in which a mask crosses a board boundary. The problem of ORing wires across a 2π boundary of a layer is solved by filling the inputs following the last wire on the last card in parallel with the first wires of the first boards. The CAM register now contains the hit pattern for one mask. This pattern is then compared to predefined patterns of good tracks stored in a CAM. In this way one can allow for inefficiencies or for faulty detector operation. If there is a match for a good track, the track counter is increased and the masks and track patterns are stored in a stack. A trigger is given to the computer if the number of tracks found exceeds a predefined threshold, otherwise a reset is given.

Checking all 2000 masks in this way would require 350 µs. From the construction of the masks one can see that the same region of the chamber is used in many masks. Consider the three innermost layers of the chamber at a given sector. If these three layers are empty it is not worth checking all possible masks for positive and negative tracks and for five different transverse momenta at this angle. In this case one can check the next sector immediately. An empty chamber would therefore require only 60×170 ns = 10 µs excluding the time for reset and synchronization. The address register MAR for the mask memory is not increased one by one. The increase depends on the contents of three drift chamber layers, the so-called reference wires. These three bits of the CAM register are used together with nine bits of the address register to compute the next mask address. The masks are therefore ordered in groups. If one of the reference wires does not give a hit the corresponding subgroup will be skipped.

1.5.6 A microprogrammed track processor with CAM and look-up tables

The processors discussed so far are fast because many steps are done in parallel by a large amount of hardware. The microprogrammed processor for the drift chamber of the TASSO experiment is a microprocessor which finds tracks within 1 ms (Schildt, Stuckenberg and Wermes 1980). In the previous processors only the wire information from the drift chambers is used. This method is sufficient if one wants to trigger on more than three tracks. If one would like to trigger on two tracks which are not required to be coplanar this coarse method gives many triggers because many wire combinations can generate a two-track trigger.

For low multiplicity events one has to use both the wire and the drift time information to find cleaner tracks. The processor reads the digitized drift times and then searches for tracks. The track reconstruction is performed in a plane perpendicular to the beam axis ($R\phi$ plane). As the magnetic field is assumed to be homogeneous all charged-particle tracks are segments of circles. In order to reach a fast tracking algorithm the circles are parametrized as indicated in Fig. 1.40. Two hits in the first and second cylinders and the interaction point are used to define a circle. The radius and the tangent of the circle are determined by the radii of the cylinders (ρ_1, ρ_2) and the angle between the two hits:

$$R = 0.5\rho_1/\sin\phi_1 = 0.5\rho_2/\sin\phi_2 \tag{1.66}$$

$$\phi_1 = \text{arc cot}\,[\rho_2/(\rho_1 \sin\alpha) - \cot\alpha] \tag{1.67}$$

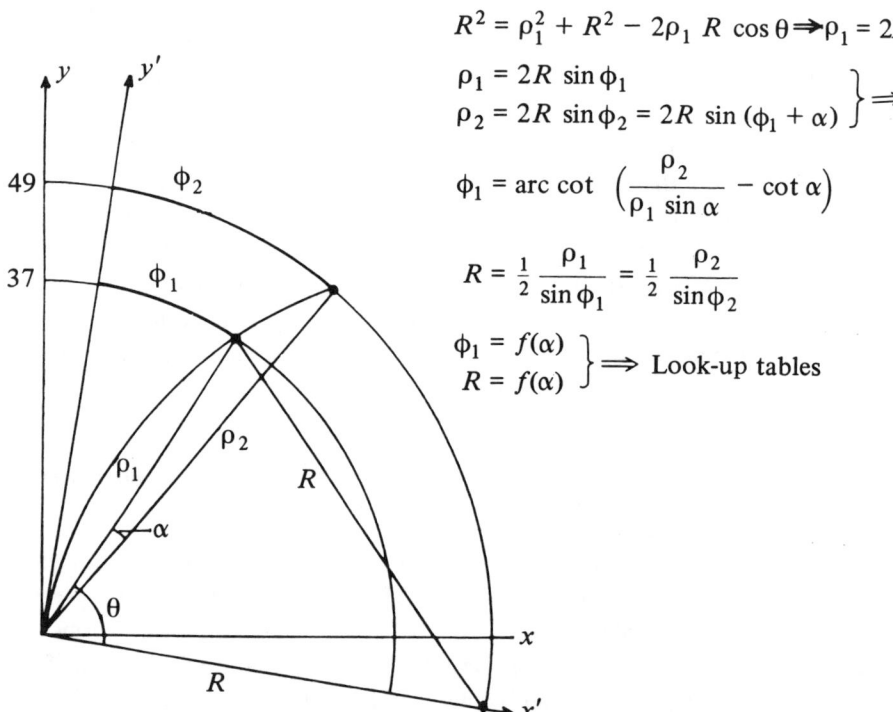

Fig. 1.40 The track parametrization for a fast track processor. With this track parametrization one gets the value α from two hits in two layers. It is then easy to compute ϕ and R with the help of look-up tables and predict the next hit.

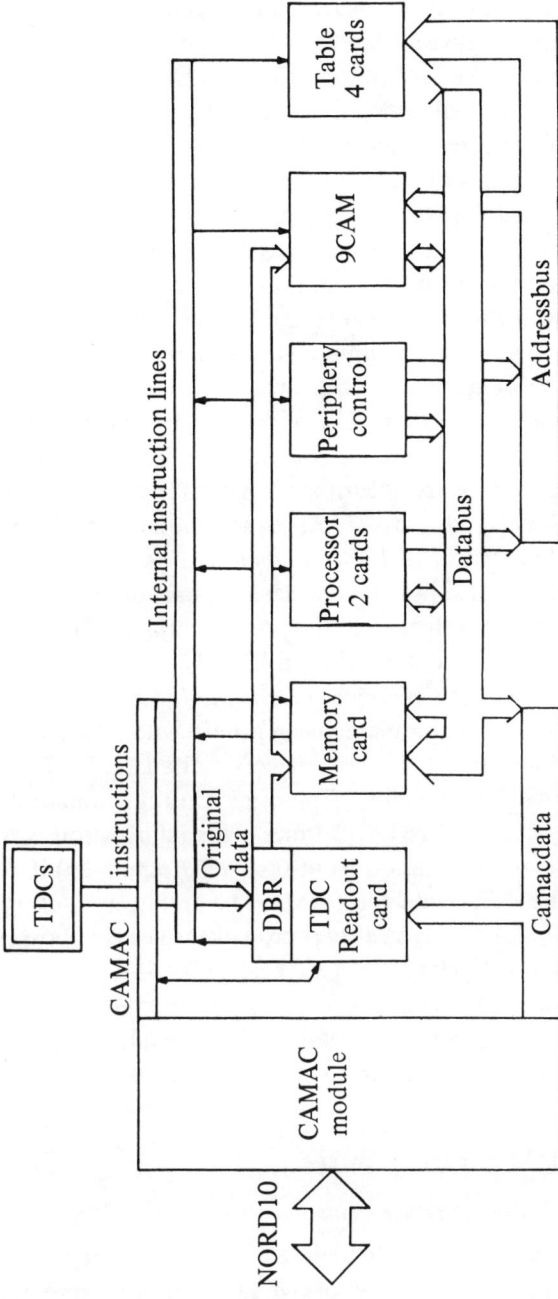

Fig. 1.41 A 'non-von-Neumann' track processor. This processor uses look-up tables and CAMs to compute tracks within 1 ms (Schildt *et al.* 1980).

Since ρ_1 and ρ_2 are constants the equations only depend on the angle α which is the angle between the two vectors pointing from the interaction point to the hits in the first and second cylinders. With a given α one can compute the starting values R_1, ϕ_1 and R_2, ϕ_2. The radius and tangent are calculated at most three times. After that the rest of the track following is just a hit search. A track is accepted if there are at least six hits in nine layers. The angle α is used only between layers $1+2$, $1+3$, $1+4$, $2+3$ and $2+4$. The maximum value of α does not exceed 0.255. Taking the resolution $\Delta\alpha = 0.00038$ due to drift time digitization into account only 665 different α values are possible for high-energy tracks. The α values are stored in five look-up tables with a maximum storage size of 1024×16 bits. The maximum value for ϕ is 0.66 or 1722 possible values, and for R it is 8191 cm or 8192 values for a resolution of 1 cm. Within a few memory cycles one gets the parameters of a possible track from the first three layers. The task is then to check whether there are at least six hits corresponding to that predicted track. This is done by the use of CAMs. A sequential search is too slow as we have seen in Subsection 1.4.1. *Association* and *matching* are done by the memory itself. It is also possible to mask off part of the associative word. Therefore one can search around the predicted point in the next chamber.

The track search is controlled by a ECL *bit-slice processor* (Motorola 1977) with a 72-bit long *microinstruction word* (see also Section 1.7). The processing unit can manipulate the registers within one cycle while an access to the look-up tables takes place simultaneously. The program is stored in a 1k × 72-bit memory. The original drift chamber data contain wire addresses and digitized drift times. This information is transformed into an integer representation from 0–177777(oct)(0–2π). Up to 40 bits per layer can be stored into the CAMs. The architecture of the processor is shown in Fig. 1.4.1. A typical two-prog event needs approximately 5000 microinstructions. A clean *Bhabha* even $e^+e^- \to e^+e^-$ requires 0.6 ms and a hadronic event 4.5 ms. It is probably a weak point of this processor that the first three layers must be efficient. If the first layers are inefficient no track parameters can be set up to start a track search.

1.5.7 Examples of triggers on energy

Analog trigger in the ASP experiment

We will now discuss the ASP detector at SLAC which surrounds the interaction region by four walls of five layers of lead glass separated by proportional chambers (Lankford 1984b). The 640 photomultipliers are

1.5 Examples of triggers

divided between digitizers and the trigger. The experiment and the trigger are shown schematically in Fig. 1.42. The trigger system forms first 80 sums of eight multipliers and then another 20 sums for each layer. These 20 layer sums each go to integration circuits and are then discriminated to define hit layers. The hit layers address a *memory look-up* which, in turn, defines allowed combinations of hits; for example the first two layers of a quadrant but not the last two. The layer sums are also summed up into a few quadrant sums which are integrated and then each is discriminated by three levels defining three energy thresholds. The resulting 12 signals address a memory look-up that counts and defines combinations of quadrant hits. The quadrant sums are also summed to form a total energy sum which is also integrated and discriminated by three thresholds.

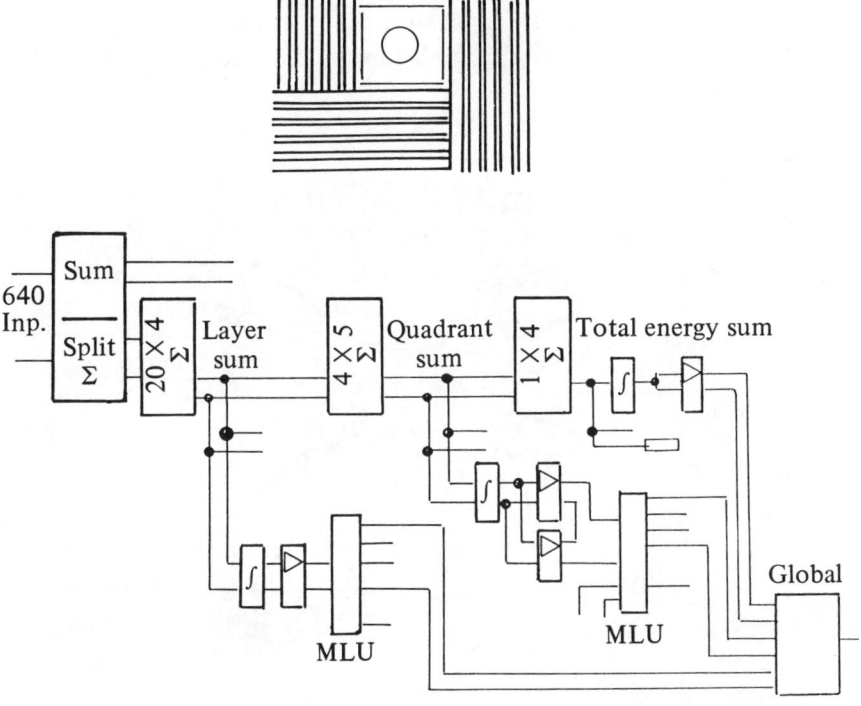

Fig. 1.42 Energy trigger for the ASP experiment with lead glass walls. The global trigger gets information from the total energy sum, the quadrant energy sum and the layer energy sum (Lankford 1984b).

Analog trigger in the UA2 experiment

In the following we will describe a modular energy trigger of the UA2 experiment at CERN. The trigger rate primarily depends on the *transverse energy* (Δy) rather than on energy. If in an experiment each calorimeter cell occupies the same $\Delta\phi\Delta y$ solid angle one can easily simplify the trigger by calibrating the calorimeter cells in transverse energy. In this case a single common threshold results in similar trigger rates independent of the position of the cell. Fig. 1.43 shows a calorimeter which consists of an electromagnetic and a hadronic compartment. To find leptonic Z^0 and W candidates only the electromagnetic part is used (Hungerbuehler 1981). The hadronic part is used in a very simple trigger such as total transverse energy. The unfolded electromagnetic calorimeter in the $\theta\phi$ plane shows the cells of $10 \times 10\,\text{cm}^2$ dimensions with a distance of 1 mm between cells. In the forward/backward region each of the 12 sectors is separated by 15 cm due to magnetic coils. The entire calorimeter consists of 480 cells with 760 photomultipliers.

Because of the geometry a shower can be shared among several neighbouring cells. Applying a simple threshold on each cell would cause

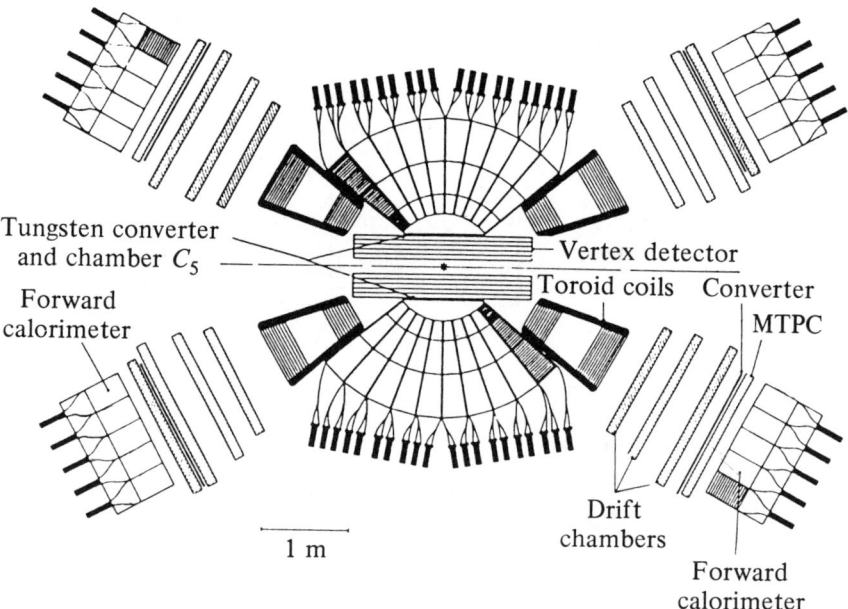

Fig. 1.43 Layout of the UA2 calorimeter experiment. This calorimeter is used to detect Z^0 and W bosons at the $\bar{p}p$ collider (Hungerbuehler 1981).

1.5 Examples of triggers

trigger inefficiencies because showers which develop in one cell will be above threshold while showers shared by two or more cells can drop below threshold. Trigger thresholds must therefore be applied to sums of neighbouring cells. Therefore in the trigger all 2 × 2 clusters are formed. Each cell in the central detector participates in four sums at each of its corners. The 480 cells are summed to 313 clusters.

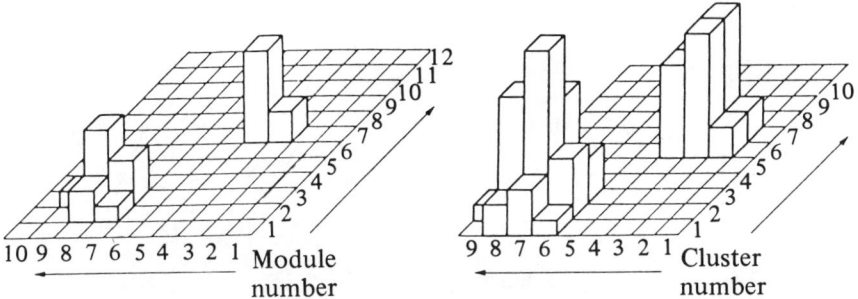

Fig. 1.44 Cluster and cell algorithm in UA2. Clusters are formed to avoid big fluctuations if a single cell or two neighbouring cells are hit.

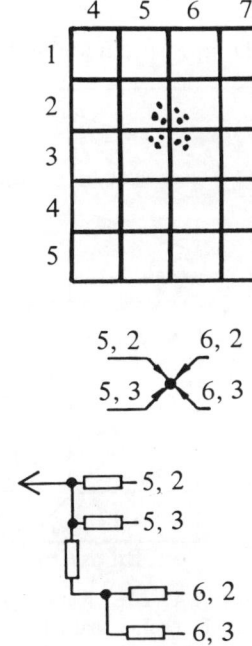

Fig. 1.45 Modules in the trigger. The trigger system is built in a modular way. Each module is responsible for one row and exchanges information with its neighbours.

78   *Real-time data triggering and filtering*

The effect of *clustering* is shown in Fig. 1.44. There are always more nonzero clusters than nonzero cells. One single cell participates in four clusters. It is now very easy to build a trigger for electromagnetic showers above a given transverse momentum by requiring at least one cluster to be

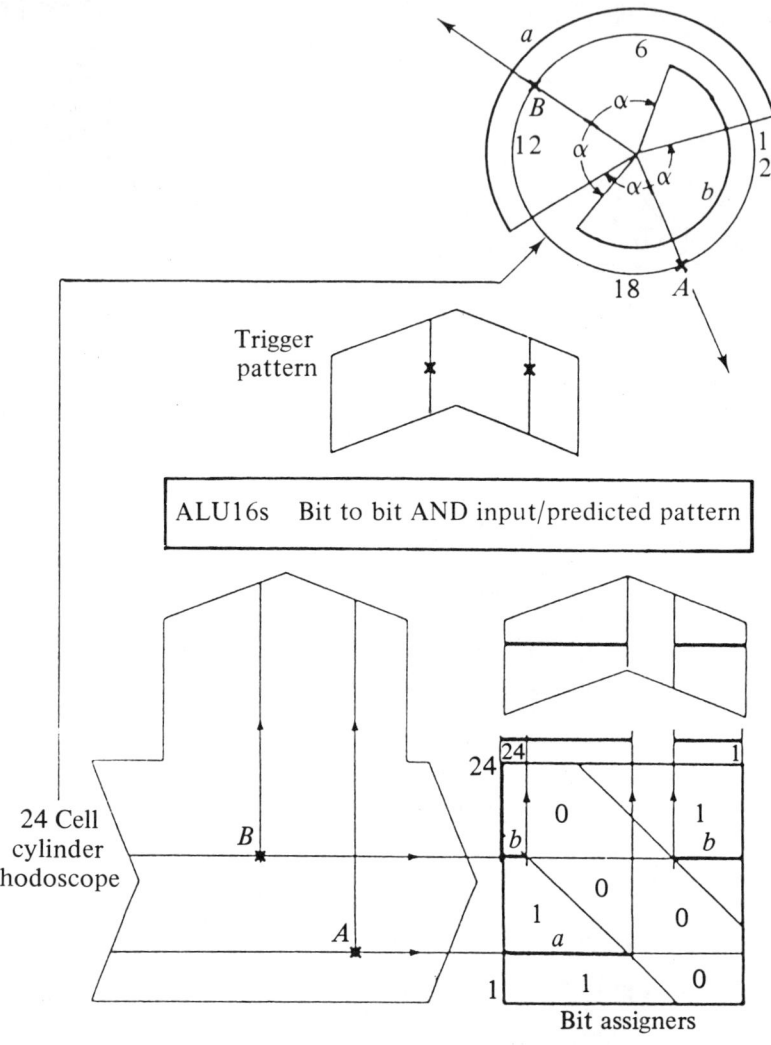

Fig. 1.46 Finding events with high e^+e^- invariant mass in UA2. e^+e^- pairs with low invariant mass generate two nearby pulses. The trigger logic requires that hits in one part of the detector need a particle in the opposite part.

1.5 Examples of triggers

above a given threshold. For some applications it is desirable to count the number of showers but the cluster algorithm increases the number of clusters. One can correct for this effect by first searching for the highest cluster and then deleting its neighbours. Two neighbouring clusters of same height are both accepted.

The trigger can be built in a rather modular way. The photomultiplier signals of both multipliers of one cell are summed and sent to an integrating sample and hold device. The output DC voltage is proportional to the charge of the multipliers; 1 mV corresponds to 1 GeV transverse energy. The voltage is kept constant to better than 20 mV for 3 µs. This introduces an error at small energies. Each module operates on a row of ten cells (Fig. 1.45). As the clusters are formed from information from two neighbouring rows the signals are split and part of the signals enters the next module. Each cluster voltage can then be compared with a reference voltage which is common to all nine clusters serviced by the module. The reference voltage is generated by a fast eight-bit Digital-to-Analog Converter (DAC) (10 ns setting time) and controlled by a synchronous counter. Each cluster exceeding the threshold sets its flip-flop. The actual trigger now uses two thresholds. The so-called Z trigger ($Z^0 \rightarrow e^+ e^-$) asks for at least two clusters above threshold. A second trigger, the W trigger, asks for one cluster at a higher threshold.

The electron pair cluster needs two clusters which are well separated in solid angle, indicative of a large invariant mass (Fig. 1.46). The trigger logic to suppress events with small invariant mass is done with the help of a bit assigner. A particle in row 2 requires the second particle to be in rows 7–20, a particle in row 5 requires it to be in rows 10–24 and a particle in row 12 requires it to be in rows 17–24 and 1–6. If we have a particle event in rows 2 and 5 there must be a signal in rows 7–24 which is not fulfilled. An event with a particle in rows 2 and 12 fulfils the trigger condition.

1.5.8 A data driven trigger on invariant mass

The last experiment showed an example of a trigger which gives a veto for low $e^+ e^-$ masses from photon pair production. Triggers on invariant mass are important for spectrometer experiments looking for *dimuon events* with high mass. These events are produced at low rates, therefore spectrometers with a large aperture are used (Greenhalgh 1984). From the trigger point of view one has to find tracks in the chambers behind the magnet and compute the invariant mass (Fig. 1.47).

$$M^2 = (P_1 + P_2)^2 = (E_1 + E_2, \mathbf{p}_1 + \mathbf{p}_2)^2$$
$$= E_1^2 + 2E_1 E_2 + E_2^2 - \mathbf{p}_1^2 - \mathbf{p}_2^2 - 2\mathbf{p}_1 \mathbf{p}_2 (1 - \Theta^2/2) \qquad (1.68)$$
$$(\cos \Theta = 1 - \Theta^2/2)$$

Neglecting the particle masses at high energies leads to $M^2 \approx \mathbf{p}_1 \mathbf{p}_2 \Theta^2$. The invariant mass of two particles at high energies is thus approximately given by $M = \Theta(\mathbf{p}_1 \mathbf{p}_2)^{\frac{1}{2}}$ whole $\mathbf{p}_1$ and $\mathbf{p}_2$ are the momenta of the two particles and Θ is their opening angle at production. The processor has to find the tracks and compute their slopes and intercepts at the magnet's centre. This information allows the computation of the bending angle making use of the bend plane and the point target approximation. The bend angle is inversely proportional to the momentum of the particle. The two intercepts at the bend plane determine the opening angle Θ in the xz plane. The Θ_y angle is

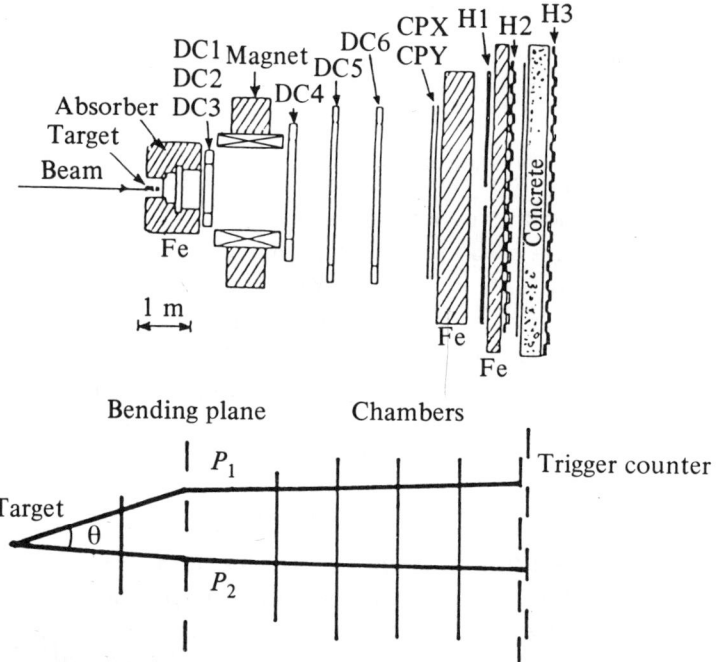

Fig. 1.47 Spectrometer to study $\mu^+\mu^-$ pairs. The target is surrounded by a dump. Wire chambers in front of and behind the magnet define the trajectories. The iron walls on the right hand side are used to trigger on muons. The lower diagram shows the part of the spectrometer which is used to determine the invariant mass.

1.5 Examples of triggers

only computed with the help of the y hodoscope behind the iron wall. The computing time needed to find tracks depends on the number of hits per chamber. Testing all combinations of two chambers needs $N_1 N_2$ tests. One can save time by defining roads and checking only those combinations within these roads. In the case of the di-muon experiment the hodoscope counters in the iron wall define a rough direction of the incoming tracks.

The quantities recorded are inverse track momentum, its intercept at the magnet centre, a two-bit number for vertical position and a sequential road counter (Fig. 1.48). Two stacks are used for the track information in two roads. After having found all the tracks in one road and the first track of the second road the final stage of the processor can start combining the track information and computing the invariant masses. The square root of the product of the momenta, the three-dimensional opening angle and the di-muon mass are calculated. The processor gives a trigger if the invariant mass is above a certain threshold. The processor needs 5–10 µs to compute the invariant mass and reduces the primary trigger rate by a factor of 10.

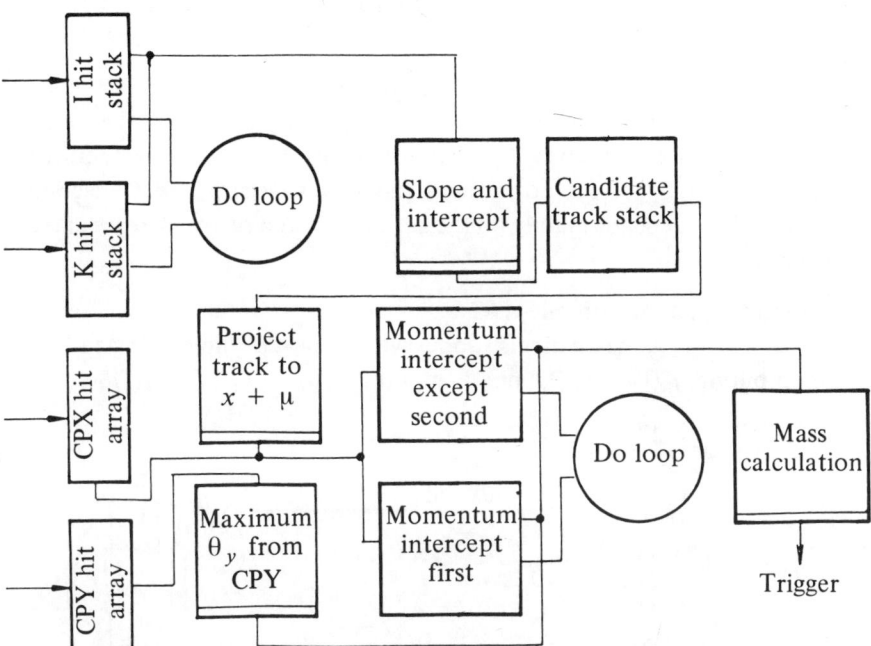

Fig. 1.48 Processor for finding invariant mass. The processor computes first the momenta of the two tracks and then, using the opening angle, the invariant mass (Kostarakis 1981).

1.5.9 Examples of triggers on interaction point

Trigger on charge division

In storage rings the position of the interaction point (vertex) is determined by the beam crossing. Transverse to the beam direction it is determined by the beam size, usually to better than a millimetre and in the longitudinal direction it is accurate within several centimetres. In many storage ring experiments, for example in TASSO, the charged particles are detected with wire chambers aligned parallel to the beam line. The only way the determination of the position of outgoing tracks can be performed is with the help of charge division. The charge of the ionizing particle is divided and appears at the two ends of the sense wiere at capacitors q_E and q_W in Fig. 1.49.

Let ρ be the resistance of the signal wire ($\rho = 190\,\Omega$) and let $Z_R = 390$ be the internal resistors needed to keep the charge then the position along the wire of length L is given by

$$z - Position = [(Q_E - Q_W)/(Q_E + Q_W)](L/2)[\rho/(\rho + Z_R)]F(t). \tag{1.69}$$

Q_E is the charge of the capacitor in front of the ADC which is loaded from the small capacitor during the gating time. The charges in the small capacitors try to equalize via Z_R. Therefore many corrections are required to determine the correct position of the vertex. $F(t)$ is the correction function which depends mainly on the gating time and on the charges in the capacitors.

(1) Subtract pedestals in all ADCs.
(2) Correct for gains. This is done with look-up tables. Instead of computing $g_E Q_E - g_W Q_W$ one computes $G(Q_E - Q_W); G = g_E/g_W$.

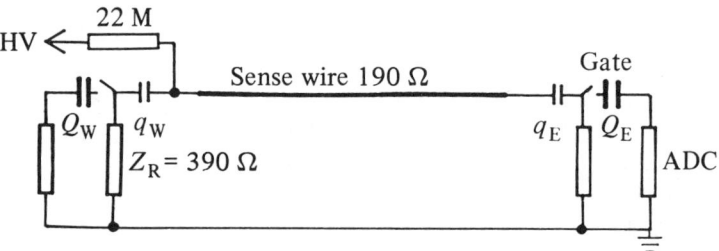

Fig. 1.49 Vertex finding with charge division. As the charges at the two ends of the wire depend on the gate widths and pulse heights a complicated algorithm is needed to evaluate the correct vertex position.

1.5 Examples of triggers

(3) A track passing near a wire gives a large pulse which needs a long time before compensation (Fig. 1.50). The correction for these pulses is small and one would like to use a short gate.

(4) Particles far away from a sense wire induce small pulses. They do not appear before the end of the drift time and tend to equalize rapidly. For these pulses the gate must be at least as long as the drift time plus some charge time for the capacitors.

(5) After having found a hit along a single wire one has to compute the vertex. Four layers are used in the vertex calculation. Hits in two neighbouring layers are averaged and the final vertex is computed by

$$z_V = (R_1 Z_2 - R_2 Z_1)/(R_1 - R_2) \tag{1.70}$$

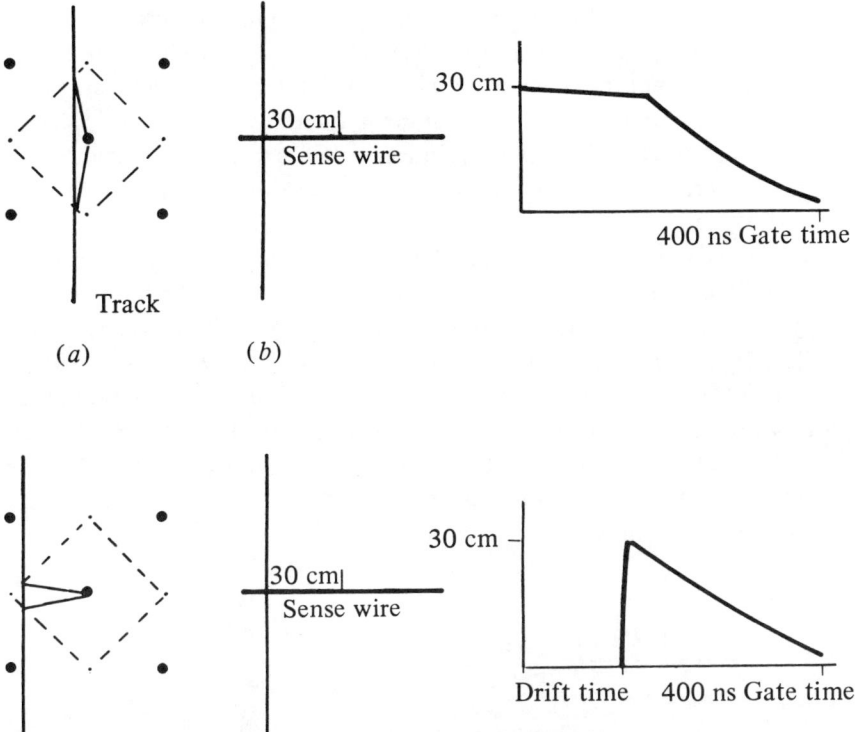

Fig. 1.50 Corrections to determine the vertex. The tracks at the wires produce different charges at different times at the small capacitors. These charges tend to equalize via the internal network. This compensation depends on the gating time and must be corrected in several steps. (a) Perpendicular view; (b) view along the wire.

where R_1 and R_2 are the radii of the chamber. As these quantities are fixed one can use look-up tables with six bits for each Z_1, Z_2 to determine the final vertex.

The time needed to find a vertex is of the order of 10 ms. This rather long time is still sufficient to stop the readout of the experiment into the main computer if this requires more than 10 ms (see Subsection 1.2.4).

1.5.10 *A trigger on interaction point for short-lived particles with a microstrip detector*

In the NA32 experiment at CERN a vertex trigger is necessary to study short-lived particles such as F or Λ_c. These rare particles contain the heavy charm quark and are produced at currently available energies at a rate which is 1/1000 that of 'normal' hadrons with light quarks. The main difficulty consists in finding a very selective signature to separate the signal from the combinatorial background in exclusive decay channel mass plots. A distinctive signature is a finite lifetime in the range 10^{-13}–10^{-11} s. This implies that tracks of decaying particles, when extrapolated each to the primary vertex, have an average impact parameter of 30–3000 μm (Fig. 1.51). One possible choice of detector is a '*Si-strip device*' which has the following advantages. The charge deposition is very localized and a large amount of charge is produced by minimum-ionizing particles. The charge transport to the readout electrodes preserves localization and the charge collection is very fast (< 10 ns). Another possible detector is a Charge Coupled Device (CCD). The great advantage of CCDs is the fact that they are two-dimensional devices. The pixel size is of the order of 23 μm × 23 μm, the effect detector thickness is 15 μm. Fifty thousand pixels are read out in 16 ms. The spatial resolution is $\sigma_x = \sigma_y = 5$ μm and the double track resolution is 40 μm.

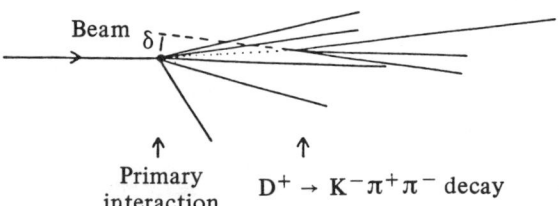

Fig. 1.51 Definition of the impact parameter δ. The spatial resolution to measure the impact parameter of short-lived particles must be of the order of micrometres.

1.6 Implementation of triggers

The experimental set-up at the target region of a big spectrometer experiment (Weilhammer 1986) is shown in Fig. 1.52. Seven microstrip detectors in the beam telescope define the incoming particle. The outgoing tracks are measured by two CCDs and eight microstrip detectors. The detectors are glued under a microscope on 14 mm thick precision quartz frames. These detector elements are then put on an optically flat granite bench to achieve high precission in the parallelism. The spatial resolution is $\sigma \approx 2.6 \,\mu\text{m}$. Depending on the decay topology the background rejection is in the range 1:4000 to 1:40000.

1.6 Implementation of triggers

1.6.1 Electronic components

In an experiment several components are read out. Taking a simple channel one can distinguish several steps (Fig. 1.53). The photomultiplier signal is connected to a discriminator which produces a standard signal if the multiplier signal is above a certain threshold. *The Nuclear*

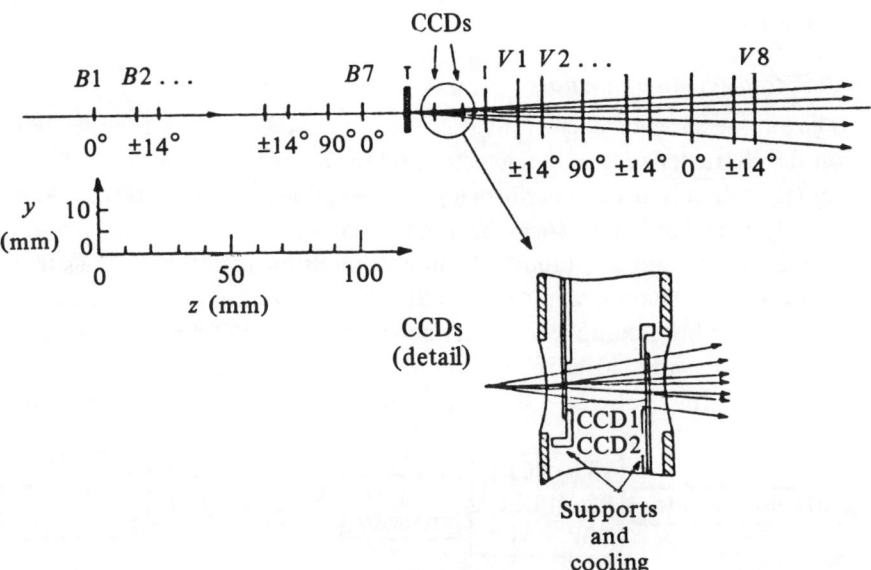

Fig. 1.52 Schematic view of the NA32 vertex detector. $B1-B7$ are microstrip detectors in the beam telescope. $V1-V8$ are microstrip detectors in the forward direction. For good reconstruction two-dimensional readout is performed by two CCDs (Weilhammer 1986).

Table 1.4. *Electronic standards*

	Logic 0	Logic 1
NIM	0 V	0.7 to -1.6 V at 50 Ω
ECL	-0.7 to -1 V	-1.6 to -1.9 V at 50 Ω
TTL	0–0.8 V	2–5 V

Instrumental Module standard (NIM) electronic makes the first decision. It operates at a speed of < 10 ns. Only simple logical operations are possible at this level. More integrated are *Emiter Coupled Logic* circuits (ECL) or *Transistor-Transistor coupled Logic* circuits (TTL) of many different types which can perform complex logic operations. Of this category we will discuss RAMs, FPLAs, CAMs and PALs. These Integrated Circuits (ICs) operate at a speed of 5–100 ns. The bit-slice processors can be used to build specific hardware. They operate on microcode. Their speed is of the order of 50–150 ns per cycle. The microprocessors operate on machine code. They can be programmed in assembler or high-level languages (PASCAL, FORTRAN, C). Data then enter a computer and are stored on tape. The typical signal levels of the various electronic standards are given in Table 1.4.

Pulseformers, discriminators

The output signal of a detector varies in length and in amplitude depending on the characteristics of the detector and the energy of the particle. These signals cannot be used to perform logical operations such as AND or OR. A pulseformer should transform the input signal into an output signal of well-defined length and amplitude. The input signals have all the vagaries that random rates, shapes, amplitudes, and cable techniques can produce. The signals of a photomultiplier can vary between 0.1 and 10 V in amplitude and between 2 ns and 50 ns in length. All these signals should produce a constant output pulse of -0.8 V and a pulse length of about 5 ns or more

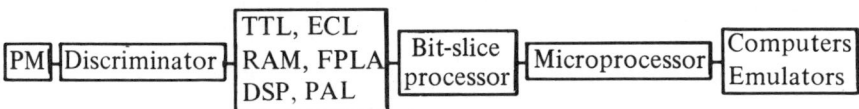

Fig. 1.53 Several electronic steps in a readout system. The speed range of the elements in a readout system varies from 10 ns to 100 ms. Typical elements are FPLAs, PALs, CAMs, bit-slice processors, microprocessors, emulators and computers.

1.6 Implementation of triggers

than 20 ns, depending on the pulseformer. The output signals should appear soon after the input signal has passed over a given threshold. This propagation time is of the order of 7 ns. To allow for good time measurement in TOF counters the time jitter of the variation of the propagation time must be below 0.3 ns. The input sensitivity of a discriminator can be changed, typically between -30 mV and -1 V. It is a challenge to the manufacturer to keep the threshold at a constant value, independent of temperature and input rates. The speed of a discriminator is given by the double-pulse resolution which is of the order of < 10 ns. A pulse which appears within the double-pulse resolution will not be registered.

When two pulses appear at the input of a discriminator with a time difference larger than the double-pulse resolution but shorter than the output pulse width an *updating discriminator* prolongs the output pulse while a *nonupdating discriminator* will ignore the second pulse. Both types of discriminators are used depending on the application.

(1) Discriminators are used within fast electronic logic to produce a standard pulse after a coincidence. The two discriminator types will then show the behaviours indicated in Fig. 1.54. To avoid oscillations for long pulses an updating discriminator must be used.

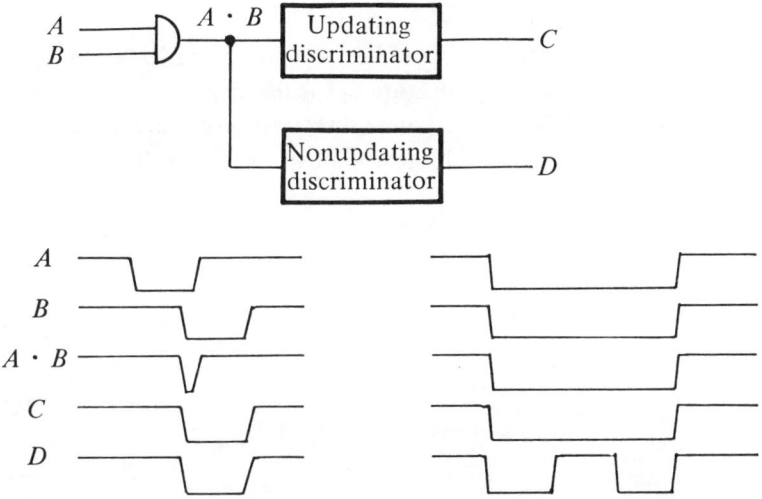

Fig. 1.54 Pulseforms of updating and nonupdating discriminators. For short pulses both discriminators behave in the same way. For long pulses the nonupdating discriminator will generate several pulses. $C =$ updating discriminator, $D =$ nonupdating discriminator.

(2) A veto counter to suppress the beam halo should be equipped with an updating discriminator to be sure that the veto works efficiently (Fig. 1.55).
(3) In experiments which count beam particles at a high rate a non-updating discriminator is advisable. Although the dead time is higher than for an updating discriminator the counting rates do not drop so much at high rates (Fig. 1.56).

A discriminator contains the following building blocks (LeCroy 1985):

(1) A *Schmitt trigger* produces a pulse of variable length but constant amplitude. It goes to logic 1 when the input pulse is above threshold and changes back to logic 0 if the input pulse decreases below 50% of the threshold.
(2) A *differentiator* produces a small spike of 2 ns length at the leading edge of the Schmitt trigger output.
(3) A *multivibrator* or pulseformer produces a pulse of a predetermined length.
(4) An *amplifier* produces a signal of given amplitude ($-800\,\text{mV}$) and cable impedance ($50\,\Omega$) with short rise and fall times (2 ns).

The main building blocks of a discriminator are shown in Fig. 1.57.

Window discriminators

Window discriminators are used if the input signal is to be between predefined limits. One application of the use of a window discriminator is shown in Fig. 1.58. We assume here an experiment in which four scintillation counters should give a hit to form a multiplicity trigger.

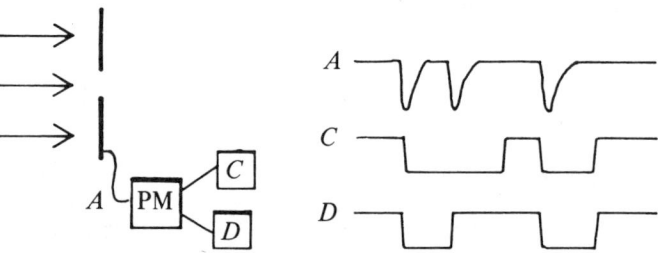

Fig. 1.55 Updating discriminator for veto counters. The nonupdating discriminator will ignore the second input pulse. In the case of an updating discriminator the second input pulse will prolong the output pulse. C = updating discriminator, D = nonupdating discriminator.

1.6 Implementation of triggers

Look-up tables

Another way to 'compute' the number of counters is with the help of a *memory look-up table* with seven input address lines and one output line (Fig. 1.59). The memory is set to zero in all cells apart from addresses 15 ($= D \cdot E \cdot F \cdot G$), 23($= D \cdot E \cdot F \cdot H$), 39($= D \cdot E \cdot F \cdot J$), 71($= D \cdot E \cdot F \cdot K$), 30 ($= E \cdot F \cdot G \cdot H$), and so on. To avoid trigger spikes during memory set-up time (5–200 ns) the trigger must be clocked or the final decision must be delayed

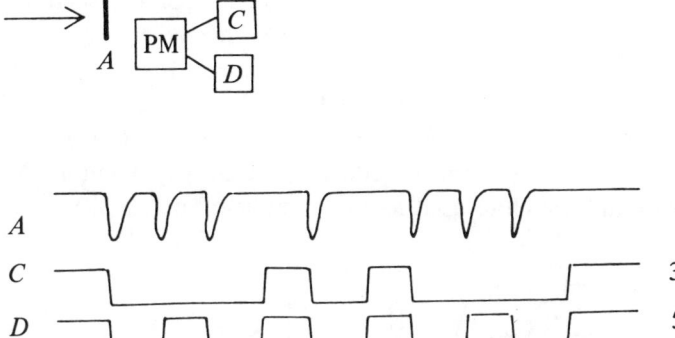

Fig. 1.56 Nonupdating discriminator for high rates. If high rates are registered a nonupdating discriminator counts more pulses than an updating discriminator. But we also see that the rate is too high for both discriminators and needs to be corrected. $C =$ updating discriminator, $D =$ nonupdating discriminator.

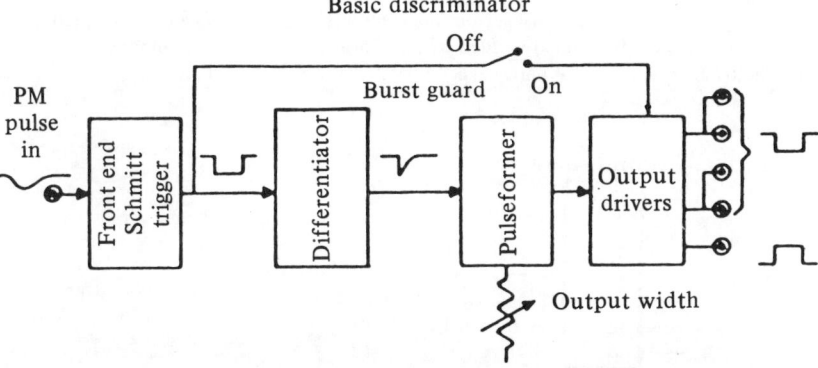

Fig. 1.57 Building blocks of a discriminator (LeCroy 1985). The Schmitt trigger limits the pulse and gives an output as long as the input is above 50% of the threshold. The differentiator is used to trigger only on the leading edge and to be independent of the pulse length.

(Fig. 1.60). Memory look-up tables are widely used to avoid the computation of complex expressions at each trigger. Instead of computing the expression or the function one stores the results for all possible arguments into a memory. This method works well if the range for an argument is not very large (1024) and if the function is smooth. Memories are also used to select a certain trigger combination out of about 12 trigger signals. This method does not work for a large number of input channels because each additional channel doubles the size of the memory. Very often it is possible to split the experiment into several small and independent sections and use cascaded memories. An example is the SLAC Memory Logic Unit (MLU) which accepts 20 input lines and outputs 5 signals. Instead of 5 Mbit RAM only a 32 kbit RAM is needed (Fig. 1.61). The access time for memories varies between 0.5 ns for GaAs memory (Bursky 1985; Mourou, Bloom and Lee 1986), 5 ns for ECL and 40–100 ns for static TTL memory. Memories are cheap and therefore used in many applications.

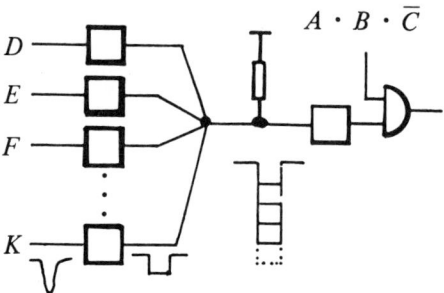

Fig. 1.58 Possible system to trigger on four counters (Fig. 1.28). The output signals (16 mA) of the pulseformers are added by a resistor. A window discriminator with thresholds between -2.8 V and -3.6 V selects triggers with four hits.

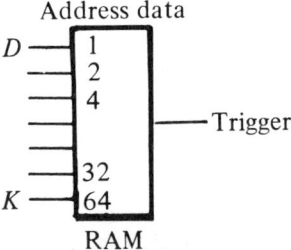

Fig. 1.59 Memory look-up table to trigger on four or more hits. Scintillation counters are connected to the address lines of a memory. Those memory cells which correspond to addresses with four address lines such as 15 ($= 1 + 2 + 4 + 8$) contain a 1.

1.6 Implementation of triggers

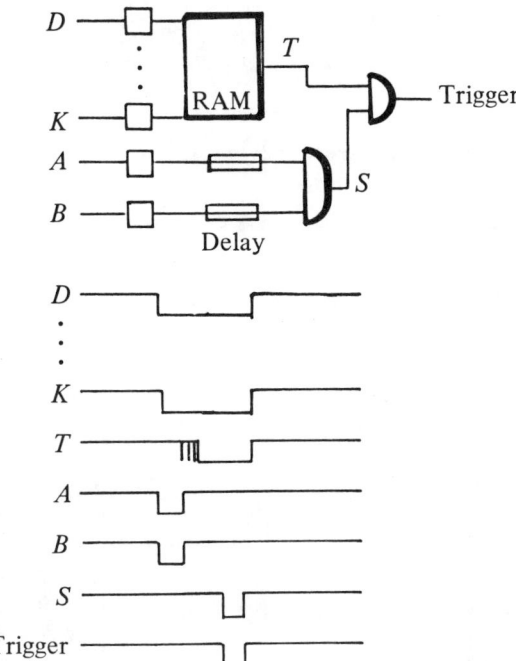

Fig. 1.60 Trigger delay for memory set-up time. A memory needs some time (5–200 ns) before the output signals are stable. The trigger signals must therefore be delayed by this length of time.

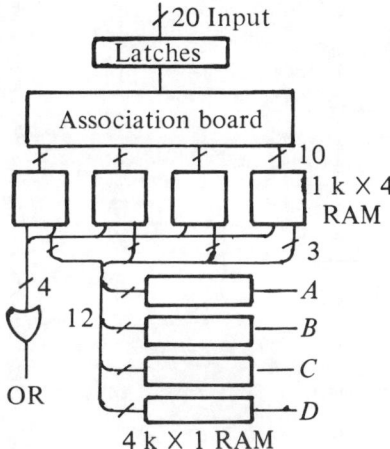

Fig. 1.61 The MLU unit. Very often one only needs a small fraction of a memory. In the MLU the 20 input bits are connected via a plug-in piggycard with 4 RAMs of 1 k. The output is either ORed or fed into another set of RAMs.

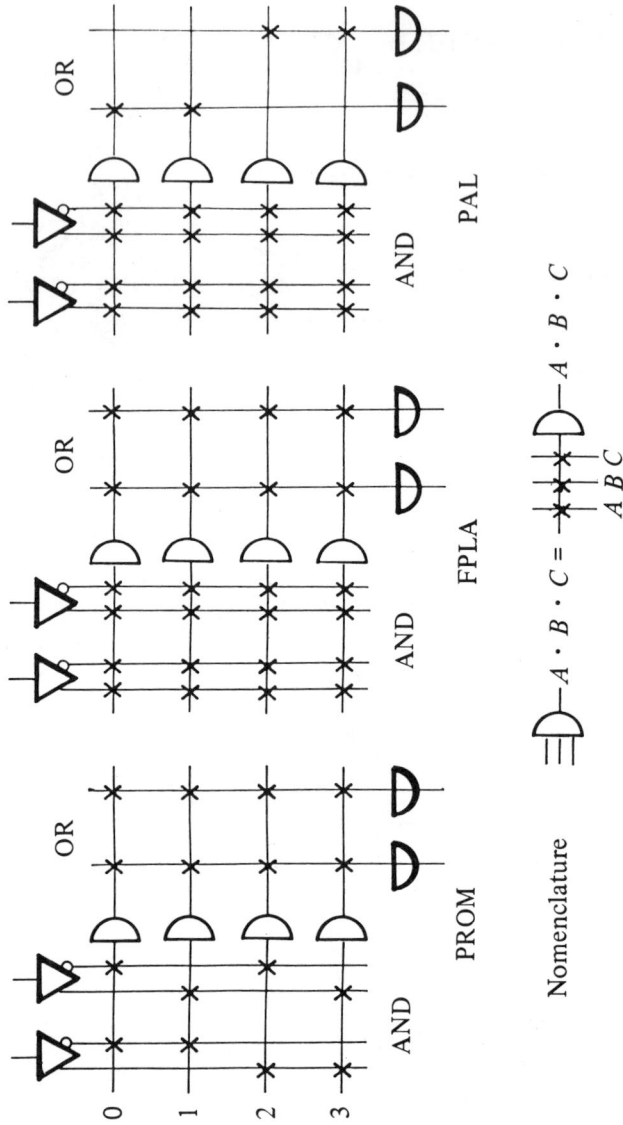

Fig. 1.62 PROM, FPLA, and PAL. A PROM has a fixed AND array to perform the address decoding. A FPLA has a programmable AND and OR field while a PAL has a programmable AND field. All devices are programmed by the customer using a PROM device to burn the fuses.

1.6 Implementation of triggers

FPLAs

We have seen several examples in which FPLA can be used to avoid having many AND and OR integrated circuits. If one wants to allow one inefficiency in a possible track road for six chambers then the trigger condition is

$$\text{Trigger} = 1\cdot2\cdot3\cdot4\cdot5 + 1\cdot2\cdot3\cdot4\cdot6 + 1\cdot2\cdot3\cdot5\cdot6 + 1\cdot2\cdot4\cdot5\cdot6$$
$$+ 1\cdot3\cdot4\cdot5\cdot6 + 2\cdot3\cdot4\cdot5\cdot6$$

Exactly this function can be performed by a FPLA or a PAL. A FPLA consists of a programmable AND array and a programmable OR array (Fig. 1.62). The customer writes the desired logical equations and derives from these how to burn the fuses in the FPLA using a PROM (Programmable Read-Only Memory) burn device.

PALs

A PAL is constructed in a similar way to a FPLA. The PAL has a fixed rather than a programmable OR array. In the example given in the previous subsection one cannot use a PAL which has only four lines ORed together because we need six ORs. In this case one has to use a PAL which is configured in a different way with six lines ORed together. In principle PROMs, FPLAs and PALs contain the same basic elements as an AND array and an OR array (Monolithic Memories 1985). In PROMs the AND array is fixed and represents the address lines, while the OR array is programmable and contains the contents of each cell. The AND array is organized in such a way that only one memory cell is enabled for all address combinations (Fig. 1.62). A FPLA has programmable AND and OR fields allowing many logical combinations. For a PAL the original OR field of a PROM is fixed and the AND field allows flexibility to perform logic equations. PALs can be programmed by a standard PROM programmer. The PAL appears to the PAL programmer like a PROM. (A PAL programmer is not a person but a hardware device used to burn the fuses and to implement the program into the integrated circuit.) During programming some outputs are selected for programming and the inputs are used for addressing. Special PAL assemblers are available which digest the logic equations and produce a binary table to burn the right fuses.

CAMs

CAMs are widely used in trackfinding processors to test whether a predicted track position matches the hits in a chamber. Within a memory

94   *Real-time data triggering and filtering*

cycle (12 ns) a CAM can deliver a yes/no decision if there are hits near the predicted position. Using conventional memories or programs one has to process all the hits of a chamber.

A CAM can be used as a normal memory to write or read information to or from certain memory locations. The operation *associate* reports on the output whether there was a match and on which address the corresponding value can be found. CAMs do not contain as many storage cells as ordinary memories. A 18-pin chip may contain eight words of two bits.

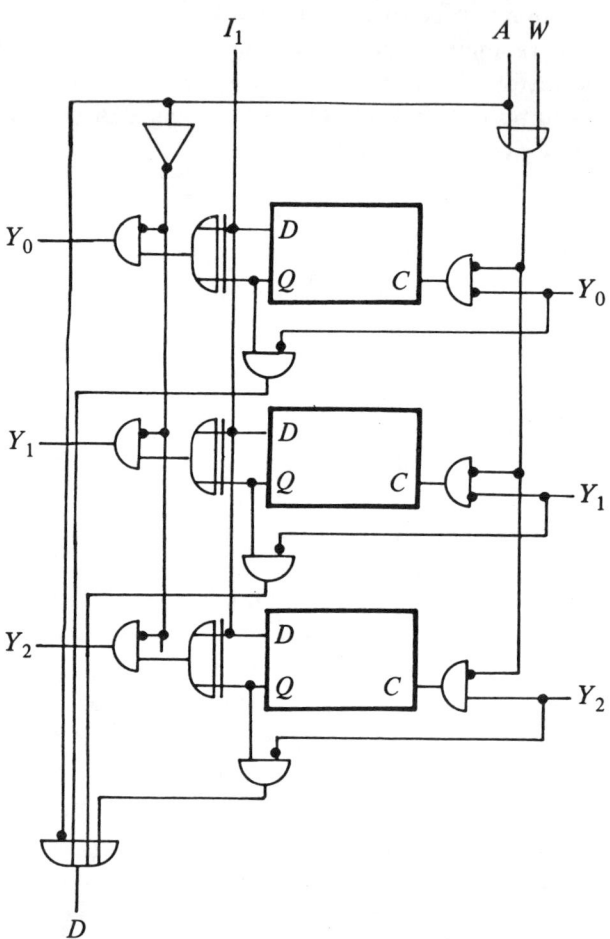

Fig. 1.63 Organization of a CAM. The CAM can be used as an ordinary memory, but, in addition, it can indicate if one of its memory cells contains a specific value. In this case the corresponding address like Y goes down. In this example negative logic is used. $Y_0 =$ means that something should be stored into address 0.

1.6 Implementation of triggers

Fig. 1.63 shows the schematics of a CAM. I_1 contains the information which should be stored into the memory or which should be used as a mask to check the memory. If the mask matches the contents of a memory cell the corresponding address line goes down indicating the address of the matching word.

1.6.2 Pipelines

In an experiment one would like to get all information such as pulse height, timing etc. to a computer for further analysis. On the other hand the output of a device is also needed to generate a trigger–a yes/no decision to take or not to take the event. The output of a device is therefore split into two branches (Fig. 1.64). One branch takes a known fraction of the output signal to an ADC, the other part is sent to the trigger. In the trigger several logical operations are performed which requires some time, in the range 50 ns to several microseconds. When the trigger appears at the gate of the ADC the original signal is already lost. One has to delay the original information and store it in a pipeline.

Delay lines

Delay lines of 50 m length can be used if the trigger does not need more than 250 ns. The signals from the photomultiplier are attenuated on the cable. This attenuation depends on the frequency. As all the photomultiplier signals have the same shape or frequency the attenuation is constant. A long

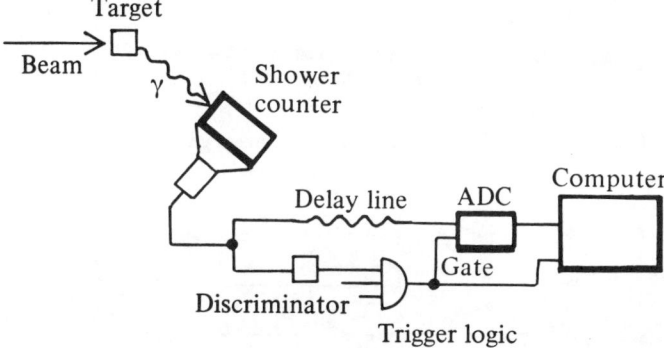

Fig. 1.64 The requirement of a pipeline. The trigger needs some time to make its decision. During this time part of the analog signal is stored in a pipeline and is available at the gate to the ADC if the trigger processing has finished. This method requires a fixed trigger time.

cable can contain the data from several events, one behind the other. For this reason such an arrangement is called a *pipeline*. A cable is an approximation of coupled inductances and capacities. Using discrete elements one gets a *lumped delay line* which can be used for delay times of 1 μs (Fig. 1.65). A disadvantage of these delay lines is the distortion of the pulse shape, but the charge of the pulse at the end of a delay line is proportional to the charge at the beginning. Delay lines need a good resistor adaption at the ends to avoid reflections in the cable.

AMU (Analog-Memory Unit)

An analog cable pipeline can take information at any time and deliver an attenuated signal at the output after a fixed time. It can only be used to store information for a short time (<1 μs). If one wants to store analog information for a longer time one can use capacities. This technique is used in *sample-and-hold ADCs*. The input signal charges a capacity. At the arrival of a trigger some microseconds later the capacity will be discharged and the signal will be digitized. If one wants to use this technique as a pipeline one has to use several capacities which are charged by the raw data and discharged some time later when there has been a trigger. This lead to the development of an *Analog-Memory Unit (AMU)* at the Stanford electronics centre (Freytag and Walker 1985). The device contains 256 analog storage cells consisting of switching transistors, a storage capacity and a differential readout buffer. The principle of this device can be seen in Fig. 1.66.

To avoid the need for many input/output connection switches to the outside of the integrated circuit one uses for each cell two input and two

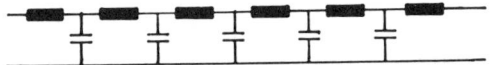

Fig. 1.65 Lumped delay line. The delay time is give by $n(LC)^{\frac{1}{2}}$.

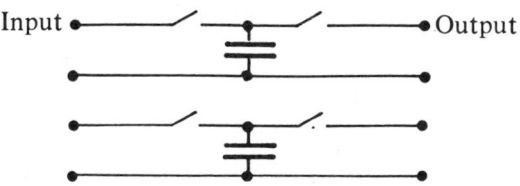

Fig. 1.66 Principle of the AMU. Several capacitors are charged if there are data available and discharged to an ADC if there is a trigger.

1.6 Implementation of triggers

output switches arranged in the rows and columns of a 16 × 16 matrix. In practice the storage capacitors are connected through Field Effect Transistor (FET) switches to the signal input. The time constant of the input combination is 1 ns, resulting in a bandwidth of greater than 100 MHz. The wide dynamic range of more than 11 bits is basically due to the large stored charge of 10^7 electrons at full scale. Readout of the analog information proceeds through a matched pair of source followers with a reference voltage accessible from the outside to inject correction levels. The device is built in NMOS technology with a feature size of 3 µm. Fig. 1.67 shows the circuit diagram of one storage cell.

CCDs

A CCD can be used to pipeline analog information with a speed of 20 MHz and a length of 512 cells. CCDs are used to delay analog signals in radar, television, and telecommunications. Each CCD contains an input stage, a

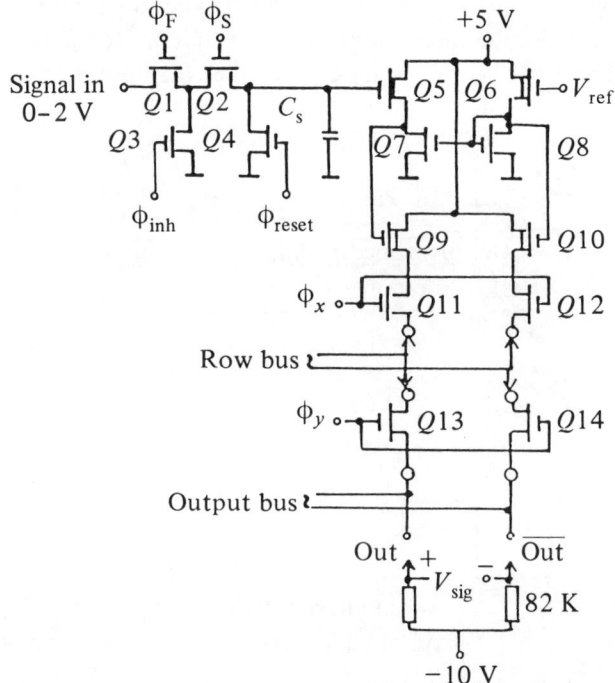

Fig. 1.67 Storage cell of an AMU. Two FET switches, one connected to the rows and the other to the columns of a 16 × 16 matrix, are used to connect the capacitors to the outside world. C_s is the capacity of storage cell, V and ϕ are voltages and Q are switches.

charge transfer register and an output stage. The charge packets are proportional to the difference between the input voltage and a reference voltage. An outer clock pulse is needed to transfer each packet to the next stage. If a CCD is operated at lower frequencies ($< 10\,\text{kHz}$) the dynamic range is reduced due to thermal charge generation in the shift registers (Barbe, et al. 1980). A schematic picture of a high speed CCD is shown in Fig. 1.68. The structure consists of a p substrate with a thick ($5\,\mu\text{m}$) high resistivity epitaxial layer on which a thin n^+ layer is implemented. The gates are isolated from the n layer by a thin oxide layer.

Digital pipeline

To avoid thermal effects within electronic pipelines one can use a digital pipeline if the input pulse can be digitized in the order of 10–50 ns (depending on the application). Flash ADCs reach this high a speed. They compare the incoming pulse with a chain of comparators and present digital information within 10 ns. This information can then be transferred to a shift register or to a memory and can be kept for a long time. The memory needs twice the speed of the input rate if it has to operate without dead time (Fig. 1.69). At the rising edge of the clock pulse one takes the

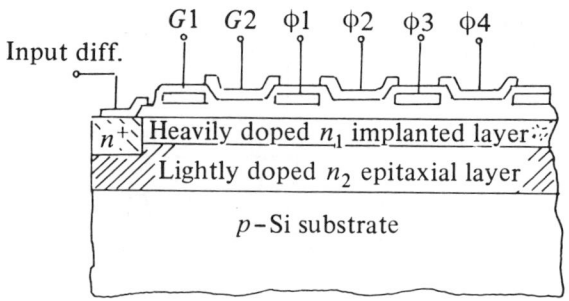

Fig. 1.68 CCD. G and ϕ are voltages.

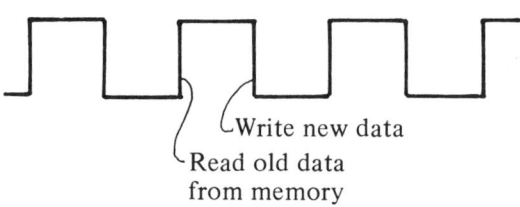

Fig. 1.69 A memory used as pipeline. On one edge of the clock, data are read from the memory and on the other edge new data are stored into the memory. If data arrive 100 ns apart a memory with a 50 ns access time is needed.

information out of the memory to the event buffer. On the falling edge new information is fed into the memory. The disadvantage of all pipelines apart from the delay line is that one cannot use pipelines to measure the exact timing information which is required for TOF measurements.

1.7 Programmable devices

1.7.1 *Bit-slice processors as event builders*

Bit-slice processors typically consist of four bit wide *Arithmetic and Logic Units* (ALU) which perform operations such as ADD, SUBtract, OR, AND, NOT or EXclusive OR. Typical examples of slice processors are the AMD 2901 from Advanced Micro Devices (Advanced Micro Devices 1978) or the Motorola 10800 in ECL technology (Motorola 1977). The cycle time of these processors is of the order of 60–70 ns. The bit slices can be cascaded to form a 16- or 32-bit processor. To avoid a delay, to transport the carry bit from one slice to the next the carry bits of all processors are handled simultaneously by a carry look-ahead generator. This device accepts up to four bits of carry propagate and carry generate signals and a carry input from the ALU and provides anticipated carries. The AMD 2901 contains 16 registers, a multiplexer to select the data sources, the function unit, and some shift registers for multiply or divide. With the help of some multiplexers one can connect the shift registers to perform shifts and rotations for full words or double words.

As an example we describe a fast processor which should read out data from an experiment via CAMAC (see Section 1.8). To perform Input/Output (I/O) the processor contains an internal bus which connects the ALU to the memory and to some CAMAC I/O registers (Fig. 1.70). The ALU can get its information from a *pipeline register*, the memory, or the I/O registers. The pipeline register gets its information from the microcode. The sequence of instructions is controlled by a sequencer which contains an incrementer for the next instruction and a stack to store the return addresses for a subroutine call. Return addresses cannot be stored in the program memory because the program often resides in a PROM. The sequencer (AMD 2911) takes the next address unless it is forced by an outside condition to jump to another program location. This outside condition can be either an arithmetic condition (zero, overflow, $A < B$) or conditions from the hardware (data ready, CAMAC Q response, (see Section 1.8) loop counter overflow). The counter can be initialized at the

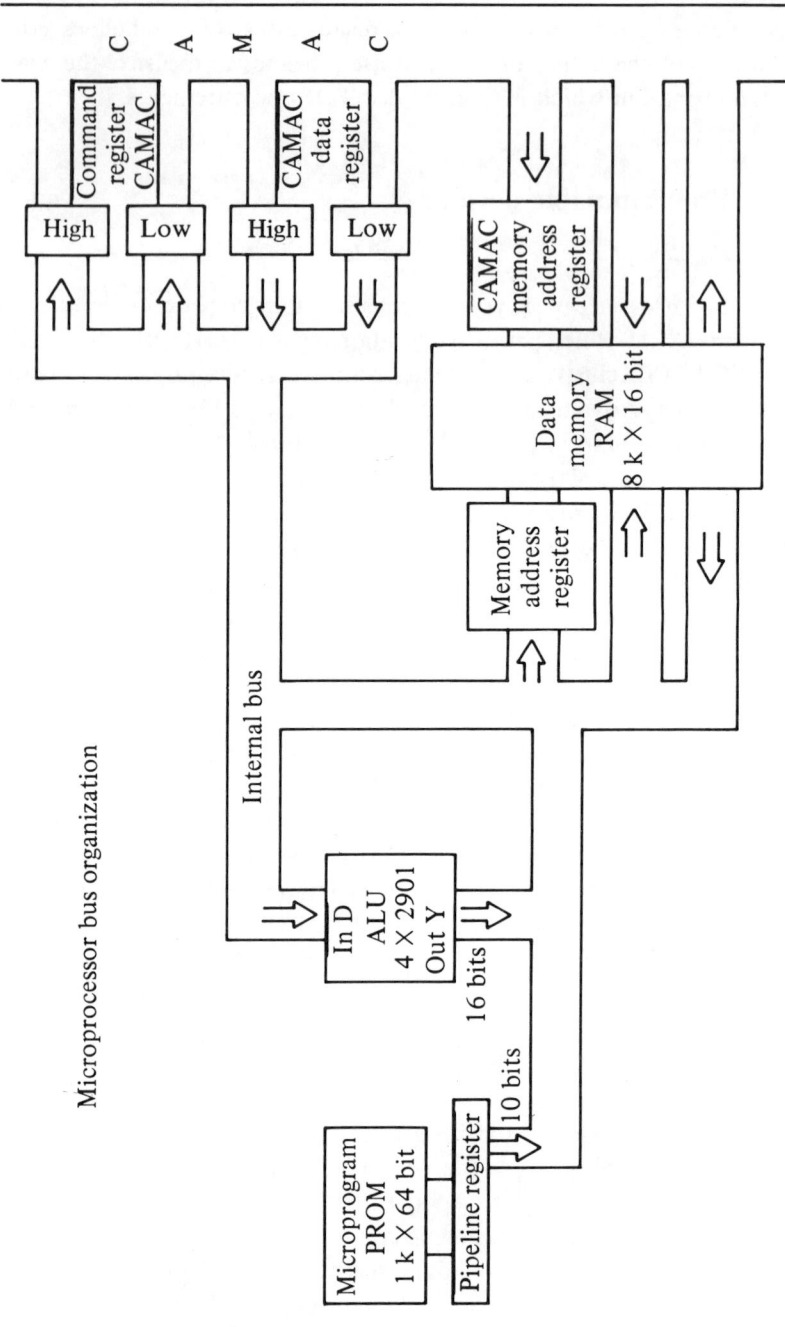

Fig. 1.70 Internal bus of a CAMAC processor with AMD 2901 bit slices. Data can be transferred to the memory, the memory address register, the ALU, and CAMAC and from the pipeline register. Such a processor is therefore very flexible.

1.7 Programmable devices

beginning of a loop. On each instruction it is incremented by 1. If it reaches 0 it gives a signal to the sequencer to finish the loop.

All the building blocks mentioned above are directly controlled by the *microinstruction word*. The length of an instruction word is therefore very long, 64–128 bits. In our example the instruction word contains ten bits for constants such as program addresses or loop counter values, four bits for the sequence instruction, five bits for the sequence multiplexer, 22 bits for the ALU, six bits for the shift/rotate multiplexers, six bits for bus source/destination and five bits for I/O operations. Each instruction performs a specific task, therefore one speaks about microinstructions or microcode. The programmer must know all the details of the processor architecture and other bits when writing a program. He or she is supported by microcode assemblers and linkers (Notz and Rehlich 1980).

The advantage of microprogrammed processors is their high speed. These processors are built for a specific application. Within the instruction itself one can ask whether data from the experiment are available or whether a CAM contains correct data. Supposing one wants to compute the following loop:

```
      S = 0
      D = 1
      DO 2 I = 1, 8
      S = (S + D)*2
    2 CONTINUE
```

This will require only two instructions in a bit-slice processor:

(1) Set $S = 0$ (register 7 = 0) and initialize the loop counter.
(2) Increment register 7 by 1, shift the result by one position to the left and test for the end of the loop counter.

The FORTRAN loop is performed by a single instruction!

```
      LRF AND ZB 7 LDCT O1771 (oct)
    2 LRU ADD ZB 7 RPCT L2 SLSS CH CNEZ
```

The first instruction makes an AND of ZERO and register 7. The result is 0 which is stored into register 7. The loop counter is set to -7 and not to -8 due to pipeline effects. The second instruction performs the loop counting, the adding by 1 and the multiplication by 2.

LRU = multiply the result by 2 and store it back into the registers
ADD = add the two arguments

ZB = the arguments for the ADD are register 7 and 0 (and the carry bit)
RPCT L2 = repeat all instruction from label 2
SLSS = set the shift multiplexers to single word shift
CH = set the carry bit to 1
CNEZ = repeat the loop until the counter is 0.

The high speed and the integration of the I/O hardware make bit-slice processors suitable for event builders, trigger processors, and data correction and filter processors. An event builder collects all the bits and pieces from different stations or crates and packs the information into its final form. While an I/O operation is being performed the processor can continue and format the previous information because the processor can perform several microcode instructions during that time.

1.7.2 Digital-signal processors (DSP)

In recent years *digital-signal processors* (*DSPs*) have undergone enormous development mainly in response to demands from the field of speech and pattern recognition, fast Fourier transforms and digital filtering. DSPs are essentially special purpose *Reduced Instruction Set Computers* (RISC), optimized for fast multiplication. With the commonly used *Harvard architecture*, i.e. separate data and instruction paths, the Texas Instrument's TMS32020 is a typical product. It has a 200 ns cycle time and can perform a 16 bit × 16 bit multiplication in one cycle. It has 288 bytes of RAM on chip and dissipates 1.5 W.

1.7.3 Transputers

Transputers are RISC computers with a very small instruction set and only a few registers. They have an internal memory and a parallel address or data bus which is 16 or 32 bits wide. Context switching from one task to another is done very rapidly. Transputers are optimized for interprocess communication, not only in their *internal* architecture, but also *externally*. Apart from the parallel bus, each chip has four bidirectional links that are used for message passing between processes running in different CPUs (Fig. 1.71). The links can operate in Direct Memory Access (DMA) mode and transmit data, while processes in the transputers continue to run. Because of these properties transputers can easily handle parallel processing which, in addition, is supported by a unique programm-

1.7 Programmable devices

ing language called *OCCAM* (OCCAM 1984). This language allows one, in a simple way, to describe parallel or sequential tasks of a program running concurrently on different transputers.

As an example one can imagine a calorimeter trigger implementation where many transputers are used to process the detector information. Each transputer receives information on a limited region of the calorimeter and performs pattern recognition on these data. Results are passed via serial links and combined by separate processes running on supervising transputers.

1.7.4 Emulators

The available computing power is a bottleneck in many experiments and prevents one from analysing the data in a reasonably short time. To overcome this problem one can try to get cheap computing power in two ways:

(1) One can buy many microprocessors to process all events. This approach is cheap but requires a recompilation of all programs and a

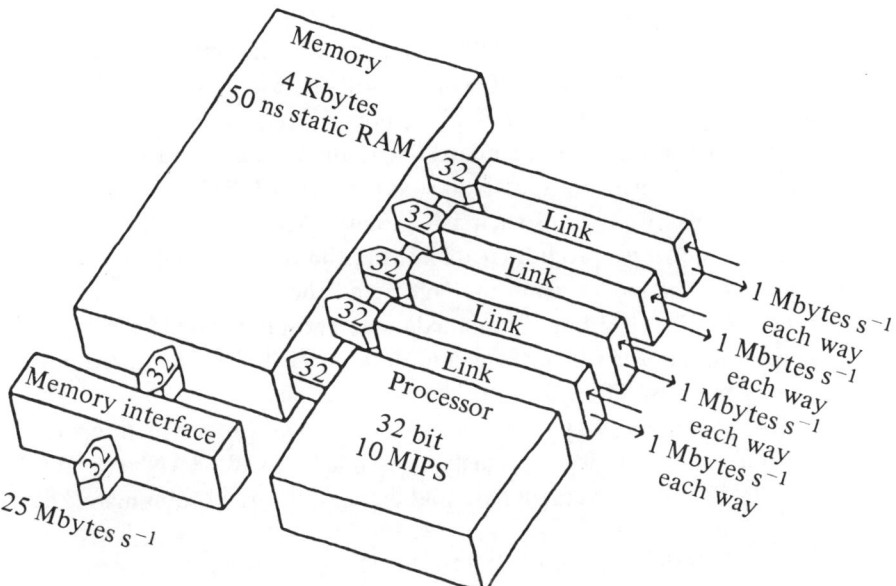

Fig. 1.71 INMOS transputer. Transputers are RISC computers with a very small instruction set. Apart from the internal bus, each chip has four bidirectional links that are used for message passing between processes running in different CPUs.

change of data formats from one floating point format to another. For FORTRAN programs compilers are available and the software effort to adapt to a new system is limited, but the user still has to rewrite some assembler and timing routines and has to learn how to run a new operating system with all the editors and commands and how to interpret error messages.

(2) One can produce a computer which is compatible with the mainframe of a big computer producer. Examples are Amdahl and Fujitsu computers which are compatible with IBM computers and the wide market of compatible personal computers. Emulators are CPUs which emulate the code of a big computer, i.e. code which has been written for one computer can run on an emulator. These emulators are not so cheap as single chip microprocessors but they have the advantage for the programmer that he or she does not need to change the code, learn a new operating system, or convert the data.

There are some *emulators* used in high-energy physics. Three of them emulate the IBM 370 code, one emulates the NORD100 (Norsk Data) code and one emulates the PDP-11 (Anthonioz, *et al.* 1981). In the following we will only discuss the IBM emulators 168/E (Kunz 1976, 1981; Bernstein 1980), 3081/E (Kunz *et al.* 1983; Fucci and Storr 1983) and the 370/E (Brafman *et al.* 1983a, b; Notz 1982; 1985a, b). The first emulator used in high-energy physics was the 168/E which is still in use in many experiments. It consists of an integer CPU, a floating point CPU, a control board, an interface, and several memory boards. The interfacing is done to IBM channels, to experiment computers like NORD100, PDP11, or VAX, or to the experiment's bus systems such as CAMAC, VME or FASTBUS.

Before transferring a program from IBM to the 168/E the program must be translated. The memories in the 168/E and the IBM are organized in different ways. The IBM has a combined memory for data and instructions. To increase speed, the instructions and data in the 168/E are stored into separate memories. The translator has to take the output of the IBM FORTRAN compiler and to split it into a data part and an instruction part. Apart from some limitations this task is possible for FORTRAN programs because the compiler places all data and constants at the beginning of the code.

The 3081/E is the successor of the 168/E. This processor is twice as fast as the 168/E. It uses two 64 bit wide buses and extra boards for multiply, divide, floating point add/subtract and an integer CPU. As in the case of the 168/E, programs for the 3081/E must be translated to separate instructions and data. Another speedup is possible with the help of software pipelining to overlap floating point and integer instructions.

The 370/E is a processor which operates in a speed range between the 168/E and the 3081/E. The big advantage for the user is that this processor uses the same architecture as the IBM. Data and instructions are not separated. Therefore the user's code and the complete FORTRAN runtime library with its I/O routines and error detection capabilities can be loaded directly without the translation step. Only the link step must be repeated to load the 370/E system in front of the user's code.

1.7.5 Parallel processing

Needs for computing power and parallel processing

An experimentalist working at a large 4π detector needs a huge amount of computing power for event processing, Monte Carlo generation and physics analysis. In addition accelerator physicists and theorists need a lot of computing power to simulate accelerators and to do model calculations, for example for lattice gauge theory.

An event of 100 kbytes of data can be processed in 10–100 s on a mainframe computer. It is not a problem for an experimentalist to find a computer with sufficient power to process a single event. The main problem is that several events are collected per second while the detector is taking data. Thus millions of events are collected per year. Event processing must keep up with event collection, otherwise there would be no feed-back to the running detector and results would be available only after a long delay.

Fast mainframe computers like the Cray-2 and the SX-2 operate at cycle times of 4.1 ns and 6 ns, respectively. With new technology like *GaAs gates* and *cryogenic Josephson junctions* one can gain a factor of 10 by using sequential computers. In order to obtain higher speeds one has to incorporate parallel processing. Parallelism in computing systems is simply doing more than one thing at a time. In principle, there is no limit to the number of concurrent actions offering improvement in computing speed.

Classification of parallel processing

Following a classification of V. Zacharov (Zacharov 1982) one can distinguish the following categories:

(1) Parallelism within *functional units*. Arithmetic, logical, and other operations can be implemented in parallel bit-serial mode by bit groups

(bit-slice processors). This category does not affect the way in which a problem is formulated.
(2) Parallelism within *processing elements.* Different operations are executed in parallel on different operands, for example. While a floating point operation takes place some integer numbers or indices are executed. Another kind of concurrency is shown for example by a multiplier which can process a stream of operands in an overlapped or pipelined fashion. Pipelining is very powerful and can give significant improvements in execution speed.
(3) Parallelism within *uniprocessing* computers. An example for a single processor is concurrent I/O or DMA.
(4) Parallelism in *many processor systems.* An obvious category for concurrency is where a computer system contains several processors. The processors may or may not share the main memory, and communicate with each other. In the most cases the separate processors are used to execute independent jobs. This gives higher throughput but the concurrency is not used to speed up the execution of individual jobs.

Another classification for parallel processing comes from the field of *vector processors*:

(1) *Single Instruction Multiple Data* (SIMD) architecture is used in vector processors to operate on data fields.
(2) *Multiple Instruction Multiple Data* (MIMD) vector processors can handle several instructions concurrently on data fields.

Pipelining for execution

In many computational instructions the instructions can be decomposed into several steps such as: fetch instruction, decode instruction, calculate the operand address, fetch the operands and execute the instruction. (Another example is: fetch a floating point number, normalize, multiply the number, normalize the result and store it back.) In a pipelined machine the instructions are decomposed into separate segments S_1, S_2, S_3, S_4 which can operate independently of each other. The advantage of pipelining can be seen in Fig. 1.72 where five instructions are performed in eight instead of twenty machine cycles.

There are, however, some problems with pipelining. The benefits of pipelining can vanish in the following cases: A computer fetches instructions one after another from the memory and then places them in a pipeline for further decoding. If a GOTO statement or a conditional branching

1.7 Programmable devices

occurs in the program then the pipeline must be cleared. A way out of this problem is to have two pipes, one is used if the condition for a branch is fulfilled and the other if it is not. Another example of contention occurs when one instruction computes and stores a result into a specific register. If this register is used in the address calculation of the following instruction the pipeline must first be emptied or delayed. Also if one instruction in a program changes the following instruction the hardware must recognize this and must clear the pipeline. Otherwise the unmodified instruction which is already in the pipeline will be used (Fig. 1.73). From the problems mentioned above one can see that it is impossible to estimate how much speed is gained for a given application because programs have GOTO and IF statements.

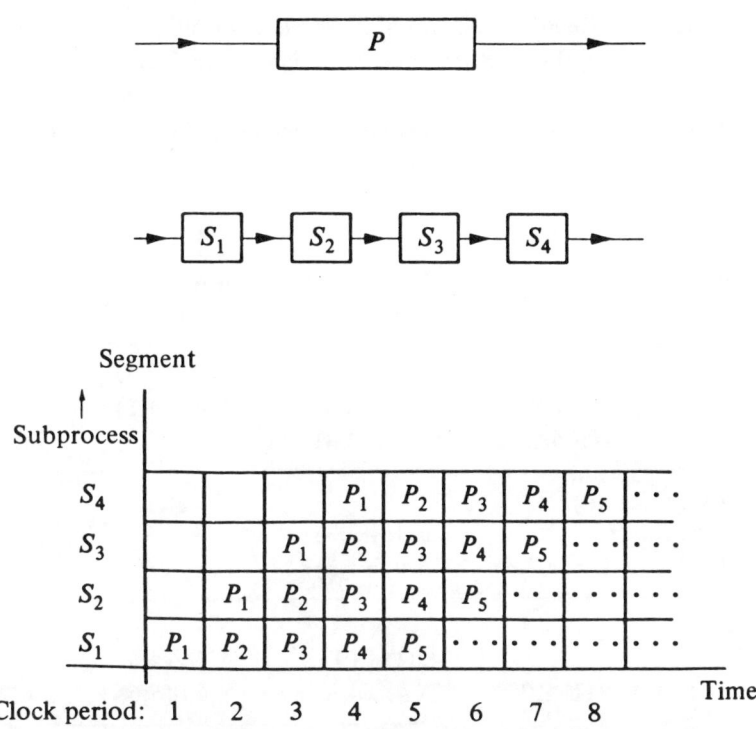

Fig. 1.72 The effect of using pipelining in a computer to speed up processing. The instructions are decomposed into small segments for fetch, decode, normalize, multiply and address calculations etc. All the segments run independently of each other and can be overlapped. In this example five instructions are done in eight instead of 20 machine cycles.

Multiple functional units

With multiple functional units one can achieve high computing speed within a single processor. Each unit is activated by a separate instruction specifying the corresponding operands. The units overlap in execution which results in problems of synchronization and bus contentions. The complexity involved in solving synchronization problems can be imagined for cases in which the input parameters for one function may result from others.

Many-processor systems

Many-processor systems use many processing elements of the same kind. In order to add two vectors of n elements, n additions could be performed in parallel in a single cycle if n parallel processors were available (vector processing). For problems which can be organized in such a manner this technique produces a substantial increase in speed over traditional serial machines. But the performance diminishes considerably for problems which are not vectorizable or if the result of each addition is examined individually. Many physical problems such as initialization and collecting

EXEC			L 1,4(2,3)	A 4,2(4,3)
ADDR.		L 1,4(2,3)	A 4,2(4,3)	S 4,0(5,6)
FETCH	L 1,4(2,3)	A 4,2(4,3)	S 4,0(5,6)	N 3,4(5,6)
EXEC			B F,4(2,3)	—
ADDR.		B F,4(2,3)	A 4,2(4,3)	—
FETCH	B F,4(2,3)	A 4,2(4,3)	S 4,0(5,6)	→ Next
EXEC			L 1,4(2,3)	—
ADDR.		L 1,4(2,3)	—	→ A 4,2(1,3)
FETCH	L 1,4(2,3)	A 4,2(1,3)	—	S 4,0(5,6)
EXEC			STC 1,(7)	—
ADDR.		STC 1,(7)	A 4,2(4,3)	—
FETCH	STC 1,0(7)	A 4,2(4,3)	S 4,0(5,6)	S 4,2(4,3)

Fig. 1.73 Problems which may occur in pipelining. The pipeline must be cleared when there are GOTO or conditional IF statements. Contention occurs if one instruction requires the result of the previous instruction or if one instruction changes the following instruction.

1.7 Programmable devices

Table 1.5. *Simulated performance of a hydrodynamic test*

Number of processors	Speedup	Efficiency
1	1.00	1.00
2	1.77	0.89
4	2.93	0.73
6	3.56	0.59
8	3.84	0.48

statistical information appear to be nonvectorizable. There is a great difference between adding two vectors of 64 elements with 64 processors and summing 64 numbers with 64 processors. In the first case one needs one cycle, while in the latter at least $\log_2 64 = 6$ cycles are required.

Another approach is to use multiple computers or processing elements which do not perform the same operation on different data at the same time but instead work on the different tasks of a calculation simultaneously. The multiprocessor systems on the market are designed to run separate problems and do not work collectively on the same program. These processors achieve higher throughput but an individual job requires as much time as on a sequential processor. The main problem in coordinated processing on a single problem lies in multiprocessor control, coordination, and interprocessor communication. If several processors write to the same memory location in a single job the result depends on the timing. The information written from the first processor is lost if a second processor later modifies the same location.

Another problem is contention between processors for memory and bus access. This can easily reduce the performance to 70% for a two-processor system even if the processors work on different programs (Maples 1984). A simulation of the expected performance for a hydrodynamic test is given in Table 1.5 (Axelrod, Dubois and Eltgroth 1983). An eight-processor system is expected to perform with only 48% efficiency. If one tries to avoid memory contentions with the help of individual memories, a large amount of interprocessor communication can degrade the system's performance.

Decomposition into a serial part and a vector part has a crucial influence on the system, as can be seen in Ware's model of multiprocessors (Maples 1984) in which one assumes that either one processor (for serial code) or all processors (for vectorizable code) are in operation. Table 1.6 shows the way in which the fraction of serial code in terms of time can degrade the performance of the system. A very small amount of 1% serial computation

Table 1.6. *Ware's model of multiprocessors*

	Relative speedup (Efficiency)			
	Number of processors (P)			
% serial code (α)	2	8	16	100
1	0.99	7.5(0.93)	14.0(0.87)	50(0.50)
5	0.95	6.0(0.74)	9.1(0.57)	17(0.17)
10	0.91	4.7(0.59)	6.4(0.40)	9(0.09)
20	0.83	3.3(0.42)	4.0(0.25)	5(0.05)

has little effect for a few processors but in a system with 100 processors it can degrade the performance by 50%.

For P processors working in parallel and part of the time in serial, the speedup $S(P)$ and the efficiency E of each processor can be calculated according to the formulae:

$$S(P) = \frac{1}{\alpha + (1-\alpha)/P} = \text{Speedup} \qquad (1.71)$$

$$E = S(P)/P = \text{Efficiency} \qquad (1.72)$$

where α is the amount of time spent in serial calculation. In a typical high-energy physics program the different events can be digested or generated independently of each other in separate processors. The I/O is sequential because data comes from a single file and is written on a single file. Also the collection of statistics and their display as histograms is a sequential task which must be done by one processor (or vector processor).

Programming aspects for parallel processing

When programming a parallel processing system some new aspects have to be considered which are not necessary in sequential programming. We will restrict our discussion to systems with several processes sharing a common file or memory, neglecting here the optimization of data structures or programs for vector processors.

As a first example let's assume that a host computer reads events from a file and transmits them to different processors for further processing. As soon as one event has finished it is transferred back to the host, together with its result, and written to the output file. This kind of operation has the consequence that *the output events are not in the same order as the input events.*

1.7 Programmable devices

In another example if we think of the generation of Monte Carlo events by a multiprocessor system, then all processors are loaded with the same Monte Carlo program and generate events, but as the programs in all the processors are identical this system will produce the same event sequence on all processors. Therefore *the random number generator must be set individually on each node after the program has been loaded.*

In the last two examples independent events are processed by independent processors. The program in each computer is a sequential program. But the host computer must handle requests from all computers and has to synchronize them. We will now consider the situation in which several processors share a single task and process concurrently. As an example let's assume that a thousand numbers must be sorted. In this case ten processors can sort hundred numbers each. Afterwards the 'presorted' numbers must be merged.

This kind of operation leads to concurrent programming (Ben-Ari 1982). Concurrent programming is motivated by the problem of constructing operating systems. Two or more sequential programs are not independent of each other. They must communicate in order to exchange data or to synchronize for external input or output. In distributed systems synchronization can be done by the sending and receiving of signals which is called '*message passing*'.

Concurrent programming leads to new programming problems which are unknown in sequential programming:

(1) *Mutual exclusion* is the abstract model of many synchronization problems. Activity A_1 of process P_1 and activity A_2 of process P_2 must exclude each other if the execution of A_1 must not overlap the execution of A_2. The most common example of the need for mutual exclusion is resource allocation. Two tapes cannot be mounted simultaneously on the same tape unit. The abstract mutual exclusion problem can be expressed as: *remainder, preprotocol, critical section, postprotocol*.
(2) *Robustness* describes the characteristic of a system that a bug in one process does not propagate to a system 'crash'. It should be possible to degrade the performance if an isolated device fails ('fail soft').
(3) *Correctness* of a program is assured when a program prints the correct answer and then stops. In concurrent programming one distinguishes two types of correctness properties: *safety* and *liveness* properties.

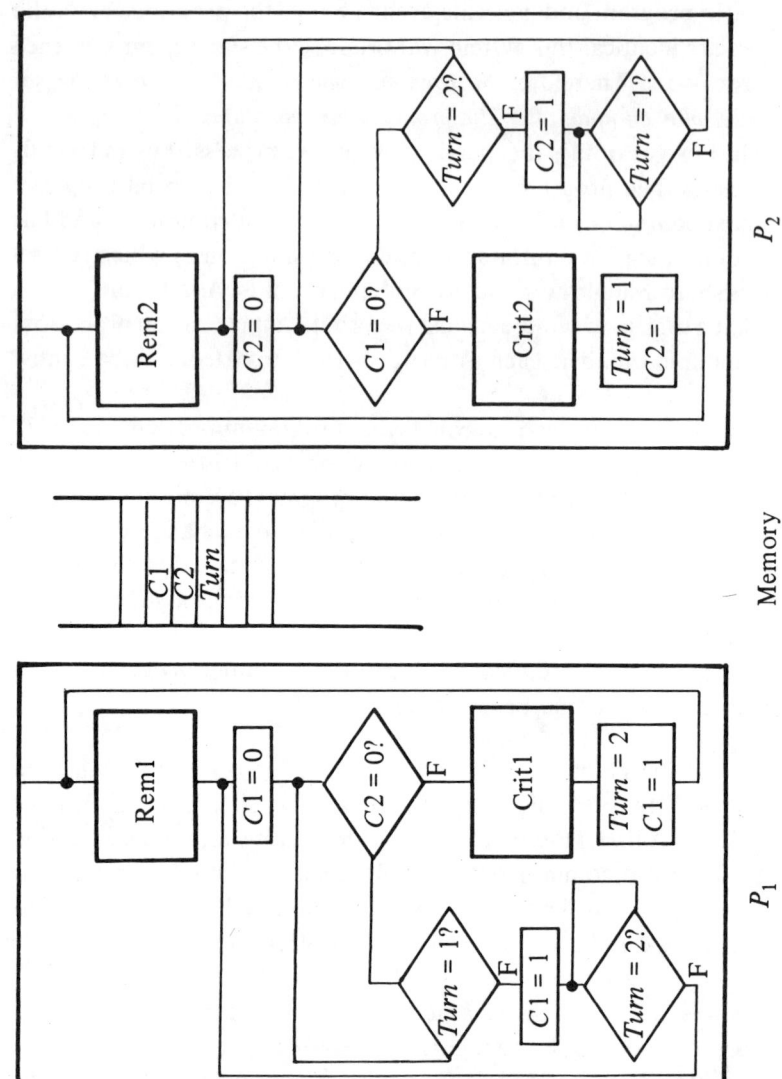

Fig. 1.74 Dekker's algorithm for two processes which should enter a critical section exclusively.

1.7 Programmable devices

(4) *Safety* properties describe the static portion of a program. Mutual exclusion is absolute and does not change during execution. In a producer–consumer problem the consumer must consume each piece of data produced by a producer.
(5) *Liveness* deals with the dynamic behaviour of a system. Liveness means that if something is supposed to happen then eventually it will happen. The most serious breach of liveness is the global form known as 'dead lock'.
(6) *Fairness* is the concept that each process wishing to process must get a fair chance relative to all other processes.
(7) *Timing* is ignored in the concurrent programming model. One makes no assumptions about the absolute or relative speed. It is a dangerous style of programming to state that mutual exclusion is not needed because process P_1 should finish its critical section before process P_2 has finished its uncritical part. Process P_2 might in future run on a faster processor. Then this assumption is no longer true and the system might crash.

We will finish this subsection with a short description of *Dekker's algorithm*. This algorithm can be used to describe mutual exclusion and is shown in Fig. 1.74. Two processors have a common memory for exchanging information. Three words in the memory are used: $C1 = 0$ if process P_1 is in its critical section; $C2 = 0$ if process P_2 is in its critical section. If $C1$ and $C2$ both are 0 (both processes want to enter the critical section) $Turn = 1$ indicates to process P_1 that it must check periodically the state of P_2 until P_2 has finished the critical portion of its program. At the end of its critical section P_2 indicates by $C2 = 1$ that it is outside the critical part. One can show that this algorithm is correct. It also fulfils the requirement of robustness. It is assumed in all algorithms that a program does not abort in its critical path.

If the critical section only contains a single word such as a counter or a synchronization flag as in the following example:

```
LOAD A
MODIFY A
STORE A
```

one can use special hardware: one uses a processor which allows read-modify-write cycles. During this cycle another processor will be delayed when it wants to access the memory.

1.7.6 The Fermilab Advanced Computer program multimicroprocessor Project (ACP)

Experimental high-energy physics involves recording and analysing the results of collisions between pairs of elementary particles. Each collision is totally independent of previous and subsequent collisions. Although an experiment can record vast quantities of data, it is possible to subdivide these data into 'events' which can be processed independently.

The primary design goal of the Fermilab Advanced Computer Project (ACP) is to maximize cost effectiveness (Nash *et al.* 1986). The system uses of the order of 100 commercial microprocessors as the computing engines. Each processing '*node*' gets an event from a '*host*' computer. When the node completes an event, the host will fetch the event from the node and deliver to it a new event for processing.

The node consists of a commercial 32-bit microprocessor, a floating point coprocessor, and a local memory which contains an entire program. A standard arrangement of nodes is shown in Fig. 1.75. The nodes reside in standard VME crates (see Section 1.8). Each crate has a single crate controller which interfaces the crate to the ACP's '*branch bus*' which connects all crates. The branch bus is optimized for high-speed 32-bit block transfer and operates at a speed of 20 MBytes per second.

The user's main software task is to split an existing program into two

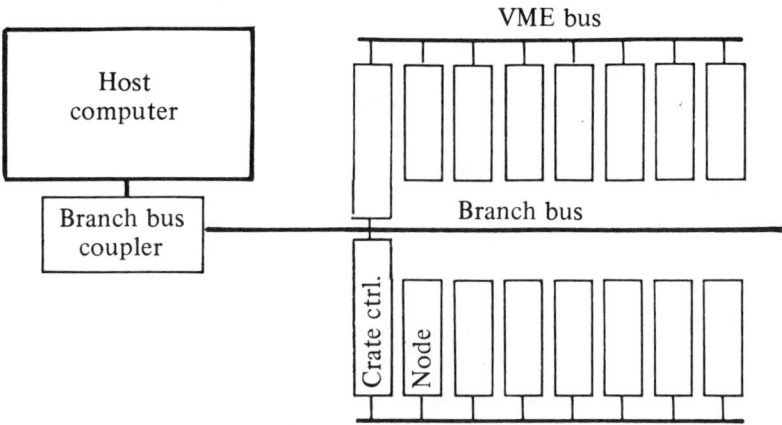

Fig. 1.75 Fermilab ACP. The system uses of the order of a hundred 32-bit 'node' microprocessors as the computing engines. They reside in standard VME crates and are connected via the branch bus to a 'host' computer.

parts: one which will run on the host and a second one which will run in parallel on all nodes. The host part contains all input and output for events and calibration data and prints out results. The node contains the actual event processing which typically consumes the majority of the CPU time.

All data passing between the host and the nodes is done by standard ACP subroutines callable from FORTRAN such as:

(1) SENDEVENT and GETEVENT which pass events between the host and a single node;
(2) BROADCAST which copies a block of calibration data from the host to all nodes;
(3) SUMNODE which sums blocks of statistical or histogram data in all nodes into a single block on the host.

For on-line filtering the ACP farm is in use at the *CDF* experiment at Fermilab and will be used in the *ZEUS* experiment at DESY.

1.8 Communication lines, bus systems

An experiment consists of scalers, ADC, TDCs, latches, registers, memories, interrupt generators, and so on. These devices must be connected to a computer for readout. In principle, one can implement all required paths by point-to-point wiring. But this approach is expensive and clumsy compared with linking the components with a few shared data paths or buses. One can design electronic modules to fit directly into the computer bus like UNIBUS, which is the computer bus for PDP11 computers. The drawback of this philosophy is that it requires that the modules are redesigned if a different computer and bus are used. This led to instrumentation buses such as CAMAC, VME, MULTIBUS II or FASTBUS (EUR 4100 1972; EUR 4600 1972; VMEbus 1985; MULTIBUS II 1984; FASTBUS 1983; FASTBUS 1985) which house plug-in modules in standard crates. The instrumentation bus is then interfaced to a computer. If the computer is replaced one then has to change the interface (Dobinson 1982).

Information exchange is initiated by a *bus master*. There may be several masters on the bus but only one is allowed to be active at a given time. *Arbitration* processes are needed to resolve *contention*. A master can exchange information with a slave by placing an address on the bus. All slaves compare this address with their individual addresses and become connected if a correspondence is found. During this connection a master can exchange information with a slave via the bus. At the end of the transfer the connection is broken to release the bus. Transactions between masters and slaves involve address and data information being transferred over the bus together with timing and control information for synchronization and specification of data flow. The set of rules needed to exchange this information is called a bus protocol (Fig. 1.76).

When one designs a bus one is confronted with the following problem (Fig. 1.77): A master *A* wants to read information from a slave *C*. What is the electrical influence of the bus gates and drivers of module *B* which remains connected to the bus? If one uses a normal AND gate to disable *B* the output of the gate is a logical 0 which is also connected to the bus like the 0 or the 1 from slave *C*. There are two ways of solving this problem:

(1) One can use inverse logic (5 V = logic 0, 0 V = logic 1 for a TTL bus). The bus lines are then connected via a *pull-up* resistor to 5 V. The device connected to the bus uses electronic circuits with open collector output. The device which wants to transfer a logic 1 then pulls the line down to 0 V.
(2) Another solution is the use of electronic circuits with *tristate* output. The output is then either logical 0 or logical 1 or it has a high resistance. Modules which don't participate in the transfer are disabled and are

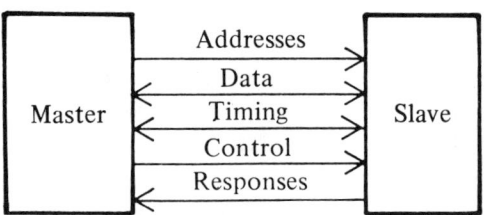

Fig. 1.76 Bus protocol. The bus protocol defines which information must be exchanged between masters and slaves.

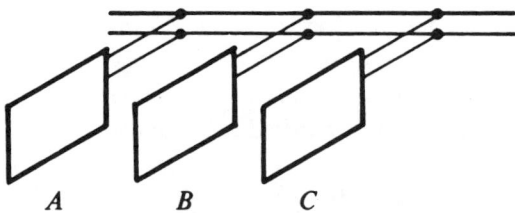

Fig. 1.77 Bus connections of slaves. With the help of tristate gates only one device is active on the bus while the other devices are passive but still connected to the bus lines.

then only connected to the bus by a high resistor which does not disturb the logic signals of the other modules.

1.8.1 Synchronous and asynchronous buses

A master puts an address on the bus and exchanges information with a slave after a connection has taken place. This is called a bus cycle. A transaction can take one bus cycle or several cycles. Many cycles are used in some systems to transfer first address information and then one or more data transfers. For each cycle it will be necessary to synchronize the transfer. There are two general ways to carry out bus cycle timing: either *synchronously* (CAMAC in the crates, MULTIBUS II, G64) or *asynchronously* (VME, FASTBUS). Synchronous transfers (Fig. 1.78) require fixed timing between master and slave. A master assumes that a slave can accept or provide information within a certain time. There is no handshake from the slave to acknowledge the cycle. The length of the cycle is determined by the master. It must be adjusted to the slowest device in the system or the master has to check the slave's ready signal at a fixed time.

Asynchronous transfers (Fig. 1.79) need a timing handshake between master and slave. The master puts valid information on the bus and waits

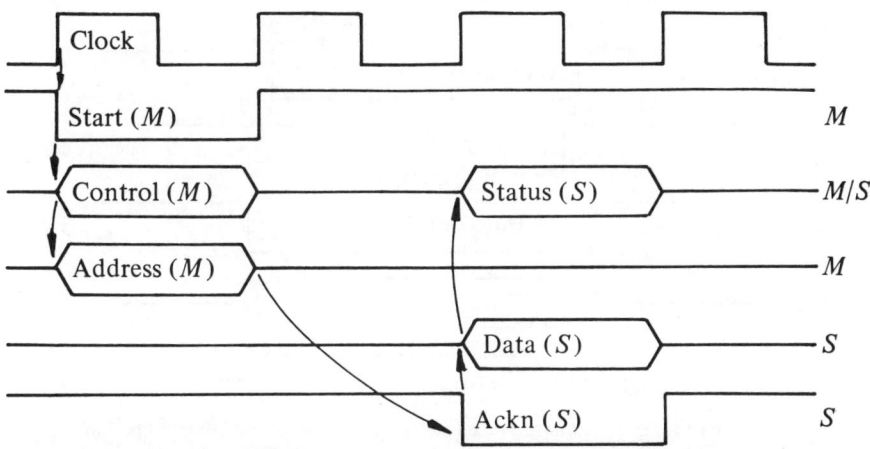

Fig. 1.78 Timing for a synchronous bus. The master asserts address and control information. The slave recognizes its address, finds the requested data, puts data and status on the bus and marks their presence with an acknowledge signal. There is no handshake between master and slave. The master assumes that the slave is fast enough to get all information. The master must be as slow as the slowest slave in the system or he has to test for, at fixed time intervals, a READY signal from the slave.

until the slave sends an acknowledgement. There is no fixed cycle time. The timing between master and slave is adjusted according to the needs of the master–slave pair. On the other hand the handshake mechanism involves a time overhead. In addition to the propagation time between master and slave there are additional electronic delays in generating the handshake mechanism. In addition, the master needs some time-out logic, in case the slave does not respond, to avoid bus hang-ups.

To summarize: a synchronous bus reaches higher speed and is preferred if all the connected devices have the same speed while asynchronous buses better suit the need when connected devices have different speeds.

1.8.2 Addressing

Address information is put onto the bus to establish a connection to one ore more slaves. This is always the first part of the transaction to

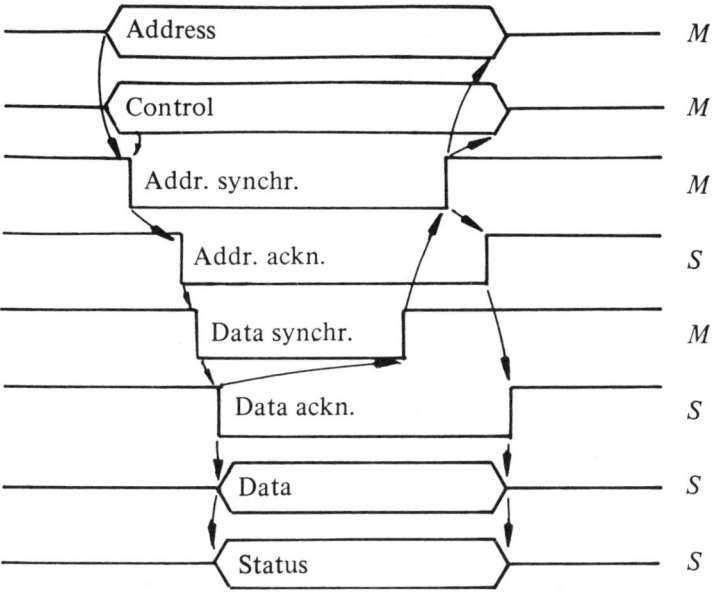

Fig. 1.79 Asynchronous bus timing. The master asserts address and control lines and after some time the address synchronization. The slave acknowledges the address, waits for data validation by the master, asserts data and status information together with the acknowledge signal. The master reads data and removes address synchronization and clears address and control lines. The slave then removes data and status information from the bus. In this way devices with different speeds can operate on the bus. Asynchronous buses are slower than synchronous buses due to the handshake mechanism.

1.8 Communication lines, bus systems

select the slave which will take part in the bus operation. There are essentially three addressing modes in use:

(1) geographical addressing,
(2) logical addressing,
(3) broadcast addressing.

In geographical addressing each slave is addressed by its position in the crate. A CAMAC crate has a crate address and each module in slots 1–23 is addressed by its slot number. An individual selection line runs from the crate controller to every station. Slave modules are enabled when this line is asserted (Fig. 1.80). Replacing a module in a CAMAC crate does not require any switch settings on the board. In FASTBUS every slot receives from its backplane connector a five-bit coded station number. Due to its position-dependent form of addressing it is not possible for two modules to have the same address and respond as slaves. Trouble shooting is easy. From the programming point of view geographical addressing is tedious. Suppose one uses a crate with two ADCs, one TDC, three ADCs, four TDCs and five pattern units. A program which should read first all ADCs, then the pattern units and afterwards the TDCs must know the positions of the devices exactly. In addition it must be changed if the positions are rearranged. In this case logical addressing is preferable. In systems with logical addressing each electronic module contains a microswitch register which represents the address of that particular module. The bus address lines are compared to the module address. In the case of a match the module sends back a connect. Another method of inserting logical address is implemented in FASTBUS. Geographical addressing is used to assign a logical address into a device register. Taking the example mentioned above the ADCs, pattern units and TDCs will have logical addresses in an increasing order so that a simple loop or DMA can readout the devices.

A drawback of logical addressing is debugging. How can one find the two modules with the same address in a system of 1000 devices? The use of

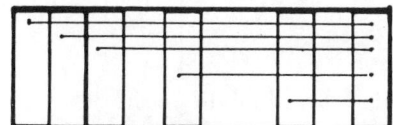

Fig. 1.80 CAMAC addressing. Each module is connected by a separate line to the master station.

Light Emitting Diodes (LEDs) on each address module opens a computer-independent way by visual inspection.

So far we have discussed how a master can establish a connection with a single slave. In many cases, however, it may be useful for a master to send information to several slaves simultaneously to reset the input for example. This is called the broadcast mode in which multiple slaves can be selected by their broadcast class.

1.8.3 Data transfers

After the selection of a particular slave, data are either written to or read from the slave. In many cases a transaction involves the transfer of many words to or from a slave. This is called *data block transfer*. Block transfers are used to read information from a disc controller or to copy an event from the slave's memory into the computer's memory. Instead of sending each transaction address, its data, the next address and next data, etc. it is much quicker to send the initial address once followed by a stream of data. The slave and master may contain internal registers to point to the next address in memory. At each data transfer the memory address registers are incremented by 1 to enable the next memory cell for the next transfer. This scheme requires a comprehensive knowledge of how data transfers are performed by the different devices. For example if the master contains a microprocessor which wants to write into some contiguous memory cells of a slave using the autoincrement mode of the slave, some types of microprocessors can give wrong result. These are microprocessors such as the TMS 9900 which do not perform just a write operation but always do a *read-modify-write* cycle. Instead of writing into locations 1, 2 and 3 the system would (Fig. 1.81): read (1), write (2), read (3), write (4), ... For these

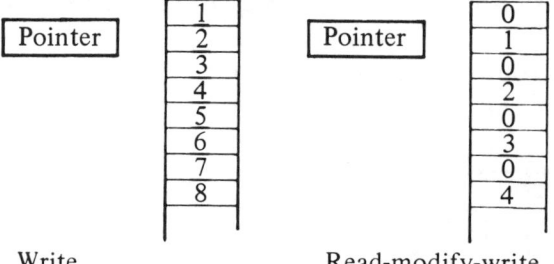

Fig. 1.81 The effect of autoincrement for different processors. The result can differ if a processor does not make just a simple write but always performs a read-modify-write cycle. In this case the autoincrement register is incremented twice.

types of processor one has to modify the slave in such a way that the memory address register is only incremented after a write operation.

1.8.4 Control lines

So far we have described operations such as read or write one or several words. We know that the instruction repertoire of a computer for transferring data is much wider. These are instructions such as store halfword, store byte, set bit, clear bit. Parts of these operations are also required in bus systems. These operations are distributed by control lines. In VME one can specify different widths of logical addresses, 16, 24, or 32 bits wide. In FASTBUS the transfer mode can vary from random data, asynchronous block transfer to synchronous block transfer. CAMAC allows many different functions in one cycle such as read, read and clear, read complement, test interrupt, clear interrupt, write, test, enable.

1.8.5 Responses

All the peripheral equipment of a computer contains status registers to inform the computer whether a device is busy, off-line, ready, or has found an error. The status register can be read by a computer in a separate cycle. A similar mechanism is needed for the slaves to inform the master that there are no more data. This information is given to the master during the data cycle and is called a response. Some of these responses are:

(1) Operation OK. If this does not appear there is no slave.
(2) There are no more data. This response is required if the master does not know at the initialization of a block transfer how much data should be read.
(3) Illegal operation such as read information from an output line printer.
(4) Parity error.
(5) Data not ready. Please wait.

Responses play an important role because of the wide variety of devices such as CAMAC or FASTBUS used as information buses. Responses permit the implementation of scanning algorithms for reading data from several modules. Fig. 1.82 shows an example where slaves are connected to several drift chambers. The readout is initialized in such a way that each module should transfer as many data as there are hits in the drift chamber. In CAMAC a slave would transfer a response $Q = 1$ if there were data and

$Q = 0$ if there were no more data. If there were no more data for $Q = 0$ the master ignores the information on the data lines and addresses the next slave.

1.8.6 Interrupts

There are many situations in which a slave has to inform the master or the computer that something has happened such as:

(1) There is a trigger. Data must be readout.
(2) There is a bus error or a parity error. Please check.
(3) The preselected time is over. Clock interrupt.
(4) The block transfer has finished.
(5) The operator has pressed a knob on the touch panel.

There are two ways for a computer to be informed that a device requires a special service. The computer can periodically read the status registers of all devices and check whether some action is required. This is called a *polling loop* or *sending a token*. A drawback of this solution is the amount of computer time needed to check all the bits and pieces. The advantage is that the program can easily use priorities and read significant status registers more often. There are no unknown interrupts due to pick-up, no illegal interrupts, and no interrupts which appear at high rates and saturate the computer. In order to make efficient use of computer time *program interrupt*

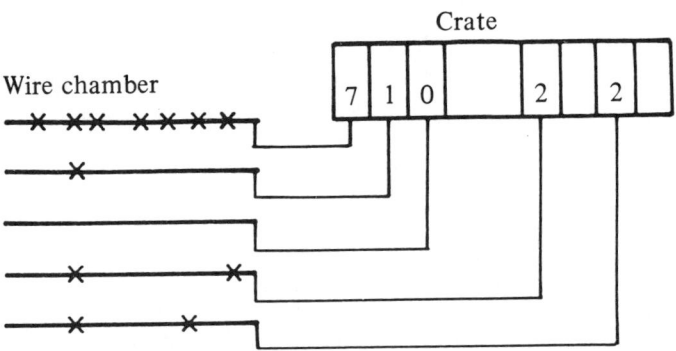

Sparse data scan

Fig. 1.82 Use of Q response to read variable length information. On the left hand side is an experiment with several wire chambers which contain some information. The crate on the right hand side should only read as much information from each chamber as necessary. This is called sparse data scan.

1.8 Communication lines, bus systems

facilities are established. An interrupt to a computer causes the program to transfer control to an interrupt service routine. This routine will analyse the interrupt and then execute a subroutine. At the end of the service the interrupted program will continue. In CAMAC the mechanism to request an attention interrupt is very simple. Each module is connected via an individual Look-At-Me (LAM) line to the controller which immediately knows which station needs service (like Fig. 1.80).

A different scheme uses a simple request line. Requests from all devices are serviced by a request handler which has to identify the request. Either it polls all slaves to identify which one has an outstanding request or it sends an acknowledgement to a special bus line which is *daisy-chained* through all the devices. When a device sees an acknowledgement it checks whether it has an outstanding request for attention. If not, the acknowledgement is passed on. If it has a request it will place an identifier on the bus which will be read by the interrupt handler (Fig. 1.83). In a system which has several computers connected to the bus there is the general problem: Which computer should be interrupted if a slave sends a request for attention? There are several schemes. Either the request handler decides with the help of the identifier to which computer the interrupt should be addressed. Or the bus, as VME, allows the transfer of interrupts on seven interrupt lines to several computers.

Another approach is used in FASTBUS, which does not use extra interrupt lines. If a slave requires attention it has to become bus master and transmit the attention message to the required destination in a special receiver block. The advantages of this scheme are that it does not require extra lines and the procedure is position-independent, but it does requires extra electronics on the slave for it to be able to become bus master.

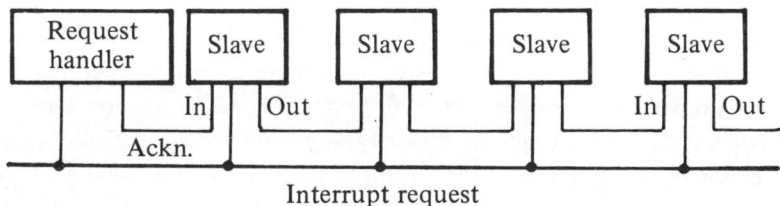

Fig. 1.83 Daisy-chained interrupt handling. The modules which are nearer to the interrupt handler have higher priority.

1.8.7 Multiple masters, bus arbitration

The use of multiple masters causes some problems in ensuring that the interrupts are transferred to the right master. Another problem is that several masters may compete for the bus. There must be a procedure to decide who may use the bus at any time. This is called arbitration. A master sends a bus request out and receives a bus grant allowing to use the bus. In centralized arbitration a single hardware unit is used with request lines to all masters. This arbitrator returns a bus grant to the master. But what happens if two masters send a bus request at the same time? One needs rules to decide who goes first. One possibility is priority arbitration which is widely used. No lower priority grant will occur as long as there is a higher priority request present. To avoid the possibility of a high priority master locking the bus one can alternatively implement a democratic arbitration scheme. Mastership is allocated in a cyclic way to all competing masters. Which scheme is used depends on the application but most buses use the priority scheme.

To avoid the extra lines to a central arbitration unit one can use alternative solutions like those we have discussed for interrupt handling. If a master wants to use the bus it asserts a bus request. The arbitrator returns a bus grant which is daisy-chained through all masters and taken by the first master issuing a request (Fig. 1.83). This form of arbitration gives highest priority to the device nearest to the master. When there are many masters this method can be slow because the bus grant must ripple through all the masters. It is also tedious if one wants to add, remove, or replace modules. One has to interrupt the daisy-chain and add more cables.

In FASTBUS a scheme was developed to allow for priority without a daisy-chain. Each master contains an internal priority vector. All masks requiring bus access use a common request line. The arbitrator then asserts an arbitration grant. The competing masters then put their vector on common bus lines which will contain the logic OR of all vectors. The masters then compare this OR on the bus with their internal vector and, if the values differ, remove the low order bits one after the other. After a certain time there is a stable vector corresponding to the winner. The masters with a vector indicating lower priority have removed all its bits. A simplified version of this system is shown in Fig. 1.84. This system has the advantage that one can remove, add, or replace modules without any recabling. We see that one has to add to the actual data transfer time the time needed by the master to get control on the bus. This time is called the *latency* of the access

1.8 Communication lines, bus systems

This time must be taken into account when a multimaster system is designed. It could happen that a single master can read data from a disc but that the same system may fail if more masters are added. At the end of the transaction a master can release the bus, but as bus arbitration may take some time it might be more efficient to keep the bus unless another master initiates a bus request.

1.8.8 Characteristics of CAMAC

The data bus is 24 bits wide. The speed is up to 3 MBytes per second. A crate has 25 slots. Eighty-two pins are connected to the backplane.

Up to seven crates can be connected by a branch highway.

Transfers are synchronous within the crate and asynchronous on the highway.

A crate controller can control the slaves in one crate. It is not possible from one controller to access a slave in another crate.

The addressing is geographical. Each controller has an address 1–7. Within a crate each slot has an extra line to the crate controller. Within each slot 16 subaddresses can be generated.

The functions define the operation. Thirty-two functions are available. F0

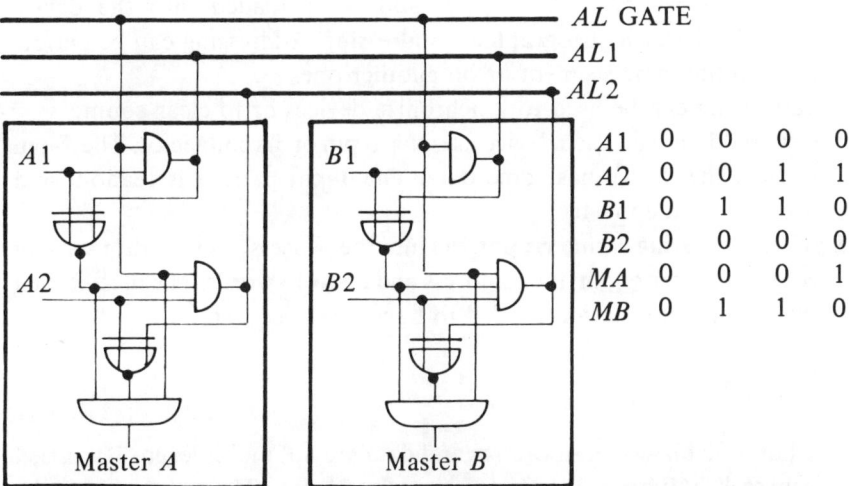

Fig. 1.84 Arbitration scheme for FASTBUS. This scheme works without daisy-chain cables. Each master puts its vector on the bus if the gate is available. The bus then contains the OR of all the vectors. The masters then remove their lowest bits if they don't match. The master with the highest vector gets the bus.

means 'read data' from the slave, F16 means 'write data' to the slave. Each module replies with a 'Q' response to indicate the status of the module.

Interrupts (LAMs) are sent from a slave on an extra line via a transformer (LAM-grader) to the crate controller. The computer can test the LAM with a F8 function command.

1.8.9 Characteristics of FASTBUS

The bus is 32 bits wide. Addresses and data are multiplexed. The speed is up to 100 MBytes per second. A crate has 26 slots. The backplane has 130 bus lines. The bus is built in ECL technology.

Normal transfers are asynchronous. For long distances one can use synchronous block transfer to avoid slowing down due to handshakes.

The basic unit is a segment. One distinguishes crate segments and cable segments. Segments can be connected by segment interconnects.

A master is capable of requesting and obtaining control of a segment. In addition he can communicate with slaves in other segments.

Each segment interconnect contains address recognition circuits. This is mainly a list of preloaded routing tables. For a bidirectional operation two lists must exist. Segment interconnects are complicated devices.

Modules can be addressed geographically (position-dependent) or logically (address-dependent). The logical address is loaded into the device address register by geographical addressing. Addressing can be performed on the same segment or on another one.

Broadcasting can be used to synchronize devices or to clear counters.

For sparse data scan each device uses the T-pin on its connector. The T-pin is set if the device has some data. The T-pin pattern is readout and decoded by the master.

Devices requesting an interrupt write into the processor's interrupt sensing control register region their address and up to 16 words. The master then has all the information needed to handle the request.

1.8.10 Characteristics of VME

The bus is 32 bits wide. Addresses and data are not multiplexed. The speed is up to 40 MBytes per second. A crate has 21 slots. The connection to the backplane is via one (single height) or two (double height) 96 pin DIN connector(s).

VME is defined on a crate basis only. In addition, a VME extension (VSB)

Table 1.7. Characteristics of data buses

Item	CAMAC	FASTBUS	VME	Multibus II	Future Bus
Bandwidth MBytes per second	3	40	30	40/48	40–60
Address width	7/25/4	32	32	32	32
Data width	24	32	8/16/32	8/16/24/32	8/16/24/32
A/D multiplexed	No	Yes	No	Yes	Yes
Board size	183 × 305	366 × 403	233 × 160	233 × 220	366 × 280
Number of connectors	1	1	1 or 2	1	
Type of connector	82 pin direct	132 pin	96/96 pin	96 pin	96 pin
Arbitration	Central	Distributed	Central	Distributed	Distributed
Interrupt	LAM pattern	Msg. passing	7 levels	Msg. passing	Message
Bus protocol	Synchr/asynchr	Asynchr	Asynchr	Synchr	Asynchr
Serial bus	extra system	Yes	Yes	Yes	Yes
Geographical addressing	Yes	Yes	No	Yes	Yes
Chip set available	No	Private	Yes	Yes	No

bus can be used to connect CPUs with local memories at high bandwidth and a VMS bus is available for long distance applications with serial links.

Data transfers are asynchronous. 8, 16, or 32-bit wide transfers are supported.

In block transfer mode the master sends a single address to a slave followed by a block of data.

Four request lines with different priorities are provided. On each level the priority is further defined by the position.

Seven interrupt lines allow the slaves to express their requests for attention.

VSB: the VSB bus has a 32-bit data path and a 32-bit address path. Up to five boards can be connected to each VSB bus.

VMS (serial bus): the VMS bus speed is 3.2 Mbits per second for short cable lengths.

1.8.11 *Standardization of data buses*

Standardization tends to be a slow process because of the time it takes to reach a consensus from initially widely disparate viewpoints. The problem is aggravated by rapid turnover of the committee membership. Recognizing the time required for this process and the desirability of having a small number of standards a committee has been set up to define the 'future bus'.

The process of defining a new standard from nothing in a committee is much slower than the introduction of a data bus by an industrial design team. These teams work in close connection with engineers and customers, and in several cases these buses are optimized for a specific microprocessor. The buses available at present were based on the assumption that microprocessors would be their main users. Some of the buses and their characteristics are summarized in Table 1.7.

2

Pattern recognition

2.1 Principles and methods of pattern recognition

Pattern recognition is a field of applied mathematics and makes use of results in statistics, cluster analysis, combinatorics, and other specialized branches. The goal of pattern recognition is the classification of objects (Andrews 1972). The range of fields from which objects might be chosen is virtually unlimited: satellite pictures, electrocardiograms, coins, printed or handwritten text, and position measurements along particle trajectories may suffice as examples of physical objects. Nonphysical objects can be found in applications such as linguistics, where certain expression patterns may provide clues as to the authorship, or authenticity of a given text.

2.1.1 Detector systems

Detector systems are normally designed to provide three independent pieces of information about a particle. The first is the energy E (or the momentum p) of the particle at a well-defined point. The second is the identification of the nature of the particle, e.g. K_0, μ^- or e^+, providing the mass m and the charge c. The third is the coordinate of one point on the track together with the direction of the track at that point. This information can be obtained from the measurement of several points along the particle trajectory. By using these three elements (energy or momentum, mass and charge, and a point plus the direction) it is possible to calculate the energy and the momentum of the particle at any point on its trajectory, and in particular at the point of interaction, the *vertex*.

If the tracks pass through a suitable static magnetic field then information about momentum can be extracted from the shape of the particle trajectory. Alternatively the energy of charged particles, and also neutral ones, can be measured by absorbing the particle in a solid block of matter, and measuring the energy dissipated. This type of apparatus is called a *calorimeter* (Fabjan and Ludlam 1982), and is dealt with in Section 2.3.

The great majority of signals for trackfinding in high-energy physics today are delivered by wire chambers, which have shown a considerable evolution over the last 20 years, as have the pattern recognition methods (Grote Hansroul, Lassalle, and Zanella 1973, Grote and Zanella 1980). In this book, there is only space to outline the most basic physical mechanisms of wire chambers, and the reader interested in more details is referred to the papers by Charpak (1978) and by Marx and Nygren (1978).

The basic unit of a wire chamber is a thin metallic wire of typically 10–50 µm diameter. In its simplest form, a detector consists of one such anode wire surrounded by a cylindrical cathode, with a voltage of several thousand volts between them. The chamber is normally filled with a noble gas such as argon, with the addition of a gas such as isobutane, at atmospheric pressure. The isobutane absorbs part of the ultraviolet radiation and in this way keeps the ionization region localized.

When ionizing radiation passes through this cylinder, the light electrons liberated will drift towards the anode wire, while the much heavier positive ions will stay behind, moving comparatively very slowly towards the cathode. Since the electric field grows inversely with $1/r$, where r is the distance from the wire centre, the electrons will gain more and more energy over their mean free path between inelastic collisions and will eventually initiate an electron avalanche through further ionization in the *multiplication region* at a distance of the order of the wire diameter away from the wire centre. The negative charge thus collected on the anode wire leads to a short negative pluse that is detected by the electronics connected to the wire which is said to have 'fixed'.

The precise operational mode of such a detector depends upon the value of the electric field potential. At low voltages, the anode wire collects all the ionization electrons, and the anode pulse height is independent of the applied potential; operated in this mode the detector is called an *ionization chamber*. At higher voltages, gas amplification occurs when additional ions are created from the primary ionization in the strong electric field around the wire. The magnitude of this effect depends upon the electric field strength around the wire, and the pulse height on the anode is proportional to the original number of ions created. For this reason, when operated in

2.1 Principles and methods

this mode the detector is known as a *wire proportional counter*. Increasing the applied potential beyond this region causes charge to build up around the wire (space charge effect) and the strict correlation between the applied potential and the pulse height is lost, i.e. the charge collected on the anode wire is no longer proportional to the original ionization. This is the *Geiger–Müller* mode of operation.

2.1.2 Proportional chambers

The mechanism described above remains practically unchanged when many parallel anode wires are arranged in one plane between two cathode planes, at a wire spacing of the order of 1 mm. The electric field is then almost homogeneous everywhere except for the region near to the wires where it resembles that in the tube. This arrangement therefore works like many independent tubes: an electron cloud created somewhere in this *chamber* will drift towards a specific wire and lead to a pulse there. A particle passing through this chamber and creating regions of ionization along its path will lead to signals on one, or several adjacent wires depending on its orientation with respect to the chamber plane. By spacing the wires at a suitable distance d, one can in this way measure the position of the impact point on the plane of the wires with a precision of about $d/3$. Detectors of over 50 000 wires (Bouclier *et al.* 1974) have been constructed with chambers operating in this proportional mode. Proportional wire planes achieve a detection efficiency for charged particles of better than 99%, the small inefficiencies arising from too low ionization, or failures in the electronics (including dead time).

Since a plane of wires measures only one of the two avalanche coordinates in the plane, the following method has been developed to measure the second coordinate as well and thus the position in space: the cathode planes are divided into parallel strips about 1 cm wide which run perpendicular to the wires, or at a suitable angle (Fig. 2.1). The avalanche induces a positive pulse on these strips which can be detected and can give the second coordinate, the third coordinate of the space point being the wire plane position. Another method consists of measuring the amount of charge flowing into amplifiers at both ends of the wire. If the wire has a reasonable resistance, the ratio of the charges will depend on the position of the avalanche along the wire. With this *charge division readout*, a precision of 1% (at best) of the wire length can be reached (see also Section 1.5.9).

2.1.3 Drift chambers

To achieve accurate results large proportional detectors need many wires, each of which needs its own amplifier. In order to reduce the cost of this and, also in order to improve the accuracy and to reduce the amount of matter in the particle's path, detectors have been constructed which exploit the same arrangement of anode wires between cathode planes in a completely different mode. To this end, the chamber is rotated by 90°, so that the particle path runs more or less parallel to the wire plane, but is still orthogonal to the wires (Fig. 2.2). Since the electron cloud drifts initially at constant speed of some $50\,\mu\text{m}\,\text{ns}^{-1}$ towards the anode wire, the time lapse between the impact of the particle and the avalanche is a direct

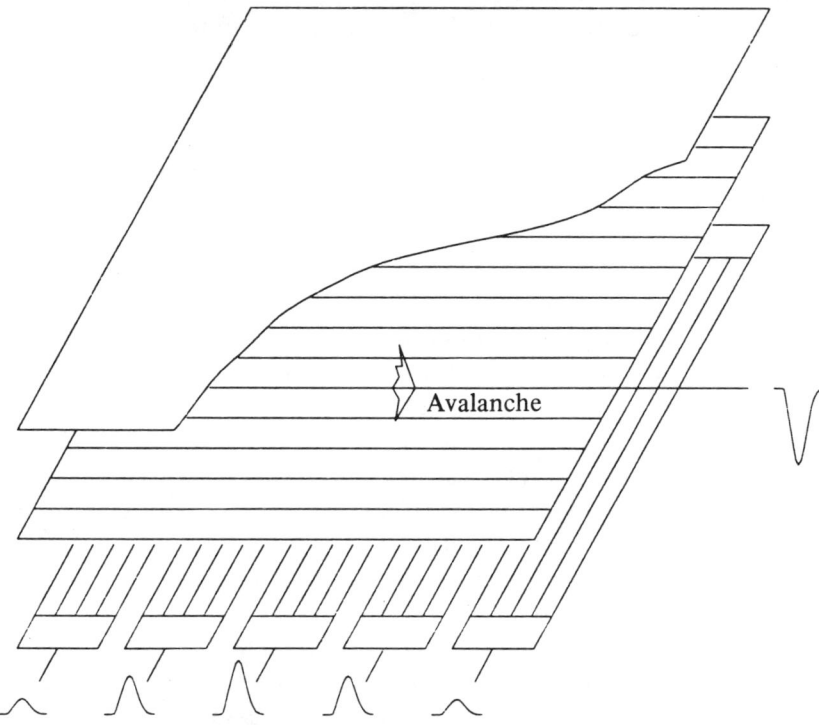

Fig. 2.1 Reading the induced pulse on surrounding electrodes. The avalanche pulse at the anode wire induces pulses in both adjacent cathode planes, one of which, in this example, is subdivided into strips which are equipped with an analog readout system. The barycentre of the pulse charge distribution gives an accurate avalanche position along the wire, and hence a second coordinate. When there is more than one avalanche in the wire plane there is no way of correlating the two projections.

2.1 Principles and methods

measure of the distance from the wire. The impact time is normally determined by separate scintillation counters. With a time resolution of the order of 1 ns, one can then measure positions with an accuracy of 50 μm. However, the apparent resolution can be considerably worse because of correlation effects (Drijard, Ekelöf, and Grote 1980). A drift space of many centimetres can be obtained, which leads to a considerable reduction in the number of wires. However, since the chambers measure only the drift distance, there is no information as to which side of the wires the particle passed. This *left–right ambiguity* represents an additional difficulty in trackfinding. The remedies employed, e.g. *staggering* the wires slightly, and offsetting drift chamber cells, cannot normally be used during the initial part of the trackfinding to reject the *image* or *ghost* tracks, but may perform this task in a subsequent step: staggering the sense wires leads to a higher error when a track is fitted to the ghost track, offsetting drift chamber cells

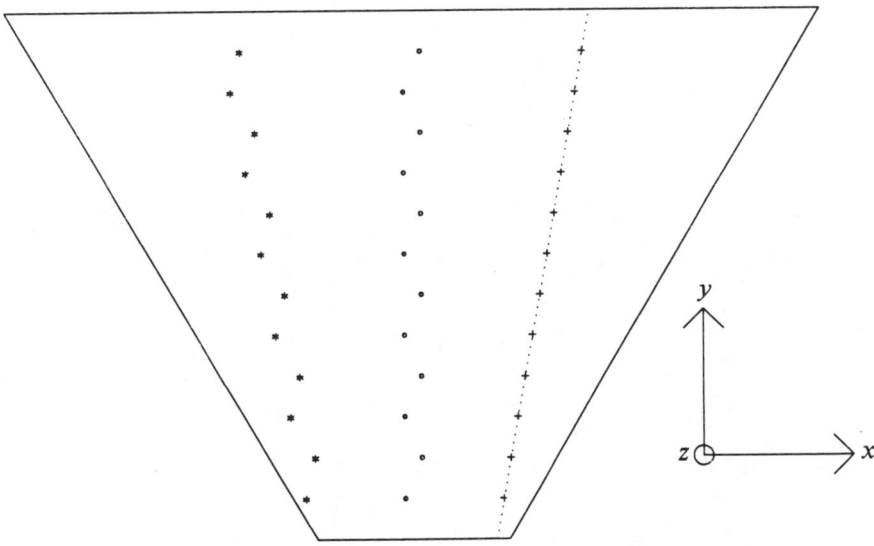

Fig. 2.2 Drift chamber cell; one cell of a cylindrical drift chamber surrounding an interaction region in a colliding beam machine. The *staggered* sense wires along the symmetry axis of the cell extend in z direction. This staggering has been very much exaggerated here in order to make it visible, in reality it is typically 200 μm whereas the wire distance is typically a few centimetres. The electric field is oriented in the x direction towards the sense wires, and is homogeneous in y and z. A charged particle passing through the righthand half of the cell leads to x position measurements along its path as well as the 'ghost' points on the opposite side which are staggered twice as much as the sense wires thus making their rejection easier.

in most cases prevents ghost track sections from adjacent, but offset, chambers being linked into one track. A typical low multiplicity event in a cylindrical drift chamber arrangement is shown in Fig. 2.3.

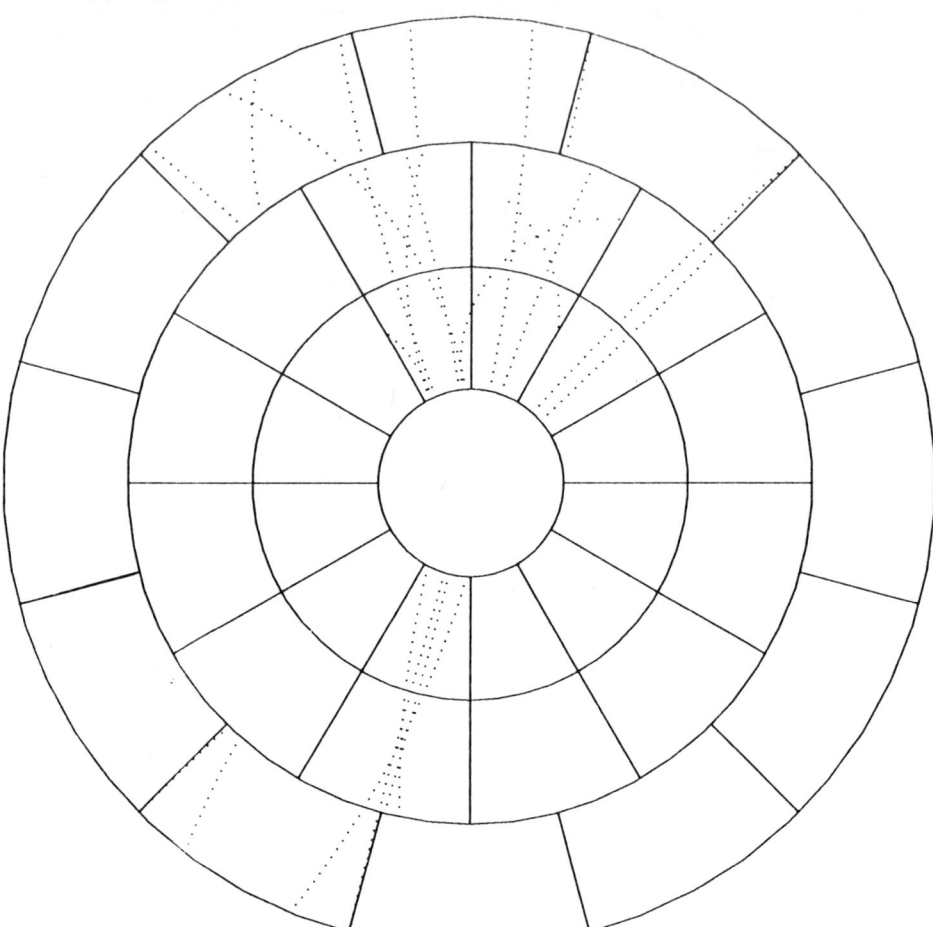

Fig. 2.3 Event in a cylindrical drift chamber detector. The picture shows a cylindrical drift chamber layout similar to the one used in the JADE detector at DESY. The solid lines represent the drift cell boundaries. The sense wires along the cell symmetry axes are not shown, but only the signals and their ghost images. The effect of staggering the outermost layer of cells with respect to the other two is clearly visible: whereas the true particle paths lead to contiguous lines of signal points, their ghost images are disrupted. The picture shows seven charged particles tracks.

2.1 Principles and methods

2.1.4 Time projection chambers (TPCs)

TPCs are in some sense an improved type of drift chamber. The main differences are the following (see Fig. 2.4):

(1) The drift space extends mainly to one side only of the sense wires, thus curing the left–right ambiguity problem. This is achieved with the help of additional field wires.
(2) In many cases, a magnetic field parallel to the electric field leads to an additional confinement of the drifting electron cloud (Marx and Nygren, 1978); this, in turn, allows for longer drift lengths of 1 m and above. With a time resolution of 1 ns one can measure the drift path length, and therefore the particle position with a theoretical accuracy of 50 μm. The longitudinal diffusion of the electron cloud tends to increase this value to typically 200 μm.

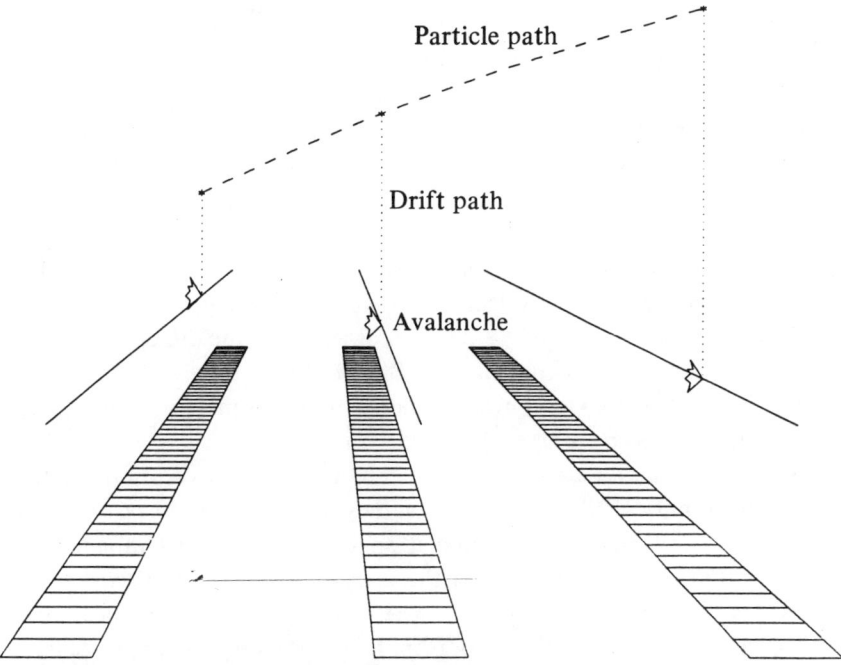

Fig. 2.4 Schematic view of a time projection readout system. The perspective view shows sense wires above rows of pads. The electric field makes the electrons drift vertically towards these sense wires where they give rise to avalanches as in the case of standard drift chambers. The induced pulses can then be read out on the pad row, where each pad is connected to its own amplifier. The scale of the picture is grossly distorted: whereas the pads are typically 5 mm by 8 mm, and sense wires are mounted 4 mm above them, the electron drift path may be 1 m or more.

(3) Opposite each sense wire, a row of small *pads* running along the wire, and oriented orthogonal to it, receives the induced pulse from the avalanche at the sense wire. The barycentre of the charges distributed over the pads can be reconstructed with a precision comparable to that of the drift length measurement.

The TPC delivers genuine space points along a particle trajectory since there is no ambiguity as to the association of signals in the two projections, the association being provided naturally by the arrival times. Because of this feature TPCs have already been used, and will be employed in most experiments which are likely to measure high multiplicity events and particle jets.

2.1.5 Pattern space

For numerical treatment objects have to be described by a finite number of parameters that are typically the results of measurements performed on the object, such as grey levels inside small areas of a photograph, frequencies and their amplitudes in a Fourier spectrum, or positions of ionized regions created by a charged particle in a wire chamber. To each of these quantities corresponds a separate dimension in the space formed by all quantities, the *pattern space* $\{P\}$. The measurement of an actual object supplies numerical values for the different measured quantities, which are grouped into a vector **x** of length n, the dimension of $\{P\}$. Each object gives rise to one point in $\{P\}$. The dimension of $\{P\}$ can be very high, e.g. 10 000 if grey level measurements are performed on a mesh of the moderate size 100 by 100.

Although in many cases $\{P\}$ is not a vector space in the mathematical sense, the totality of all measurement values is often called the *measurement vector* for convenience. $\{P\}$ being a space of some fixed dimension n requires that all n components of the measurement vector **x** exist for all objects in $\{P\}$. There are two possible reasons why this might not be the case in a given problem. Firstly, the coordinates are missing because of detector inefficiencies or other similar reasons, i.e. they should, in principle, exist, but are not available for a given object. In this case they may be predicted from neighbouring values of the same object, or through some other procedure.

Secondly they are missing because they are not defined, not even in principle. As an example consider a track which does not reach the rear part of a detector, either because it leaves the detector through the side, or

2.1 Principles and methods

because it is curved backwards by a magnetic field. In such a case, if one wants to use pattern space transformations such as feature extraction (Subsection 2.1.11) one is obliged to split the full set of objects into subsets, e.g. by subdividing the phase space in the example above such that several pattern spaces, each with its own dimension, can then be treated separately.

2.1.6 Training sample and covariance matrix

In many different places in this book, a *training sample* is used to tune an algorithm, or to serve as a comparison when geometric acceptance, efficiencies, and other parameters have to be calculated. A training sample has to be used when there is no mathematical description of the objects studied, and can be used for convenience even if a mathematical description exists. It consists of a set of objects, which in the case of high-energy physics are usually particle trajectories, but can also be single points, point pairs, track segments, or even complete events. A typical training sample consists of several thousand objects. Its careful selection is very often the condition for the success of a method to be applied. It has to be representative for the problem under study. To give a trivial example: if the aim is to find tracks down to $1\,\text{GeV}\,c^{-1}$ in a detector inside a magnetic field, it is no use employing a training sample with tracks of $5\,\text{GeV}\,c^{-1}$ and above. Even a sample of trajectories with uniform populations at 1, 2, 3, and so on up to $50\,\text{GeV}\,c^{-1}$ will give a strong bias to high momenta. In this case, it is much better to provide a uniform population in intervals of $1/p$, i.e. in equal steps of curvature, since this is a crucial parameter for trackfinding. A less trivial example is that of event selection: the physics to be studied in a given experiment may just correspond to a small and badly populated region of the total phase space, if again a uniform distribution of the training sample over the phase space is used. This could mean that although the total reconstruction efficiency over such a nonrepresentative training sample may be 99.5%, the probability of reconstructing one of the events really looked for could be close to zero.

Closely related to the training sample is the covariance matrix. In the case of a known probability density distribution of stochastic variables x_i, one element of the covariance matrix **C** is defined as

$$c_{ij} = \langle (x_i - \langle x_i \rangle)(x_j - \langle x_j \rangle) \rangle = \sigma_{ij} \tag{2.1}$$

where '$\langle \cdots \rangle$' denotes the expectation values, and σ_{ii} is the standard deviation of variable i. The more common case is, however, the one in which the probability distributions along the different axes are not known, in which case σ_{ij} is replaced by its estimator s_{ij} which is defined as in Equation (2.1), but now with '$\langle \cdots \rangle$' denoting the 'mean over the training sample'.

The covariance matrix is frequently composed of three normally independent, and consequently additive contributions. The first contribution arises from the correlation between the different measured quantities, i.e. it expresses the fact that the measurements contain redundant information, and the measurement vectors do not fill $\{P\}$, but only a subspace of it (the *constraint surface*). The second contribution results from statistical measurement errors, and the third from the 'statistical' behaviour of the physical object, e.g. a particle undergoing multiple scattering in matter.

The different contributions to **C** can be obtained separately from different training samples, provided such training samples are available, i.e. from Monte Carlo generation: if real measurements are used in a training sample, the effects are, of course, all mixed up. In feature extraction (Subsection 2.1.11) and parametrization (Subsection 2.2.3), a training sample without any errors is used, in order to be able to see the effect of the nonlinearity of the constraint equation. For goodness-of-fit checks, i.e. χ^2 calculations, the full covariance matrix has to be used.

In certain circumstances it may be advisable to provide more than one covariance matrix: such is the case when tracks have to be fitted over a large momentum range, when, for low momenta, the multiple scattering and, for high momenta, the measurement errors are very often the dominant contributions to **C** (Gluckstern 1963; Mecking 1982).

2.1.7 Object classification

A *class* of objects is formed by all those objects fulfilling certain criteria which define the class. The goal of pattern recognition is to classify measured objects, i.e. to find a decision function or, in the more general case, a procedure f such that

$$c = f(\mathbf{x})$$

is the number of the class to which the object that is described by **x** belongs; c is sometimes considered as a point in the *classification space* $\{C\}$.

In high-energy physics pattern recognition we mostly deal with classes that do not share objects. In this case, because to each object there belongs

2.1 Principles and methods

a point in pattern space, the task of pattern recognition consists in finding a set of *hyperplanes* (planes of dimension $n - 1$) that divide $\{P\}$ into disjoint regions of classes. If, in addition, the different classes form clusters in $\{P\}$ that are well separated from each other (Fig. 2.5), one may find separating hyperplanes that are rather smoothly curved, or even linear. The following

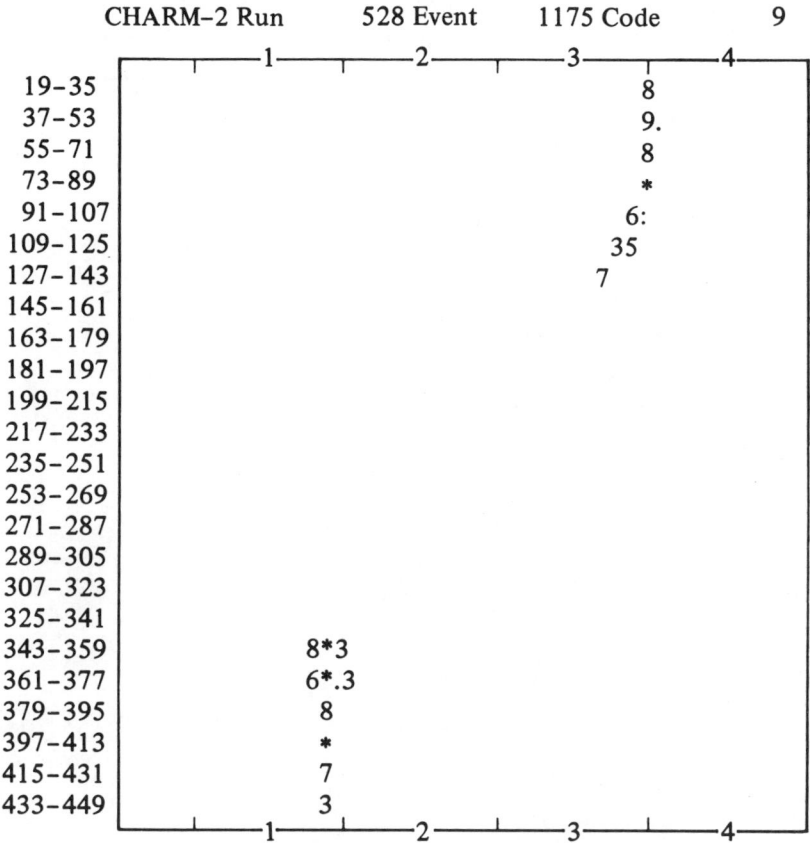

Fig. 2.5 Two clearly separated object classes in $\{P\}$. The figure shows an event in the top view of the detector of the CHARM-2 neutrino experiment at CERN. The picture is very much compressed longitudinally in that several planes are grouped into one line (numbers at left). The first active plane in the detector is plane 21. The picture shows a double event: a beam halo muon has entered the front of the detector (at the top) and stops after about 120 planes after having passed through about 6 m of glass absorber. Further upstream (near the bottom of the picture), a neutrino has interacted in a charged-current event, creating a hadronic shower with an outgoing muon. The total length of this part of the detector is 35 m, the width is about 4 m. The different print symbols here represent groups with different numbers of wires.

example from high-energy physics may serve to illustrate the concepts of pattern space and dividing hyperplanes: A detector provides measured space points along particle trajectories. If we treat these space points as *objects* and the tracks into which they have to be assembled as *classes* then $\{P\}$ is simply the three-dimensional Euclidean space. The tracks form *clusters* in $\{P\}$. They do not normally share objects (= points) if we exclude the vertex region. Thus, we can define hyperplanes in the form of tube-like structures around each track (this is often called a *road* in the two-dimensional case), and for straight tracks we can even find linear hyperplanes separating them.

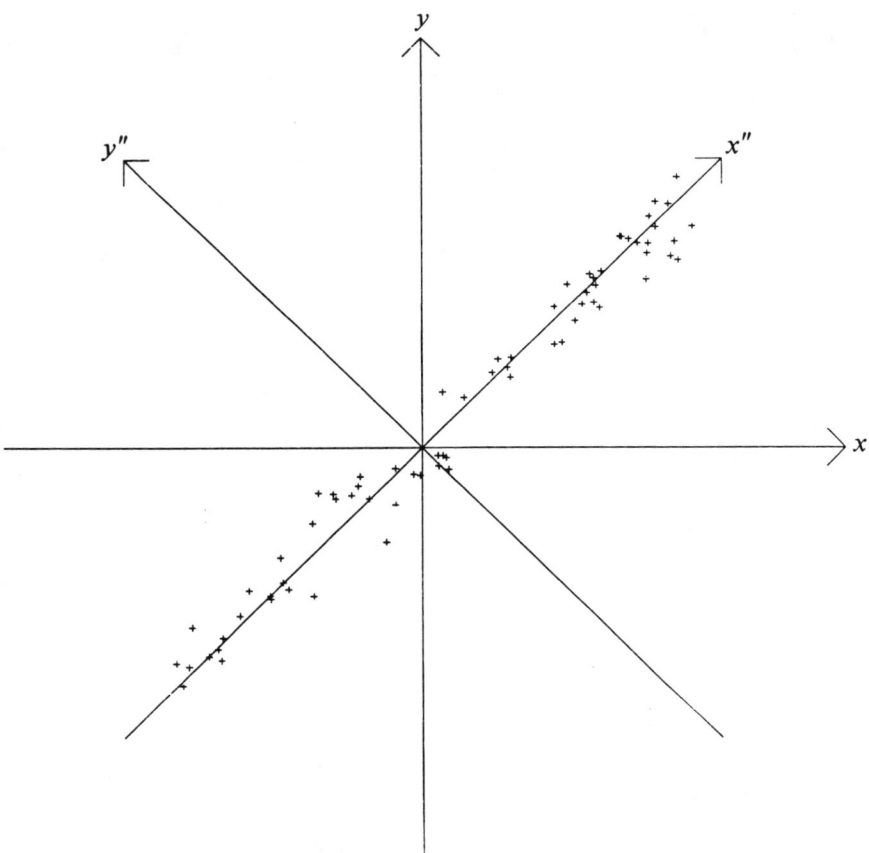

Fig. 2.6 Strongly correlated coordinates in $\{P\}$. The figure shows points with strongly correlated x and y coordinates. In a different coordinate system (x'', y'') the correlation almost does not exist, and the absolute values of the y'' coordinate are much smaller than the corresponding y values.

2.1 Principles and methods

2.1.8 Feature space

The pattern space formed by the *raw measurements* is often not very convenient to work in. The first reason for this lies in the frequently rather heterogeneous nature of the vector **x** which consists of components that are sometimes difficult to compare, like comparing apples to oranges. In addition, these components may have very different significance for the pattern recognition task. Consider for example a vector **x** of which the first 10 000 components are the grey levels of a satellite photograph scan on a 100 by 100 mesh, and component 10 001 is the time at which the picture was taken. Clearly this last coordinate is very different from the others and will

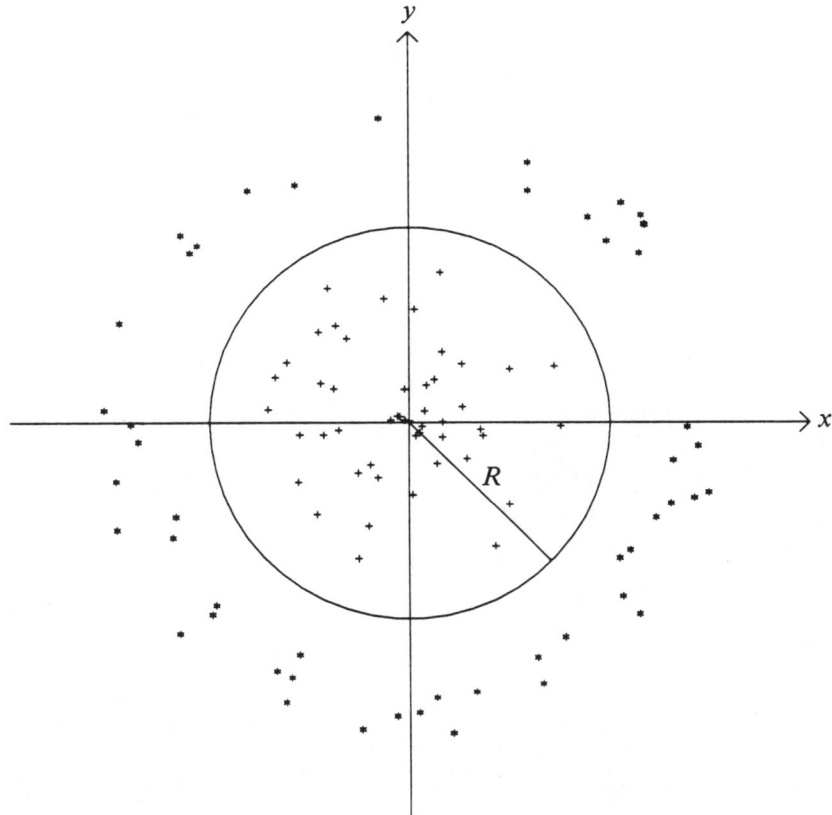

Fig. 2.7 Two object classes divided by a curved hyperplane. The two object classes are represented by points inside a circle (class 1) and outside the circle (class 2) in a Cartesian coordinate system. The hyperplane $x^2 + y^2 = R$ is a circle.

probably have a much higher importance for the pattern recognition task than any of the others alone.

The second reason is the sometimes very high dimensionality of $\{P\}$, which can easily reach values of 10^6 or more. Not only is the manipulation of vectors of this size tedious and slow even on the most modern computers, but such a high number of dimensions implicitly means that the coordinates must be highly correlated (Fig. 2.6), since the object measured will only have a rather small number of macroscopic or otherwise significant features. High correlation between coordinates is equivalent to saying that the object classes do not fill n-dimensional space regions in $\{P\}$ (*bubbles*), but that they occupy subspaces of much smaller dimensions.

Finally the third reason can best be illustrated with the help of Figs. 2.7

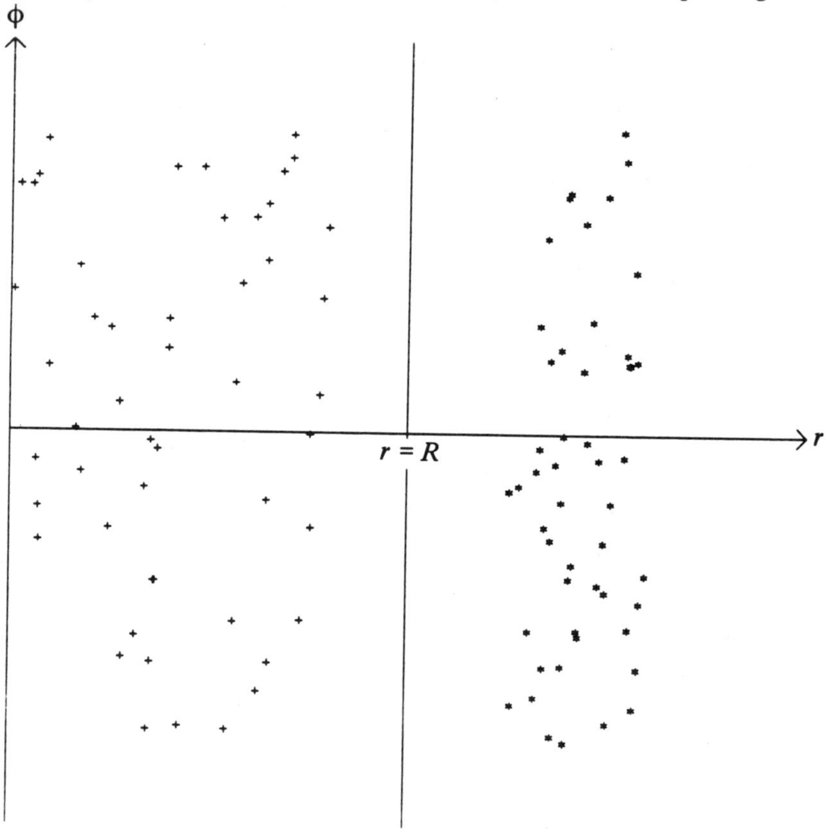

Fig. 2.8 The two object classes of Fig. 2.7 in feature space. The same points as in Fig. 2.7 are now plotted in a coordinate system with the axes r and ϕ. The hyperplane $r = R$ is now linear.

2.1 Principles and methods

and 2.8. A simple transformation of coordinates from x, y to r, ϕ with

$$r = (x^2 + y^2)^{\frac{1}{2}}, \qquad \phi = \sin^{-1}(y/r) \qquad (2.2)$$

converts the dividing hyperplane, a circle in Fig. 2.7, into a linear one in Fig. 2.8. This is an example of a simplification of the problem without a reduction of the dimensionality (2). All of this leads to a transformation of the pattern space of dimension n into the feature space $\{F\}$ of dimension m:

$$\mathbf{z} = R(\mathbf{x}), \qquad \mathbf{z} \in \{F\}, \qquad \mathbf{x} \in \{P\}$$

with an arbitrary transformation R. If R is linear, it can be represented by a matrix multiplication (apart from a possible translation):

$$\mathbf{z} = \mathbf{T}\mathbf{x}$$

with the transformation matrix $\mathbf{T}$ of dimension (n, m).

With the help of the three spaces introduced, the task of pattern recognition is then most elegantly formulated as the task of finding the transformations in

$$\{P\} \Rightarrow \{C\} \quad \text{or} \quad \{P\} \Rightarrow \{F\} \Rightarrow \{C\}$$

2.1.9 Classes, prototypes, and metric

Classes of objects are normally defined by enumerating all of their members, such as in the classification of printed letters with one letter per class, or by specifying a set of *prototypes* in cases where there are too many of them, such as in the case of handwritten letters. The objects themselves can be defined in pattern space or in feature space, and there either directly as vectors, or as functions thereof. Some or all of the parameters in these functions may depend on the object itself; such is the case if an object, for example a particle track, is required to form a helix in space of which only the direction of the principal axis is known, this being the direction of the magnetic field vector.

In most cases of automatic pattern recognition, some kind of metric is needed to decide whether an object belongs to a certain class or not. A metric provides a distance function d, obeying the rules:

$$d(\mathbf{x}, \mathbf{y}) \geq 0$$
$$d(\mathbf{x}, \mathbf{y}) = d(\mathbf{y}, \mathbf{x})$$
$$d(\mathbf{x}, \mathbf{y}) \leq d(\mathbf{x}, \mathbf{z}) + d(\mathbf{z}, \mathbf{y}) \quad \text{(Triangle inequality)}$$
$$d(\mathbf{x}, \mathbf{y}) = 0 \quad \text{iff} \quad \mathbf{y} = \mathbf{x}$$

where **x**, **y**, **z** are vectors in the space considered. Of course, the Euclidian metric

$$d(\mathbf{x}, \mathbf{y}) = [(\mathbf{x} - \mathbf{y})^T(\mathbf{x} - \mathbf{y})]^{\frac{1}{2}} \tag{2.3}$$

obeys these rules. The metric must be chosen such that the distance between two objects is in some way proportional to their degree of being different.

As a counterpart one can define a similarity measure between two objects by requesting that a function s exists such that

$$0 \leq s(\mathbf{x}, \mathbf{y}) \leq 1$$
$$s(\mathbf{x}, \mathbf{x}) = 1$$
$$s(\mathbf{x}, \mathbf{y}) = s(\mathbf{y}, \mathbf{x})$$
$$[s(\mathbf{x}, \mathbf{y}) + s(\mathbf{y}, \mathbf{z})]s(\mathbf{x}, \mathbf{z}) \geq s(\mathbf{x}, \mathbf{y})s(\mathbf{y}, \mathbf{z})$$
$$s(\mathbf{x}, \mathbf{y}) = 1 \quad \text{iff} \quad \mathbf{x} = \mathbf{y}$$

If d is a metric distance, then $s = e^{-d}$ is a metric similarity. If d is bound, i.e.

$$d_m = \max(d(\mathbf{x}, \mathbf{y})) < \infty$$

then

$$s = 1 - d/d_m \tag{2.4}$$

or

$$s = 1 - d^2/d_m^2 \tag{2.5}$$

would be alternative measures of the similarity (Schorr 1976).

Both the distance and the similarity functions can be used to decide whether a measured object belongs to a given class or not. A slightly unusual example of this is the difference between a measured track **x** and its prototype $\mathbf{x}_f$ which has been found through a linear least squares fit to a track model:

$$\chi^2 = (\mathbf{x} - \mathbf{x}_f)^T \mathbf{W}(\mathbf{x} - \mathbf{x}_f) \tag{2.6}$$

with the weight matrix $\mathbf{W} = \mathbf{C}^{-1}$, where **C** is the covariance matrix (Subsection 2.1.6) known *a priori* either from theoretical considerations, or as defined by a training sample of tracks with errors:

$$\mathbf{C} = \langle (\mathbf{y} - \mathbf{y}_f)(\mathbf{y} - \mathbf{y}_f)^T \rangle \quad \text{(dyadic product)} \tag{2.7}$$

The statistical distribution of χ^2 is, of course, well known, and cuts can be applied at different confidence limits.

2.1 Principles and methods

2.1.10 Template matching

It may happen that for a given classification task, the coordinates of the object can only assume the values 0 and 1, such as in the following example of numerals, where they correspond to the two grey levels *white* and *black*. Each object defining the class then has an associated vector consisting of 0s and 1s, called the *template*. Similarly, a measurement of an object will give such a vector **x**. If the class is described by a total of m prototype templates $\mathbf{y}_m$, a comparison of **x** with all of them has to be performed in order to find the one that it fits best. This procedure is called *template matching* (Young and Calvert, 1974) and consists simply of a vector multiplication:

$$n_m = \mathbf{y}_m^T \cdot \mathbf{x} \qquad (2.8)$$

where m runs over all templates. Grouping the template vectors into a matrix **Y**, the complete procedure comes down to one single matrix

0000000000	0000000000	0000000000	0000000000
0001111000	0000001000	0001111000	0001111000
0010000100	0000011000	0010000100	0010000100
0100000010	0000101000	0100000100	0000000100
0100000010	0001001000	0000000100	0000000100
0100000010	0010001000	0000000100	0000000100
0100000010	0000001000	0000001000	0000001000
0100000010	0000001000	0000010000	0000001000
0100000010	0000001000	0000100000	0000000100
0100000010	0000001000	0001000000	0000000100
0100000010	0000001000	0010000000	0000000100
0010000100	0000001000	0100000000	0010000100
0001111000	0000001000	0111111100	0001111000
0000000000	0000000000	0000000000	0000000000

Fig. 2.9 Templates for four numerals. The four numerals 0, 1, 2, and 3 are outlined by print characters '1' in a 10 by 14 mesh with the print background made of characters '0'. The four numerals can be seen best by holding the page at some distance, or at a flat angle to the line of sight.

multiplication:

$$\mathbf{n} = \mathbf{Y}^T \mathbf{x} \tag{2.9}$$

where the maximum component of **n** defines the match. Template matching is clearly very well suited to being implemented on vector processors, and has already been applied to trackfinding in high-energy physics (Georgiopoulos, Goldman, Levinthal, and Hodous 1986). It is also a longstanding method for crude and fast trackfinders in real-time systems (see Subsections 1.5.3, 1.5.5).

In Fig. 2.9, four numerals are outlined in a 10 by 14 squared mesh. The template matrix **Y** can be directly read off the figure: each column of **Y** is constructed by placing the ten columns of 0s and 1s of one digit below each other, such that the total column length becomes 140.

This method can be generalized in two ways. The first consists in replacing the condition that the variables have to assume the values 0 or 1 by requesting them to lie in some interval, or around a given value with a known error distribution for each variable; in this latter case, the weighted sum of squares of differences between the measured and the prototype values can be used as a matching criterion.

The second generalization can be applied in cases where the number of possible classes is too high to establish a complete dictionary for direct template matching. In this case, one may try to find a solution in two steps, by grouping classes into *superclasses* and performing the template matching on those. Once the *superclass* has been found, one could then continue by using the smaller directory valid for the classes in this superclass only, or by applying another method such as a combinatorial search. This approach has already been used in high-energy physics trackfinding (Georgiopoulos *et al.* 1986).

Another obvious way of reducing the size of the dictionary is to use the object description in feature space rather than in pattern space. For example, in order to distinguish and recognize different wave types such as sinoidal, rectangular, or triangular, it is sufficient to build a dictionary in feature space with the first five or so Fourier spectrum components of each wave form rather than to describe each wave in object space by a great number of points on it.

2.1.11 Linear feature extraction

If an object such as a particle trajectory is described by an n-dimensional vector **x** in pattern space, and the number of free parameters

2.1 Principles and methods

(in this case the number of integration constants of the equation of motion) is f, then the object class occupies a f-dimensional subspace in $\{P\}$, as defined by the $n - f$ constraint equations between the coordinates. These equations are, however, very often not known explicitly, e.g. if the particle moves in an inhomogeneous magnetic field, and are, in addition, frequently non-linear, in which case the subspace of the object class is curved.

As an example, consider the points on a circle of radius r_i around the origin to be the objects of class number i. The constraint equation, and accordingly the hyperplane it defines, is not linear:

$$x^2 + y^2 = r_i^2$$

One may, however, subdivide this hyperplane into regions which can be approximated by linear hyperplanes, in this case straight line segments forming a regular polygon. The actual subdivision will be dictated by the nearest neighbouring classes, and by the precision one has to achieve. This procedure is always possible in regions where the hyperplane is a unique function of the constants of integration. In an actual application, it may be sufficient simply to subdivide the phase space region under study into smaller regions of equal size. Attempts to use curved hyperplanes have not been successful.

Let us suppose now that a class of objects is defined by a set of prototypes, a *training sample* (Subsection 2.1.6). Then the 'best' linear feature extraction algorithm, which leads to the 'best' linear approximation of the curved subspace, can be defined by requesting that it minimizes the sum of the squares of the distances of all vectors in the training sample to the linear subspace. This can be written in mathematical terms as follows: Let $\mathbf{q}_i (i = 1, \ldots, n)$ be an arbitrary orthonormal basis in the n-dimensional pattern space. Let

$$\mathbf{a} = (\mathbf{x} - \langle \mathbf{x} \rangle) \tag{2.10}$$

denote the shifted vectors of the training sample, where '$\langle \ldots \rangle$' stands for the 'mean over the training sample' as before. Then the object vectors $\mathbf{a}$ can be written as

$$\begin{aligned} \mathbf{a}_i &= \sum_{i=1}^{n} a_{ji} \mathbf{q}_i \\ &= \sum_{i=1}^{m} a_{ji} \mathbf{q}_i + \sum_{i=m+1}^{n} a_{ji} \mathbf{q}_i \end{aligned} \tag{2.11}$$

The 'best' linear feature extraction now obviously consists in finding the basis $\mathbf{q}_i$ which minimizes the mean quadratic error if the expansion above is truncated at m, i.e. which minimizes

$$S(m) = \left\langle \left(\mathbf{a}_j - \sum_{i=1}^{m} a_{ji}\mathbf{q}_i\right)^2 \right\rangle$$
$$= \left\langle \left(\sum_{i=m+1}^{n} a_{ji}\mathbf{q}_i\right)^2 \right\rangle \qquad (2.12)$$

Because the basis $\mathbf{q}_i$ is orthonormal, multiplication of Equation (2.11) with $\mathbf{q}_i^T$ gives

$$a_{ji} = \mathbf{q}_i^T \cdot \mathbf{a}_j \qquad (2.13)$$

leading to

$$S(m) = \left\langle \left(\sum_{i=m+1}^{n} a_{ji}\mathbf{q}_i\right)^2 \right\rangle$$
$$= \left\langle \left(\sum_{i=m+1}^{n} (\mathbf{q}_i^T \cdot \mathbf{a}_j)\mathbf{q}_i\right)^2 \right\rangle$$
$$= \left\langle \sum_{i=m+1}^{n} \sum_{k=m+1}^{n} (\mathbf{q}_i^T \cdot \mathbf{a}_j)\mathbf{q}_i^T \cdot \mathbf{q}_k(\mathbf{a}_j^T \cdot \mathbf{q}_k) \right\rangle$$
$$= \left\langle \sum_{i=m+1}^{n} (\mathbf{q}_i^T \cdot \mathbf{a}_j)(\mathbf{a}_j^T \cdot \mathbf{q}_i) \right\rangle \quad \text{since } \mathbf{q} \text{ orthonormal}$$
$$= \sum_{i=m+1}^{n} \mathbf{q}_i^T \langle \mathbf{a}\mathbf{a}^T \rangle \mathbf{q}_i$$
$$= \sum_{i=m+1}^{n} \mathbf{q}_i^T F \mathbf{q}_i \qquad (2.14)$$

with the dyadic product matrix

$$F = \langle \mathbf{a}\mathbf{a}^T \rangle \qquad (2.15)$$

If $\mathbf{e}_i$ is a normalized eigenvector of F corresponding to the eigenvalue e_i, then

$$\mathbf{e}_i^T F \mathbf{e}_i = \mathbf{e}_i^T \langle \mathbf{a}\mathbf{a}^T \rangle \mathbf{e}_i = e_i \qquad (2.16)$$

holds. From this follows

$$\langle (\mathbf{e}_i^T \cdot \mathbf{a})(\mathbf{a}^T \cdot \mathbf{e}_i) \rangle = e_i$$

or

$$\langle (\mathbf{e}_i^T \cdot \mathbf{a})^2 \rangle = e_i \qquad (2.17)$$

2.1 Principles and methods

Since '$\langle \cdots \rangle$' stands for 'mean over training sample', e_i is equal to a sum of squares and therefore is not negative. If the vectors **a**, i.e. the original track vectors, span the full pattern space, then the eigenvalues are all positive, since

$$e_i = 0 \quad \text{would mean} \quad \langle (\mathbf{e}_i^T \cdot \mathbf{a})^2 \rangle = 0$$

leading to

$$\mathbf{e}_i = 0 \tag{2.18}$$

because of the condition above, namely the **a** spanning $\{P\}$. Equation (2.18) however, contradicts the assumption that $\mathbf{e}_i$ is normalized.

In conclusion, in the case in which the tracks span the full original pattern space $\{P\}$, **F** is positive definite, otherwise it is semidefinite with its rank equal to the dimension of the subspace spanned by the track sample. Consequently, since **F** is at least semidefinite, minimizing Expression (2.14) means minimizing each term separately. Introducing Lagrange multipliers λ_i then requires that the terms

$$\mathbf{q}_i^T \mathbf{F} \mathbf{q}_i - \lambda_i \mathbf{q}_i^T \cdot \mathbf{q}_i + \lambda_i \tag{2.19}$$

have to be minimized (remember that $\mathbf{q}_i^T \cdot \mathbf{q}_i = 1$). Setting the derivatives of this expression with respect to the components of $\mathbf{q}_i$ to zero gives the minimum condition for $S(m)$:

$$\mathbf{F} \mathbf{q}_i = \lambda_i \mathbf{q}_i \tag{2.20}$$

i.e. $\mathbf{q}_i$ is an eigenvector of **F** with eigenvalue λ_i. The minimum of $S(m)$ is then simply

$$S(m) = \sum_{i=m+1}^{n} \lambda_i \tag{2.21}$$

By taking the eigenvectors of the largest m eigenvalues as the new basis, one has now found the 'best' subspace obtainable with linear feature extraction. The average distance of any track in the sample to this subspace is given by $S(m)$. This method is named after Karhunen-Love (Young and Calvert 1974). It is also known as *principal component analysis*.

In the general case a certain number, m (greater than or equal to f, the number of free parameters), of the components of **y** will be significant, and the remaining $(n - m)$ components will have small absolute values (Brun, Hansroul and Kubler 1980).

The result is that the eigenvectors of the *dispersion* matrix

$$\mathbf{F} = \langle (\mathbf{x} - \langle \mathbf{x} \rangle)(\mathbf{x} - \langle \mathbf{x} \rangle)^T \rangle \tag{2.22}$$

form the rows of the transformation matrix **T**, such that the new coordinates become

$$\mathbf{y} = \mathbf{T}\mathbf{x}, \quad \mathbf{y} \in \{F\}, \quad \mathbf{x} \in \{P\}$$

As an example, the eigenvalues and eigenvectors for the transformation (x, y) into (x'', y'') of Fig. 2.6 have been calculated to be:

eigenvalues:
 5.4774, 0.0385
eigenvectors:
 0.720613 0.693337 (1)
 0.693337−0.720613 (2)

where the eigenvectors have been written row-wise such that the above matrix is the transformation matrix. The slope of the x'' axis with respect to the (x, y) system comes out as 0.962.

2.1.12 Minimum Spanning Tree (MST)

The MST is used in cluster analysis, and as such is of some interest for trackfinding, if we consider tracks to be clusters of points. It has been applied successfully both to trackfinding (Cassel and Kowalski 1981), and calorimeter analysis in high energy physics (CHARM-II experiment, CERN). To understand it, a minimum knowledge of the expressions used in graph theory is necessary.

A *graph* consists of *nodes* and *edges*. A node can represent any object, and is often graphically represented by a point (see Fig. 2.10). An edge can be symbolized by a line connecting two nodes, and expresses the fact that some well-defined relation between these two nodes exists. As an example, consider the objects to be space points, and the edges connecting the nodes belonging to the same track. If a positive *weight* (this could be a metric distance or a metric similarity, see Subsection 2.1.6) is assigned to each edge, the graph is called *edge-weighted*. For space points, the distance between them may serve this purpose. An *isolated point* is a node without edge. A *connected graph* is a graph without isolated points. In a *fully connected* or *complete* graph, all nodes are directly connected with all other nodes. A *path* between two nodes is a sequence of edges connecting them. A *loop* or *circuit*

2.1 Principles and methods

is a closed path, i.e. a path connecting all nodes in it to themselves. A *tree* is a graph without circuits. A *spanning tree* is a connected graph without circuits (Fig. 2.11). Finally, a *minimum spanning tree* is a spanning tree for which the sum of the edge weights has a minimum for a given graph. If all edge weights are different, the MST is unique.

A MST can be constructed in the following way:

Start with any node, call it the initial tree T_0.

If a tree of order k exists already, add to it that node outside T_k for which the minimum edge weight connecting it to T_k (i.e. to some node in T_k) is minimum for all nodes outside T_k.

When there are no more nodes outside T_k that can be connected, the MST is complete.

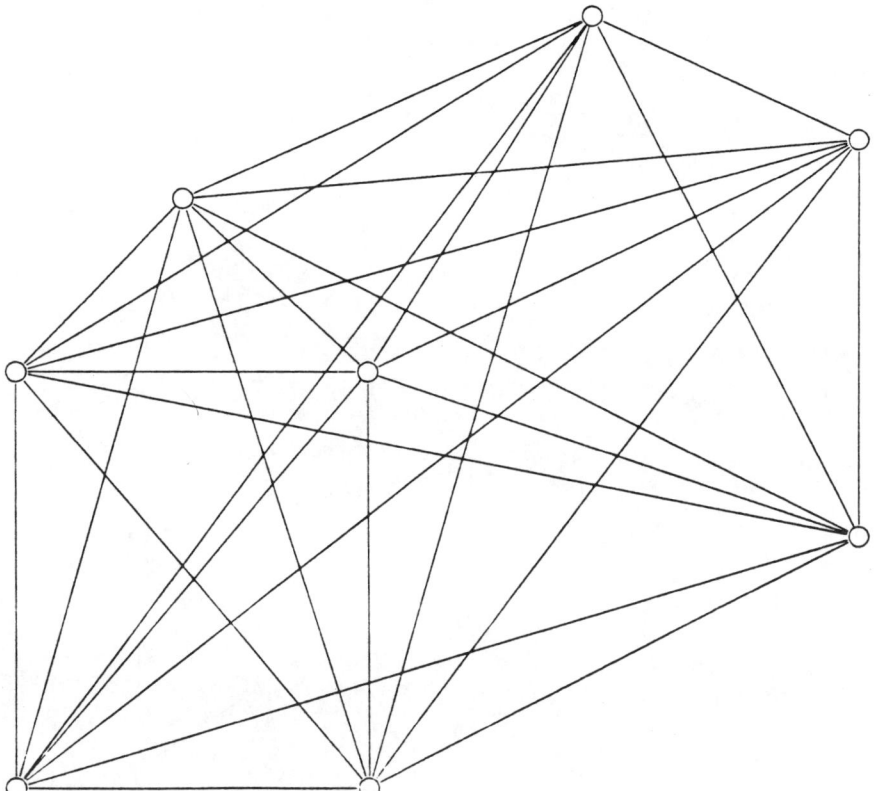

Fig. 2.10 A fully connected graph. The *nodes* of the graph are represented by small circles, the *edges* by connecting lines.

152   *Pattern recognition*

If the graph contains n nodes, and all edge weights have to be calculated once (e.g. point distances), the computing time for this calculation is, of course, proportional to $n(n-1)/2$.

By using a special algorithm to implement the above procedure, called the *Prim single fragment method* (Zahn 1973), the number of edge-weight comparisons to be made becomes proportional to n^2, so that the whole MST construction time is then also proportional to n^2. In applications such as trackfinding this time can be further reduced by setting an upper limit for the edge weight, i.e. limiting the distance between consecutive points on a track.

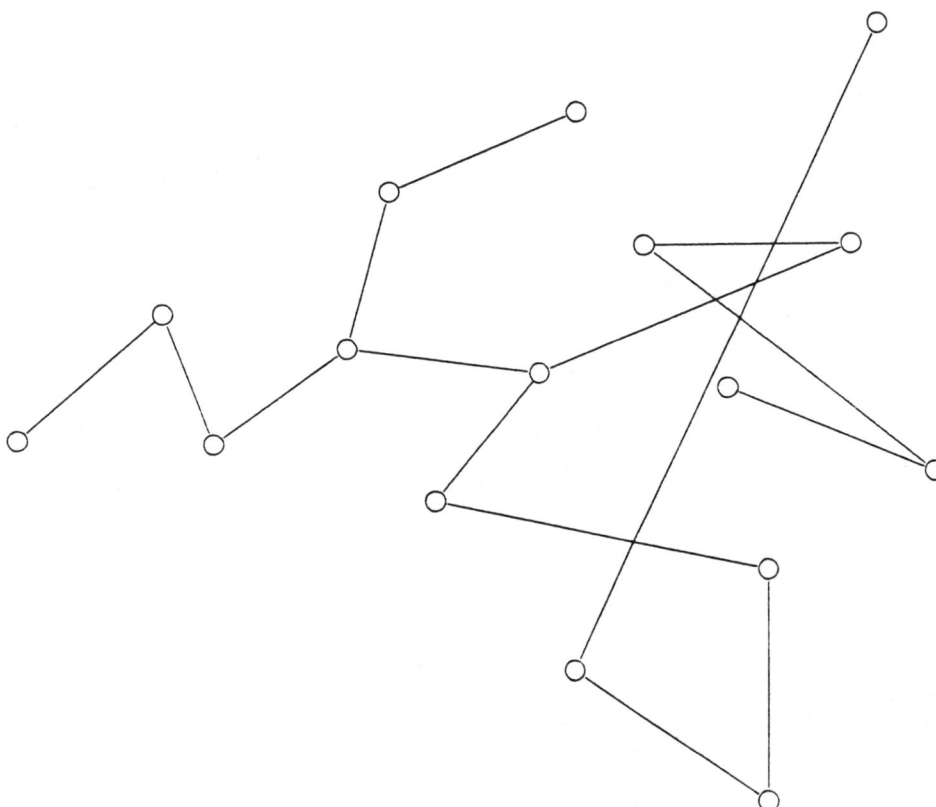

Fig. 2.11 A spanning tree. Each node is connected to at least one other node. From any given node there is only one possible path to any other node. This is equivalent to the fact that no closed paths (*loops* or *circuits*) exist.

2.1.13 Compatibility graph

A further graph theory method has been used to resolve problems in high-energy physics trackfinding. It provides the solution to the following problem, which arises from the uncertainty with which points are assigned to tracks, and the subsequent result that the same point may be assigned to more than one track: For a given graph of nodes and edges, find the subset of a maximum number of nodes that are not linked with each other, and are therefore said to be *compatible*. In high-energy physics trackfinding, this problem arises when signals, or complete track segments are found that can belong to more than one track (candidate) at the same time. If such tracks are represented by nodes, and are called *incompatible* if

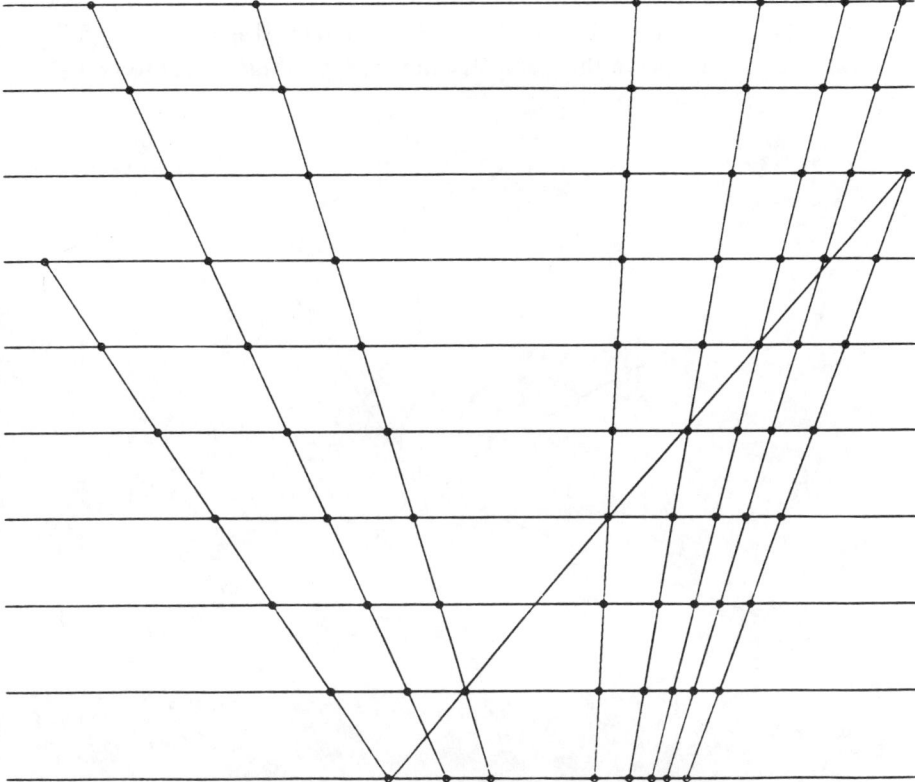

Fig. 2.12 Tracks from a vertex plus a spurious combination. Straight tracks are plotted in an arrangement of parallel detector planes. If the track finding program does not use the vertex as an extra point on the track, it is likely to find the 'ghost' track which crosses the others. Such a case can be treated with graph logic (see Fig. 2.13).

they share e.g. track segments, and this fact is expressed by an edge connecting each pair of such tracks, then we have exactly the situation described above, since it can be argued that the most plausible underlying event structure is the one that allows for a maximum number of compatible tracks (see Fig. 2.12). In this figure, straight tracks coming from a common vertex have been found in a set of ten planes. In addition, the hypothetical trackfinding algorithm has found an additional 'ghost' consisting of spurious combinations of hits from the different correct tracks. If two tracks are called incompatible when they have at least one signal in common, and if incompatible tracks are connected, then the graph in Fig. 2.13 is created from the tracks in Fig. 2.12. Obviously, in this case the maximum number of compatible tracks can be obtained by deleting the 'ghost' represented by the node in the centre. By linking compatible tracks rather than incompatible ones, the problem is then to find all maximum size complete (fully connected) node sets in this complementary graph. Algorithms to find all

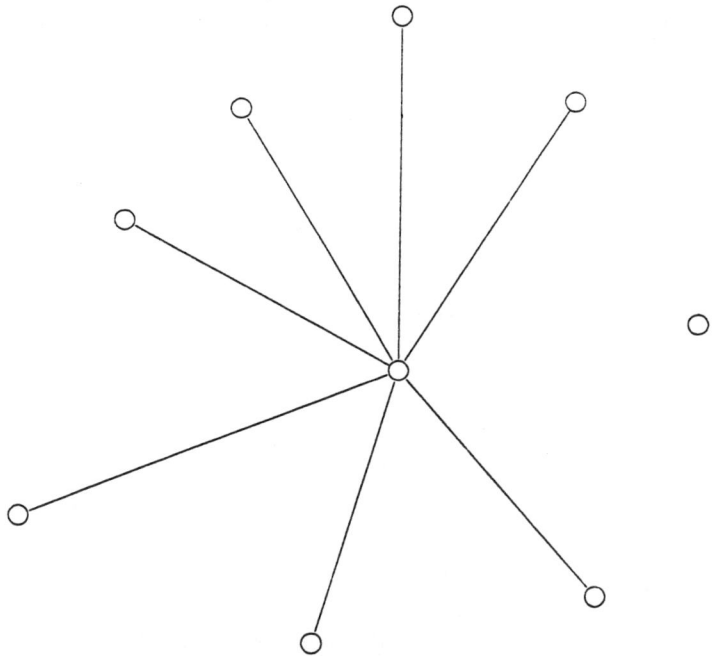

Fig. 2.13 The tracks of Fig. 2.12 in graph representation. Each track is now represented by a node. Tracks are called *incompatible* if they share at least one signal in Fig. 2.12. Incompatible tracks are connected by an edge. The requirement is to remove a minimum number of nodes until all tracks are compatible. Obviously, this leads to the removal of the 'ghost' track (in the centre of this graph) which is incompatible with most others.

maximum complete subsets of a graph exist, but are too lengthy to be explained here in detail (Das 1973). A FORTRAN subroutine using the algorithm by Das can be obtained from CERN (CERN Program Library V401).

2.2 Numerical techniques for trackfinding in high-energy physics

2.2.1 Circles, polynomials, and splines for curve approximation

On many occasions, the particle track is conveniently approximated by a simple curve during the process of trackfinding. This arises when the particle moves in an inhomogeneous magnetic field, in which case there is no analytical description of the track, or even when the projected track is an ellipse or a circle, namely when the speed of the calculation is important.

The simplest and fastest approximation to a curved track is the parabola:

$$y = ax^2 + bx + c \tag{2.23}$$

which can actually be expressed as a linear function in three of the y values (the y values could be the measured coordinates along a track):

$$y = a_1 y_1 + a_2 y_2 + a_3 y_3 \tag{2.24}$$

with

$$a_i = [(x - x_j)(x - x_k)]/[(x^i - x_j)(x_i - x_k)] \tag{2.25}$$

and $i \neq j, k, j \neq k$.

This formula is a special case of Lagrange's formula, Equation (2.26), for polynomials of any degree, the correctness of which follows immediately from the fact that two polynomials of degree n are identical if they share $n + 1$ points:

$$y = \sum_{j=0}^{n} \frac{\pi_j(x)}{\pi_j(x_j)} y_j \tag{2.26}$$

with

$$\pi_j(x) = \prod_{\substack{k=0 \\ k \neq j}}^{n} (x - x_k) \tag{2.27}$$

Pattern recognition

Thus, if a hit has to be predicted in a detector plane as function of three points already on the track, Expression (2.24) can be used, where the coefficients a_i, which depend only on the plane positions and are therefore constant, have been stored in a table for all combinations of three planes that may be used.

The simplicity of Expression (2.24) can be compared to the expression for a circle: the centre (x_c, y_c) and the radius r of a circle through three points are given by

$$x_c = (b_1 - b_2 + c_1 a_1 - c_2 a_2)/(c_1 - c_2) \qquad (2.28)$$

$$y_c = b_1 + c_1(a_1 - x_c) \qquad (2.29)$$

$$r^2 = (x_1 - x_c)^2 + (y_1 - y_c)^2 \qquad (2.30)$$

with

$$a_i = (x_{i+1} + x_i)/2, \quad i = 1, 2 \qquad (2.31)$$

$$b_i = (y_{i+1} + y_i)/2, \quad i = 1, 2 \qquad (2.32)$$

$$c_i = (x_{i+1} - x_i)/(y_{i+1} - y_i), \quad i = 1, 2 \qquad (2.33)$$

and the points have to be numbered in such a way as to avoid division by zero. Because of this last condition the above formula is inconvenient and 'risky' to use, since it will, of course, become imprecise when the values in the denominator approach zero. This can be avoided by performing a coordinate system shift followed by a rotation in such a way that (x_1, y_1) coincides with the origin, and y_3 becomes zero:

(1) Shift:

$$x'_i = x_i - x_1, y'_i = y_i - y_1, i = 2, 3 \qquad (2.34)$$

(2) Rotate:

$$d = (x'^2_3 + y'^2_3)^{\frac{1}{2}} \qquad (2.35)$$

$$c_\phi = x'_3/d \qquad (2.36)$$

$$s_\phi = y'_3/d \qquad (2.37)$$

$$\tilde{x}_i = x'_i c_\phi + y'_i s_\phi, i = 2, 3 \qquad (2.38)$$

$$\tilde{y}_2 = y'_2 c_\phi - x'_2 s_\phi \qquad (2.39)$$

(3) Find centre, radius:

$$\tilde{x}_c = x_3/2 \qquad (2.40)$$

2.2 Numerical techniques for trackfinding

$$\tilde{y}_c = [\tilde{y}_2 - \tilde{x}_2(\tilde{x}_3 - \tilde{x}_2)/\tilde{y}_2]/2 \tag{2.41}$$

(having checked that $\tilde{y}_2 \neq 0$)

$$r^2 = \tilde{x}_c^2 + \tilde{y}_c^2 \tag{2.42}$$

(4) Rotate and shift centre to original system:

$$x_c = \tilde{x}_c c_\phi - \tilde{y}_c s_\phi + x_1 \tag{2.43}$$

$$y_c = \tilde{y}_c c_\phi + \tilde{x}_c s_\phi + y_1 \tag{2.44}$$

Up to now, only analytical functions have been mentioned. A class of very useful nonanalytical functions are the spline functions that consist of pieces of polynomials (and are thus piecewise analytical) linked at certain points call *knots*. At the knots, the derivatives of a spline function exist only up to a certain limit k which is specific to the spline in question, such that the derivative $k + 1$ becomes a step function. Actually, all spline curves can be expressed as indefinite integrals of step functions. Since the spline curves are not analytical (for a function to be analytical, it must have infinitely many derivatives in each of its points of the contiguous open interval where it is defined (Dieudonné 1979)), they are much more readily adapted to follow a path given by a series of points than a polynomial. As an example, both a 12th order polynomial, and a cubic spline are fitted to points (marked by little crosses) lying on a curve composed of two Gaussian functions in Fig. 2.14:

$$f(x) = e^{(x+3)^2/3} + 2e^{(x-3)^2/2}$$

The figure speaks for itself. Spline curves are used in many graphical applications, such as the drawing of road maps.

In high-energy physics trackfinding, two types of splines have mainly been used, simple third order or *cubic* splines, and B-splines. Third order splines have been fitted both to the particle track, and to its second derivative; in the latter case their explicit double integration, leading to so-called *quintic splines*, delivers an approximation to the track as given by the equation of motion in an inhomogeneous magnetic field (Wind 1974).

Spline functions are mathematically defined in the following way: Given n knots

$$x_1 < x_2 < \cdots < x_n$$

and the values

$$x_0 = -\infty, x_{n+1} = +\infty$$

a spline curve $s(x)$ of degree m is defined for all x values through the following properties:

(1) In each interval $[x_i, x_{i+1}]$, $i \in [0, n]$, $s(x)$ is given by some polynomial of degree m or less.
(2) $s(x)$ and its derivatives of orders $0, 1, \ldots, m-1$ are continuous everywhere.

Splines are conveniently expressed with the help of truncated power functions x_+^m:

$$x_+^m = x^m \text{ for } x > 0 \qquad (2.45)$$

$$x_+^m = 0 \text{ for } x \leq 0 \qquad (2.46)$$

Any spline function s of degree m can be uniquely expressed as

$$s(x) = p(x) + \sum_{j=1}^{n} c_j (x - x_j)_+^m \qquad (2.47)$$

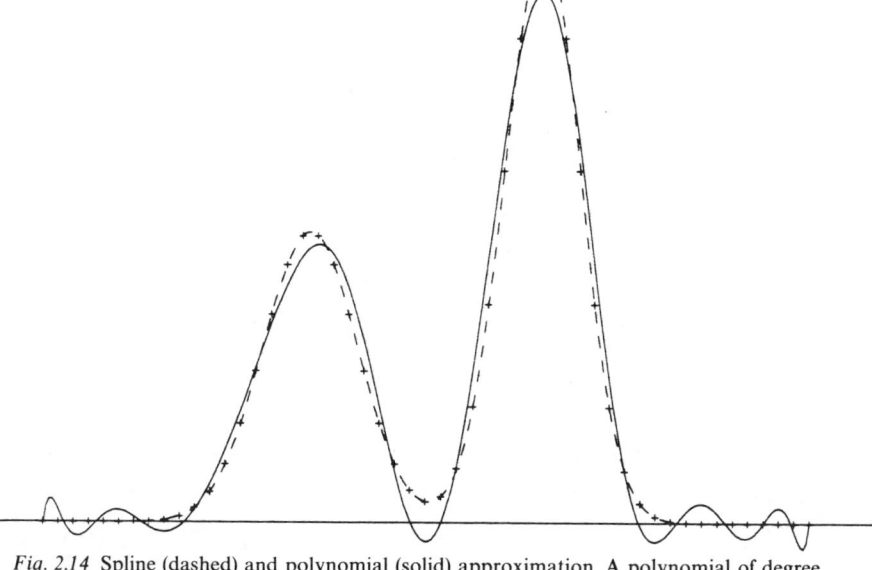

Fig. 2.14 Spline (dashed) and polynomial (solid) approximation. A polynomial of degree 15 is fitted to 51 equidistant points on the curve which is given by the sum of two Gaussian curves. A cubic spline through the same points as nodes is calculated. The polynomial approximation is particularly bad where the curve is flat over an extended area since such a behaviour is, of course, impossible for a polynomial curve.

where $p \in \prod_m$ (class of polynomials of degree m or less). A *natural* spline is a spline of odd degree $2k-1$ with $p(x)$ of degree $k-1$ (rather than $2k-1$):

$$s(x) = p(x) + \sum_{j=1}^{n} c_j(x-x_j)_+^{2k-1}, p \in \prod_{k-1} \tag{2.48}$$

For a natural spline, the coefficients satisfy the relation

$$\sum_{j=1}^{n} c_j x_j^r = 0, \; r \in [0, k-1] \tag{2.49}$$

If a function is to be approximated by a natural spline, where function values are given at the knots $x_1, \ldots, x_n$, the number of unknowns $n+k$ matches the number of constraint equations from Equations (2.48) and (2.49). This is probably the reason why such splines are called 'natural'. It can be shown (Greville 1969) that the system of linear equations is nonsingular, i.e. that the spline representation is unique.

The 'popularity' of splines becomes obvious if one tries to find the 'smoothest' interpolating function g for n data points given at $a = x_1 < x_2 < \cdots < x_n = b$. 'Smoothest' means here that the kth derivative of g varies as little as possible in the interval $[a,b]$, i.e. that

$$\int_a^b [g^{(k)}(x)]^2 \, dx = \text{minimum} \tag{2.50}$$

If $C^k[a,b]$ denotes the class of functions for which k derivatives exist on $[a,b]$, and only functions $g \in C^{k-1}[a,b]$ are considered, such that $g(x)^{(k)}$ is piecewise continuous in $[a,b]$, then it can be shown (Greville 1969) that the smoothest interpolating function for a given k is precisely the natural spline of order $2k-1$, as long as $k < n$. For $k = n$, the solution is, of course, the uniquely defined polynomial of degree $n-1$, and for $k > n$ there are infinitely many polynomials of degree $k-1$.

From this it follows that third order natural splines or *cubic* splines are the best lowest approximation to a curve looking 'smooth', since the case $k=1$ leads simply to a connection of the data points by straight lines. For cubic splines, the explicit formulation of $s(x)$ as a sequence of pieces of polynomials still looks relatively simple:

$$\begin{aligned} s(x) &= a_{00} + a_{01}x & x \leq x_1 \\ s(x) &= a_{i0} + a_{i1}x + a_{i2}x^2 + a_{i3}x^3 & x_i \leq x \leq x_{i+1}, i \in [1, n-1] \\ s(x) &= a_{n0} + a_{n1}x & x \geq x_n \end{aligned} \tag{2.51}$$

There are $4n$ unknowns, and $4n$ constraint equations: $1 + 2(n-1) + 1$ from the function values at the knots, n from the continuity of the first, and n from the continuity of the second derivative at the knots. The matrix of the system of linear equations has band structure, which has computational advantages.

The formulation of splines with the help of truncated power functions is not satisfactory mainly for two reasons:

(1) Expression (2.48) is biassed in the sense that for increasing x more and more terms enter in the expression. Consequently, the matrix of the system of linear equations resulting from Expression (2.48) has the form of a band matrix plus a triangular matrix, which is inconvenient for large n.
(2) Even worse, it can be shown (Cox 1982) that this system of linear equations is inherently ill-conditioned, which means that it has to be solved using very high precision in the computer or, better, it has to be avoided.

One would prefer a basis other than truncated power functions that is in some sense 'symmetric' and leads to well-conditioned systems of equations. The so-called *B-splines* form such a basis. They are defined with the help of divided differences (Abramowitz and Stegun 1970): Given $k+1$ abscissa values $x_0, \ldots, x_k$, and a function $f(x)$ assuming values f_k at x_k, the first divided difference is defined as

$$[x_0, x_1]f = (f_0 - f_1)/(x_0 - x_1) = [x_1, x_0]f$$

and the jth divided difference becomes

$$[x_0, \ldots, x_j]f = ([x_0, \ldots, x_{j-1}]f - [x_1, \ldots, x_j]f)/(x_0 - x_j) \quad (2.52)$$

This can be written in terms of functional values:

$$[x_0, \ldots, x_k]f = \sum_{j=0}^{k} f_j/\pi_j(x_j), \quad (2.53)$$

where $\pi_j(x)$ is as in Equation (2.27). For a given set of knots x_k with $x_k \leq x_{k+1}$ for all k, the normalized B-spline $N_{n,j}(x)$ is defined by

$$N_{n,j}(x) = (x_j - x_{j-n})[x_{j-n}, \ldots, x_j](. - x)_+^{n-1} \quad (2.54)$$

where the expression with the "." is defined as follows: $[\cdots]f(., x)$ means that the divided difference is to be taken with respect to the variable at the '.', e.g.

$$[x_0 - x_1](. - x) = \{(x_0 - x) - (x_1 - x)\}/(x_0 - x_1)$$

2.2 Numerical techniques for trackfinding

which is equal to 1. This is a special case of the theorem that the nth (divided) difference of any polynomial of degree n is a constant.

Applying this theorem, it can be seen from Equation (2.54) that a B-spline extends over a number of intervals which is one higher than its degree. For $N_{n,j}(x)$ this means that it is only different from zero for $x_{j-n} < x < x_j$. This can be seen as follows: for $x > x_j$, the truncated power functions are all equal to zero. For $x < x_{j-n}$, the truncated power functions are all equal to polynomials of order $n - 1$, hence their nth divided difference is zero. When a function is interpolated with the help of B-splines, the matrix of the corresponding linear system has consequently band structure.

2.2.2 Interpolation and extrapolation

The choice of the trackfinding method may depend on the precision with which additional points can be predicted if a part of the track has already been found. A related problem is the linking of track segments that have been found independently.

Interpolation is more precise than extrapolation. For polynomials this can be seen most easily from the Lagrange formulation:

$$y(x) = \sum_{j=0}^{n} \pi_j(x) y_j / \pi_j(x_j) \tag{2.55}$$

with $\pi_j(x)$ as in Equation (2.27) and where n is the degree of the polynomial y, and y_j are the values at x_j.

If just one of the y values, say y_k, is wrong by an error Δy_k, then the error in y for all x is simply

$$\Delta y = [\pi_k(x)/\pi_k(x_k)] \Delta y_k = p(x) \Delta y_k$$

with some polynomial $p(x)$ of degree n. All $n + 1$ zeros of $p(x)$ are at x_j, $j = 0, \ldots, n$. Therefore, for $x \in [x_0, x_n]$, $p(x)$ oscillates between its $n - 1$ extrema, which, however, for arbitrary x_j can be made to assume any value, since $\pi_k(x_k)$ can become very small. On the other hand, for equidistant x values, one would hope that the absolute value of $p(x)$ remains below some convenient limit (preferably $= 1$) for $x \in [x_0, x_n]$.

This is not exactly the case, although the maximum absolute values are independent of the x step $h = x_1 - x_0$, and behave fairly well as can be seen from the following list (the degree of p is given in brackets):

$$\max(p(x)) = 1(2), \ 1.056(3), \ 1.151(4), \ 1.257(5),$$
$$1.357(6), \ldots, 6.5(10), \ x \in [x_0, x_n].$$

This means that for polynomials of low degree, the interpolation error is of the order of the measurement errors.

On the other hand, for x outside $[x_0, x_n]$, $p(x)$ will grow faster than d^n if $d = \min(|x - x_0|, |x - x_n|)$, since there the differences $(x - x_j)$ all have the same sign. This means that the extrapolation error will grow rapidly even for polynomials of low order, and has to be watched carefully.

The interpolation of a circle with $y_1 = y_3$ leads to errors of the same order of magnitude as the measurement error, whereas the extrapolation error is roughly proportional to d^2/s^2, where d is the distance to the centre of the segment, and s the segment length (Pimiä 1985), as long as s is small compared to the radius.

The interpolation and extrapolation errors (introduced by measurement errors) of splines and other functions more complicated than polynomials are normally difficult, if not impossible to evaluate in a closed form. The approach taken to this problem is in most cases a Monte Carlo simulation of all possible tracks under study, including all errors such as multiple scattering, δ electrons, and other known reasons for measurement errors, and the establishing of *road* widths with the help of histograms.

2.2.3 Parametrization

Leaving aside quantum-mechanical effects which are very small, the track of a particle in vacuum is completely defined by five parameters, e.g. the five integration constants of Equations (3.40a) and (3.40b). The trajectory can then be expressed as a function of one single independent variable, such as the track length s, or the coordinate along an axis of a coordinate system in space, say x. On the other hand, for a plane $x = $ constant, we can consider the values (y, z) on the trajectory for this x value to be functions of the five parameters, englobing in this way not just one trajectory, but all trajectories passing through the plane $x = $ constant,

$$y \equiv y(y_0, z_0, y'_0, z'_0, 1/p)$$

and z accordingly. Finding an expression of this type is called *parametrization*. A different parametrization will normally be necessary for each x value.

As a simple example, consider circles through the origin in which the x axis is a tangent. Then, for a given x_1, the y values can be expressed as a function of the only free parameter left, the radius r (the formula is valid for circles bending upwards or downwards, if the sign of curvature is included

2.2 Numerical techniques for trackfinding

in r):

$$y_1(r) = r[1 - (1 - x_1^2/r^2)^{\frac{1}{2}}], \ r \geqslant |x_1|$$

In high-energy physics data analysis, parametrization has been applied in three different forms:

(1) To calculate the coordinates of a track in a given detector plane as functions of the five kinematic quantities. This is useful both for Monte Carlo simulation, and for track fitting, and is, of course, very much faster than tracking each particle individually through the magnetic field, and then calculating the detector hits.
(2) To parametrize the track hits in some of the 'planes' as functions of track hits in other 'planes'. This is used in trackfinding to predict coordinates once a part of the track has been found (Lassalle, Carena, and Pensotti 1980). This is described in more detail in Subsection 2.4.2.
(3) To express the five kinematic quantities as functions of the hits (= coordinates) in a given detector. These values are normally not good enough to be final, but may represent good starting values for an iterative fit procedure.

In all three cases, the most obvious merit of the method is its speed. Indeed, if the magnetic field is rather well behaved, and, more importantly, if the phase space region in question is small (Aubert and Broll 1974), one can hope to express the desired quantities by polynomials of relatively low order. This high speed makes the method very valuable in real-time applications, see e.g. Subsection 1.5.6. An extra benefit comes from the fact that the magnetic field need no longer be used explicitly once the parametrization has been performed.

The most serious drawback of the method lies in the fact that normally it requires much study and optimization effort. The choice of the phase space region, and the selection of the *training sample* are particularly delicate operations. For example, it is important to choose more points in pattern space near the borders of the trajectory cluster than near the centre. Furthermore it should be clear that any given parametrization is only valid for a specific detector arrangement, and that each change in the detector set-up, and each different subset of planes that must be treated (owing to detector inefficiencies) require a new parametrization. However, missing points can sometimes be replaced by their predictions (see Subsection 2.4.2).

Normally, a linear least squares fit is used for convenience to parametrize a function of several variables. This can be done with the help of the program MUDIFI (CERN Program Library E5001) which uses for the fit products of monomials, Legendre, or Chebyshev polynomials in each variable. The program keeps only functions that reduce the sums of squares of residuals significantly when added into the fit. In this way, a great number of polynomials can be tried in an automated fashion.

From the remarks on feature extraction it should be clear that it is advisable to perform such a multidimensional fit preferably after a feature extraction transformation, such as performed by the program LINTRA (CERN Program Library E5002). LINTRA and MUDIFI are designed to work in sequence. In particular, LINTRA will produce the FORTRAN code of a routine performing the Karhunen–Loeve transformation (Brun, et al. 1979, Brun, Hansroul, and Kubler, 1980).

Feature extraction followed by parameterization should normally always be considered and even tried out when a detector has to be studied, be it only to gain some further insight into the track sample and the response of the detector.

2.3 The task of trackfinding in high-energy physics

Given a set of position measurements in a detector, the task of trackfinding is to split this set into subsets (= classes) such that:

(1) Each class contains measurements which could be caused by the same particle.
(2) One class (which may possibly be empty) contains all measurements that cannot be associated with particles with sufficient certainty. (These may stem from accidental signals, from distorted measurements, from ambiguities in the association with tracks, or from weaknesses of the track model.)

This definition is modest enough to represent a realistic goal (Grote 1981). It reduces trackfinding to a cluster analysis problem, where a cluster is described as follows (Andrews 1972):

A cluster is loosely defined as a collection of vectors or points that are close together.

This very general definition stresses the decisive role of the metric used to define clusters. Points which are 'near' when a specific metric is used ('near' could be defined relative to the mean value of all pair distances) may be 'far' from each other when a different metric is used.

2.3 The task of trackfinding

The measurements along a particle trajectory have to fit to a track model, which is given by a function in space obeying the equation of motion. The human brain is very good in recognizing clusters, e.g. a sufficiently dense sequence of points along a smooth curve, even in the presence of *kinks* (sudden changes in the direction), vertices, overlapping tracks, and background, but it is rather poor in fitting these points to a given curve, except for a straight line, whereas a computer excels in this.

Many trackfinding methods are based on these complementary capabilities: first, they try to find *track clusters* as efficiently as the eye does, and then they test them rigorously with a fit to a track model, making use of all the *a priori* knowledge of the detector performance and resolution, the particle motion, and both systematic and statistical errors. This has as the consequence that most practical methods basically proceed in two steps. First a subset of measurements is selected, forming a *track candidate*. Second, a decision function is used to check whether or not the track candidate is an acceptable track.

A further consequence of the unmatched superiority of the human cluster finding performance is the requirement that the tracks should be easy to recognize in a suitable graphical presentation of the coordinates. This means that the number of points per track, and the point density along the track should be high enough to make this possible. This is of the greatest importance for testing the software with real data. If this goal is not met, the quality of the trackfinding can only be checked with Monte Carlo events, and its efficiency with respect to measured data of real events will always remain in doubt. Fortunately, this fundamental principle is now generally accepted.

The two-step procedure of track candidate definition and track candidate checking is often split further in order to make it faster. In an initialization phase a certain number of measurements are selected and are either rejected or accepted as a possible start of a track. If they are accepted then additional measurements are processed, in several steps, each time considering a possible rejection, until a final check is performed on the fully assembled track. The gain in speed over the basic two-step procedure, which first selects a full track candidate and then applies a decision function, is three-fold.

Firstly, each reduction in the number of points per track candidate brings a considerable reduction in the number of such candidates, since the number of combinations grows with the power of this value: if there are m planes and n tracks, then the total number of combinations is of course n^m. Secondly, the track candidate allows the prediction of an interval in which correct points can be found. This interval should be large enough to allow all points, that can possibly belong to the track, to be found, even if this may lead to the temporary acceptance of wrong points. Otherwise, if the interval is too narrow to include all points, one will bias the trackfinding procedure in a way which is very difficult to evaluate. Thirdly, the application of the final decision function can be very time consuming, for example if an iterative fit to a track model has to be performed. This results in a sequence of decision functions from simple and fast to precise but slow that improves the speed if really at each intermediate test a good fraction of the wrong candidates is rejected.

Of course, great care is needed to avoid the rejection of good track candidates by any of the approximate and simple tests, because these will normally all be biassed in their acceptance of tracks once they start rejecting good tracks.

For this reason as well, one should carefully avoid the rejection of points at this level. If more than one point is found inside the prediction interval, it is advisable to split the current track candidate into as many pieces as necessary in order to include one point in each of them, rather than take for example the point nearest to the prediction (except when the correct track model is being used, and when confidence in the track candidate is very high). The multiplication of track candidates can normally easily be dealt with: in most cases, the wrong track cannot be extended further if a track following method is used. However, it may sometimes lead to several highly incompatible tracks at the level of the final fit, and only then should one chose one of those that has passed the final acceptance criterion (Subsection 2.1.13).

If, for a given method, all possible tracks have been found, a certain number of points will normally remain unassigned to any track. These form the *background*. If several different methods, or the same method with different cuts, are applied a good measure of the efficiency is essential. Such a measure, as will be shown later, is specific to each experiment (Subsection 2.3.5). It will typically be based on a calibration sample (measured events) or a training sample (Monte Carlo events).

2.3.1 Point removal

Once a track has been found, it is very tempting to remove its points from the pool of points before continuing the track search, in order to have a smaller number of combinations to consider. This can, however,

2.3 The task of trackfinding

only be done safely if the redundancy is high enough, so that a track will still be found if one or two of its points are missing, since two (curved) tracks may, of course, have points in common. Even so, a slight bias against the second track is introduced by removing some of its points. Of course, the removal of all points of a *wrong* track will normally lead to one or several good tracks being lost completely; because of these risks, a suitable compromise between removing and not removing tracks is normally reached in doing both, insofar as the points are removed (marked in some way) from the track candidate initialization, but not from the set of points that can be added to an existing candidate through prediction. This procedure reduces the number of combinations to be tried, and therefore speeds up the whole procedure, but allows the same point to be found on more than one track.

2.3.2 Track quality

In the case of a conflict between two or more tracks in the sense that they are incompatible, and one of them has to be chosen as the (only) correct one, but also in a more general context, it is desirable to have some measure of the quality of a track. Clearly, a simple and well-defined number such as the χ^2 value of the final fit comes to mind, provided that the fit is really unbiassed. However, this is a risky, and in most cases arbitrary choice, because the χ^2 values are, of course, distributed according to well-known laws, and by no means does a small χ^2 indicate with certainty that a track is 'better' than another one with a higher χ^2 value, as long as this value is 'possible' inside the acceptance range chosen. This remains true even if confidence levels are compared rather than the χ^2 values which depend on the number of degrees of freedom. If, in addition, the track model is not absolutely correct, short tracks will on average have higher confidence values than longer ones; preferring tracks with higher confidence values would thus lead to the opposite of the desired effect, since a long track, in general, deserves more confidence than a short one.

On the other hand, the number of points and the absence of *holes* (hits predicted but not found in a chamber) are a rather safe measure of quality, provided the final acceptance test has been passed successfully. This leads quite naturally to a hierarchical search for tracks in which the longest tracks are looked for first, either by starting on the outside of the detector (as seen from the interaction region), or, in the case of interpolation, by spanning the

168   *Pattern recognition*

initial track candidate over the longest possible interval. Of course, once the longest tracks have been found and 'removed' in the above sense, shorter tracks will normally have to be looked for in chamber subsets.

This consideration, taken together with the earlier one about the dangers of track candidate elimination in an early stage of the trackfinding, leads to the following situation: in all cases where tracks are found in separate pieces, be it because the detector consists of independent parts, or because the trackfinding algorithm leads to a segmentation of the track (Subsection 2.3.4), it is highly advisable to keep even weaker incompatible partial tracks until the final step, in which the complete tracks are assembled and tested with the final check, normally a fit to some track model.

2.3.3   Working in projections or in space

Particle detectors that measure genuine space points along a particle trajectory have emerged over the last years (Subsection 2.1.4). For such detectors, if the coordinates of the space point are really measured with comparable accuracy, the choice is obvious: the trackfnding should be done in space. This will become clear in the following, since in this case only the advantages remain: the disadvantages of working in space all arise from space points being reconstructed from measurements in different projections.

However, the number of detectors giving only one coordinate per measurement (or two coordinates with very unbalanced precision) is still high, either because old chambers are used, or because one of the two coordinates is error prone or imprecise or both, which is frequently the case (Pimiä 1985). In such a case, a detector 'chamber' supplies (apart from the plane position) only one coordinate, whereas two are needed to reconstruct the impact point of a particle in the 'plane' of the wires. If several planes with parallel wires are placed behind each other, they provide one projection of the track onto a plane orthogonal to the wires. For a reconstruction of the track in space, at least two different projections are necessary, and extra projections provide further information, which can be invaluable if the chambers are less then 100% efficient, or if the wires in different chambers are more or less orthogonal, in which case it is necessary to correlate the two projections of a track. The different projections can be provided either by high-voltage strip readout, or by charge division (Subsection 2.1.1), or by wire planes with different wire orientations.

To reconstruct the trajectory in space, as is necessary in the great

2.3 The task of trackfinding

majority of cases, there are basically the following choices:

(1) Combine several local projections into space points, and perform the trackfinding on those space points.
(2) Find the tracks in the different projections independently, and subsequently match the tracks in the different projections.
(3) Find the tracks in only some of the projections, and match them with the help of the other projection(s). In this case, one needs fewer points in the projections used only for matching, since the tracks are not reconstructed there.

Space points can be reconstructed from a wire and one or two high-voltage strip readout coordinates on the corresponding cathode plane(s). This point will give the avalanche position at the wire hit, regardless of the particle's impact angle with the wire plane, since the induction spot is always opposite the avalanche. This means that the particle does not necessarily cross the strips which give a signal. Space point reconstruction is normally preferred to trackfinding in the wire plane and high-voltage projections independently. Since the induction spot size is typically 1 cm or more, signals from nearby tracks frequently overlap in the cathode plane, and, as a result, the trackfinding algorithms may find fewer tracks there than among the more precise anode wire signals.

When the different projections are all provided by similar wire planes, it is possible to proceed either by reconstructing 'space points' and then associating these into tracks, or by finding the tracks in the projections independently, and then matching them. In the first case four wire planes are necessary to define each 'space point', no three of which may have identical wire directions, since one needs four straight lines to define a unique line in space (which cuts all four). This approach not only provides a space point, e.g. the centre of the track segment thus constructed, but also the trajectory direction at that point, which is certainly an advantage (Eichinger 1980). Of course, if the impact angle (the angle of the particle track with the wire plane normal) can be limited to small values, three wire planes are sufficient; three wire planes are also necessary for the resolution of ambiguities.

Accordingly, space point reconstruction in most cases requires a higher number of wire planes than the reconstruction of tracks in projections. This is because in most cases the point density along the track must not fall below a certain threshold which is typical for each detector and experiment. This threshold is a result of the generally accepted request that all detectable tracks should be visible to the human eye when plotted in a suitable way. If, say, 10 points per meter are required, this can be achieved by 22 planes (10 with vertical, 10 with horizontal, 2 with inclined wires for matching) in the case of separate projections, whereas for space points 30 or even 40 planes would be required.

This argument is reinforced by the fact that the particle detection efficiency e (the probability that a charged particle crossing one wire plane leads to a detectable signal) is smaller than 1. This means that for the typical e values of around 99%, three to four times more points are lost in space than in projections. Missing points, however, cause normally a lot of trouble in the trackfinding.

One factor in favour of space points is, however, the disturbing occurrence of overlapping tracks in projections, and their virtual absence in space. If for two tracks $y_1(x), z_1(x)$ and $y_2(x), z_2(x)$ we define an overlap in the interval $[x_1, x_2]$ by

$$|y_2 - y_1| < \delta_y \text{ and/or } |z_2 - z_1| < \delta_z \text{ for } x \in [x_1, x_2]$$

then the 'and' must hold for tracks defined in space, whereas the 'or' applies to tracks in projection. From this it follows immediately that the probability of having overlapping tracks is high for projections, particularly if the tracks are curved, and small in space. For all practical purposes, overlaps in projections have to be considered as the rule, whereas in space they may be treated as exceptions.

Finally, one may observe that the computing time for both alternatives is roughly equivalent: whereas in the second case, space points have first to be constructed, the trackfinding takes almost twice as long per track in the first case, since each track has to be found separately in the two projections. It is clear from this argument that reconstructing the tracks in more than two projections is a bit wasteful: if three or four equally good projections can be provided, it is preferable to work in space.

In summary one might then say that for simple event topologies, working in projections is sufficient and cheaper, whereas the more expensive approach of working in space is recommended for complicated event topologies in which tracks crowd in certain regions of the detector. Of course, the second approach may simply be almost unfeasible technically. Such is often the case in cylindrical detectors at collider experiments, where the only way in which wires can be mounted without great effort is parallel to the beam direction. For this type of application, time projection chambers are an almost ideal solution from the trackfinding point of view, since they provide genuine space points (Subsection 2.1.4).

2.3 The task of trackfinding

2.3.4 Treating track overlaps

The most common, and most recommended approach is to exclude regions where two or more tracks overlap from the primary trackfinding. (In order for these areas to be found, the primary trackfinding will already have to iterate once, i.e. recognize that an overlay exists, then restart the trackfinding with this knowledge.) Supposing for the moment that we know how to exclude such regions, the only other cause for finding more than one hit in a track road is noise, i.e. spurious points that do not belong to any recognizable track.

Such noise can lead to two types of errors:

(1) The track search can be initialized incorrectly by including a noise point in the initial track segment or track candidate.
(2) A noise point may be taken instead of the correct point during subsequent stages of the track-finding. This may happen if, either the correct point is missing, or the point prediction algorithm does not work properly, or both. When a noise point is taken instead of the missing correct one, normally very little harm is done if the point prediction is reasonably precise. In the case in which more than one point may be accepted as a correct solution to the prediction algorithm, one has to open a *branch* for each of these points, and follow both tracks, which in the case of a noise point will normally lead either to a (very short) branch which can be pruned, or to two highly incompatible tracks, differing in just one point, if the trackfinding manages to continue with the correct track even after having included a wrong point in the track. Such incompatibilities (and even more complicated ones) can then be solved with the methods described in Subsection 2.1.13.

As has been pointed out before, simply taking the point nearest to the prediction is risky, and can only be done with some confidence if the correct track model is used for point prediction, and not an approximation, such as a parabola, or a spline curve.

Let us now consider the problem of recognizing regions where tracks overlap. There are basically two possibilities:

(1) Just two tracks cross at some sufficiently large angle, and the point prediction is so precise that (on one or both of the two tracks) only one extra point from the other track is found during prediction, precisely at the crossing point. In this case, one can apply the same algorithm described above for noise points, and both tracks will safely be found

(this will only be the case when the detector provides undistorted signals even for a region where two nearby charged particles are detected by the same detector element, e.g. a wire).

(2) In all other cases, the prediction in neighbouring chambers will lead to the same problem of multiple solutions, i.e. if the *incoming* track is interpolated or extrapolated over several adjacent chambers, there will be more than one point in the prediction interval for most or all of them, whether this now stems from noise or from tracks. In this case, it is easiest to stop the incoming track right there, and start a new one in a clean region.

In this way, some of the tracks will only be found in segments, and these segments have to be joined together in a subsequent step. This can be done using similarity or distance criteria based on these segments, and a method such as the MST (Subsection 2.1.12), by using inverse similarities as edge weights, with typical similarities such as overlapping parts, similar curvature, direction, and inverse distance of the centre points (Pimiä 1985).

Once two or more segments have been joined into a safe track in this way, which normally implies a fit to the correct track model, one can then interpolate this track in the overlap regions and simply pick up the 'nearest to prediction' points. Most importantly, in this way one can clean the overlap area so that one is able to find short, or very curved, tracks after the complete cleaning operation, tracks which do not extend far enough into originally clean areas to be found there.

2.3.5 *Efficiency*

When attempting to optimize an algorithm, it is desirable to have a quantitative measure of its performance. In the case of pattern recognition this measure is called the *efficiency*. There are several possibilities for defining an efficiency:

(1) For a number of events with a fixed number n of reconstructable tracks each, the average efficiency $\langle e \rangle_n$ is $\langle f \rangle/n$, where f is the number of tracks found correctly in each of these events. This number normally decreases slowly with increasing n. Sometimes an overall efficiency is quoted, which is based on an 'event mix' with varying multiplicities, either $\langle \langle f \rangle/n \rangle$, or $\langle f \rangle/\langle n \rangle$. The first of these expressions gives higher values than the second one.

(2) The fraction of events with *all* tracks found correctly can be used. This

2.3 The task of trackfinding

represents a very severe measure, and will decrease rapidly with $\langle n \rangle$.
(3) The fraction of events in which at least a minimum fraction, say 90%, of the tracks has been found correctly.
(4) Any of the above metrics, but in which certain tracks which are notoriously difficult to find are excluded such as tracks below a certain momentum, or in particularly tricky detector regions.

All of these approaches, however, are far from providing a satisfactory definition of efficiency. To be of any use, such a definition has to include in some way the number of incorrect tracks constructed, i.e. those tracks which are made from points belonging to different tracks in reality, or containing spurious points which cannot be reasonably associated with any track: the higher the number of these incorrect tracks, the lower the efficiency must become. Otherwise, the best method, with a guaranteed efficiency of 100%, would simply consist of listing all possible point combinations, among which the correct tracks must be present.

On the other hand there is a real need for a quantitative measure of the efficiency which should be defined for each specific case beforehand. It is only in this way that different algorithms can be compared in an unbiassed fashion. The number of correct tracks has to be known, of course. This is the case in Monte Carlo generated events, but for real data it must be found in an independent way, often through scanning of an event sample by eye. This should preferably not be performed by the experimentor, or even a physicist in the same collaboration, but rather by an independent person, since 'a physicist's eye scan is always biassed by wishful thinking'. Alternatively, if the number of tracks in a measured event can be concluded from some independent source, e.g. from scintillator pulse heights, this may also serve to define the trackfinding efficiency.

Except for events with a constant multiplicity per event, global figures such as the overall efficiency defined above should be avoided, and instead the more meaningful dependence on n should be quoted. Such an efficiency could then, for example, be defined as

$$\langle e \rangle_n = \max(0, \langle f - qw \rangle)/n \tag{2.56}$$

where n is the fixed multiplicity, f is the number of tracks found correctly, w is the number of additional wrong tracks, sometimes called 'ghosts', and q is the 'punishment' factor for ghost tracks. Of course $q = 0$ can be accepted if the ghost tracks can be removed subsequently by a method which is not part of the trackfinding.

It should finally be pointed out that the computing time used by a given method is also important, and sometimes forces compromises to be made between a good, but awfully slow method, and a fast, but less efficient one. In most cases, however, the efficiency is the more important consideration, and one will normally try to speed up a good method by extending the algorithm which gives the desired results rather than use a less efficient method, or, as Weinberg (1972) has put it, 'Any program that works is better than any program that doesn't'.

2.4 Methods of trackfinding in high-energy physics

2.4.1 A classification

The different trackfinding methods can be classified as *global* or *local*. A method is called *global* if all objects (points) enter into an algorithm in the same way. This algorithm then produces a table of tracks, or at least a table in which the tracks can be found more easily than among the original data. The algorithm can be considered as a general transformation of the totality of the event coordinates. The computing time of a global method should, in principle, be proportional to n, the number of points in the event.

A *local* method, on the contrary, is one that selects one track candidate at a time, typically by starting with a few points only (track candidate initialization), and then makes predictions as to further points belonging to this track candidate, e.g. by interpolation or extrapolation of the current track model based on the track candidate found so far. If additional points are found, they are added to the candidate, otherwise the candidate is dropped after a certain number of attempts, depending on the degree of detector inefficiency the algorithm wants to allow for. Since local methods invariably have to make fruitless attempts in order to find track candidates, and thus use the same point in different combinations, the computing time increases more rapidly than linearly with the number of points.

Pure global methods are independent of the order in which points enter the algorithm, local methods are not, since the treatment of each point depends on the initialization, and the 'trackfinding history' inside an event in general. However, a good trackfinding algorithm gives the same tracks even if the order of the raw data from each chamber is randomized (and therefore the measurements enter the trackfinding algorithm in a different order).

2.4.2 Local methods

Track following method

This is typically applied to tracks of the *perceptual* type, where the track can be more or less easily recognized by the human eye from the displayed coordinates. This requires the measurement of highly redundant data, and is nowadays aimed at by practically all experiments.

An initial track segment is first selected, consisting of a few points (up to three or four), and this segment is normally chosen as far away from the interaction region as possible, since there the tracks are, at least on the average, more separated than anywhere nearer to it; in addition, in this way the longest tracks will be found first.

Next step, a point is predicted by extrapolation into the next chamber in the direction of the vertex. This extrapolation may be of 'zero' order simply by choosing the nearest neighbour, first order (straight line), second order (parabola), and possibly higher orders, or other track forms such as circles or helices (if working in space). In all cases the aim is to have a fast point prediction algorithm, representing the track locally by the simplest model possible. Speed is of highest concern in real-time applications, see e.g. Subsection 1.5.4. For chambers which are sufficiently close, the parabola extrapolation will be sufficient in many cases with magnetic fields, since it preserves the sign of curvature as does a real particle track. In addition it is very fast, since a parabola through three points is given by a linear expression in the three dependent coordinates, Equation (2.24), where the coefficients depend only on the plane positions, and can therefore be calculated once and for all for the prediction of hits in any of the chambers.

A rather sophisticated example is the track parametrization which has been applied successfully in the OMEGA detector (Lassalle, *et al.* 1980). From a track sample generated at the target and tracked through the detector, Karhunen–Loeve feature extraction gives the significant coordinates as well as the insignificant ones which act as constraint equations. The transformation is applied to sets of successive planes. Since each constraint equation contains the track coordinates in these planes, the coordinate in one particular plane is fully defined when the coordinates in all other planes are known. A track coordinate in a given plane can then be calculated from the part of the track already constructed in the other planes, normally those preceding it. Because sometimes tracks have to be followed in a direction away from the target, and because chamber inefficiencies have to be dealt with at the track initalization stage, an average of three sets of planes have to be parametrized per plane, and the coefficients kept, in total several hundred sets (the detector consists of about 100 planes). This is an example of parametrizing track coordinates as a function of other coordinates. Although the equations are linear, the tracks may be curved as has been pointed out before. The prediction errors are smaller than the intrinsic detector resolution which is given by the wire spacing.

The tracks are followed in space, although each detector plane measures only one coordinate. During the track following phase, i.e. after the initialization, missing hits are simply replaced by their predictions thus avoiding having to parametrize an impractically high number of different sets of planes. Of course this can only be done for two to three successive missing signals at most. It can be argued that in reality this track following does not use extrapolation but interpolation, since owing to the choice of the track sample for the parametrization, the target region is implicitly used as a constraint for the track coordinates. To put it differently, the fact that the tracks have to go through the target region reduces the phase space that has to be parametrized and thus increases the success of the method.

Another example of track following can be found in Mess, Metcalf, and Orr (1980), where muon tracks are followed through a calorimeter, this task being made more difficult by considerable multiple scattering and the presence of hadron showers. The target calorimeter in this experiment consists of 78 slabs of marble each 8 cm thick with a plane of proportional tubes at the back of each slab. The wires are oriented vertically and horizontally alternately, there are no inclined wires. Behind the target calorimeter is placed an end-calorimeter consisting of 15 circular plates of magnetized iron each 15 cm thick interspersed with proportional tube planes.

The muon trackfinding is hampered by the presence of a shower, and multiple scattering in the marble and iron. It proceeds as follows: firstly, those shower signals are removed for which this can be done safely without removing (too many) points of the muon track. This removal can only be partial because the difference between the shower signals and that of a minimum ionizing particle is not as high as expected due to space charge effects. Secondly, the track is initialized combinatorially in the magnet system where conditions are cleanest, and then followed back into the target calorimeter, where a concurrent search for other tracks is made. The tracks are fitted to straight lines in the field-free regions, resulting in some of them being found only as pieces because the particle has undergone intense bremsstrahlung somewhere on the track. Finally these pieces are joined together. The trackfinding efficiency is over 99%.

2.4 Methods of trackfinding

In summary, the track following method is in one way or another concerned only with the local track model, since it always looks only at the next few points, using the most recently found ones to extrapolate the track. This allows a simple, and hence fast, track model. On the other hand, once the distances become too big, the approximate model will not be precise enough; and because of the measurement errors, even an absolutely correct tracking based on a few recently found points is problematic, since most detectors deliver sufficiently precise track parameters only when the full track is used in a fit.

The track following method uses the combinatorial initialization of candidates. Once a few points have been added to a candidate, it is very likely to be a good track, thus keeping the overhead of following wrong candidates rather small. Accordingly, the computing time is normally proportional to a number between n and n^2 (n is the number of points). In the case of the TRIDENT trackfinding program in the OMEGA detector at CERN (Lassalle, et al., 1980) the computing time is approximately proportional to $n^{1.5}$.

Track road method

In this case, there is no extrapolation as in the track following method, but a much more precise interpolation between points is used to predict extra points on the track. By using initial points at both ends of the track (and one point in the centre of curved tracks) a simple model of the track is used to predict the positions of further points on the track, by defining a *road* around the track model. This track model may be (almost) precise, such as a circle in the case of a projection onto a plane which is orthogonal to the field vector of a homogeneous magnetic field, or a straight line in a field-free region; or it may be approximate, in which case the width of the road has to be established by Monte Carlo tracks. In principle, the better the model the narrower the road can be, but rarely can one use the theoretical road width of, say, three standard deviations of the detector resolution. This may be due to systematic errors in the position, to signal clusters, to signals being hidden by background signals, etc. The track road method is slower than the track following method; but sometimes it is the only workable method available, particularly in the case of widely spaced detector planes (Fröhlich, Grote, Onions, and Ranjard 1976), where the redundancy in the coordinate measurements can be very low, i.e. tracks with as few as three space points may have to be accepted. Most modern detectors provide a density of measurements that is high enough to permit the use of faster methods. However, even when there is sufficient redundancy the road method can sometimes be superior in performance and speed when compared with the track following method, for example in drift chambers with left–right ambiguity. Berkelman, in a description of the CLEO drift chamber (Berkelman 1981) argues rather convincingly (although without proving it) that in this case knowledge of the precise curve, a circle, can only be used profitably if the initial track candidate spans the full track range. The CLEO chamber consists of nine concentric cylindrical drift layers with wires parallel to the cylinder axis. It has been used in an electron–positron collision experiment at the Cornell Electron Storage Ring (CESR). In the trackfinding, a pair of coordinates (a signal plus its mirror image) is chosen combinatorially in an outer and an inner cylinder, and the origin serves as third point. A first approximation of the curvature is calculated from the raw data. If it is acceptable, a drift time correction is applied, and four roads are established (for the four possible combinations of a signal pair in an outer cylinder and a signal pair in an inner cylinder) inside which points are searched for and added to the track candidate. After this, the candidate with the highest number of signals is accepted as the correct one. Where there is more than one suitable candidate, the error of a least squares fit is used to decide between them. In a further step the origin is dropped as a constraint, leading to eight possible candidates from the three initial point pairs. Berkelman believes that his method is less vulnerable to overlap confusion and therefore much faster in execution than the track following model of the type applied at TASSO (Cassel and Kowalski 1981) which is described in the discussion of the application of the MST in Subsection 2.4.3.

Since (in a magnetic field) a road has to be initialized by three points, and since combinations in different planes have to be chosen either because of detector inefficiencies, or because not all tracks may reach the last plane, and since most initial combinations of three points will be wrong, the computing time of this method is typically proportional to a factor between n^2 and n^3 (n is the number of points). In the case of the MARC track finding program for the Split Field Magnet (SFM) detector at CERN, the time was proportional to $n^{2.3}$ (Fröhlich, et al. 1976).

Track element method

Here, a track candidate is constructed in two steps: firstly, short track elements are made up of points, normally inside 'natural' subdivisions of the detector, such as drift chamber cells. From this track candidate, zero order (nearest neighbour), first order (straight line), or second order (parabola) extrapolations or interpolations are used to define track elements, each of

2.4 Methods of trackfinding

which is then condensed into a *master point* (the weighted average of the cluster) plus a direction. Secondly, these master points are combined using track following or other track finding methods.

The great advantage of this method is its speed, compared to using all (up to several hundred) points per track directly. It is therefore appropriate for detectors with a very high point density, and was, historically, frequently used for bubble chamber analysis. In addition, the left–right ambiguity of drift chambers can be solved at the track element level. The reduced number of points, and their wider spacing are compensated by their higher precision, and the fact that they have a direction associated with them (Eichinger 1980).

This method has been applied in the JADE detector at DESY (Olsson *et al.* 1980). There, the cylindrical drift chamber surrounds the beam tube inside the homogeneous field of a solenoid magnet. The drift chamber sense wires are thus parallel to the beams, and to the magnetic field lines. They are staggered by 200 µm to either side in order to be able to distinguish signals from their mirror images. Three concentric rings of chambers provide up to 48 points for an outgoing track. The detector looks similar to the one sketched in Fig. 2.3.

Charge division on the wires delivers a z coordinate, but with two orders of magnitude less precision. Consequently, the trackfinding is performed in the xy plane of the drift time signals only, which is orthogonal to the wires. Track elements are made out of four or more points in each of the 96 detector cells separately. To each segment a parabola is fitted, and the mirror image tracks are rejected at this stage because of their poor χ^2 values. Track segments are then connected through quadratic extrapolation into full tracks.

2.4.3 Global methods

A totally different approach is used in the *global methods*, which are also applicable to a wider range of problems in cluster analysis. In these, all points are considered together, and a procedure exists for classifying all tracks simultaneously.

The complete combinatorial method

This method works in the following way:

(1) Split the set of all position measurements into all possible subsets.

(2) Fit a track model to each subset (= track candidate) and call it a track if it fits the model.

Although this method can be applied directly in some simple cases, where there are only very few coordinates in total, for most practical problems it is too time consuming. Consider five tracks in ten planes, producing 50 measurements, and assume that 1 ms is required to fit the track model to each subset. Even if we ignore possible multiple hits of tracks in the same plane, the processing will take about three hours of computing time, which is normally prohibitive.

Histogramming method

In this case, one defines a set of n different functions of the point coordinates and enters the function values in a histogram of n dimensions (one dimension for each function), where in practice n is normally 1 or 2, as will be explained later. An n-dimensional histogram has to be visualized as an n-dimensional array of n-dimensional cells; each cell contains a counter which is increased by a given weight when the coordinates of a measurement, consisting of an n-tuple of values, define a point inside that cell. When these n-tuples of function values have been entered, and if the method works correctly, tracks form *clusters* or 'peaks' in the histogram; these have then 'only' to be found, and the problem is solved.

A simple example may illustrate this method. Suppose that the interactions always take place at the same point, and that there is no magnetic field. In this case, the tracks will form a *star* around the interaction point (at 0, 0 in a suitable projection). This could, for example, be the case in a colliding beam machine in the projection orthogonal to the beam direction. If we introduce an x, y coordinate system, and use as a function any of

$$y/x, \sin^{-1}[y/(x^2+y^2)^{\frac{1}{2}}], \tan^{-1}(y/x) \tag{2.57}$$

then the tracks will appear as peaks at specific function values.

Another application of the histogram method is possible if the tracks are circles through the origin (interaction point) as in Fig. 2.15. In this case, the inverse (conformal) transformation

$$u = x/(x^2+y^2)$$
$$v = -y/(x^2+y^2) \tag{2.58}$$

will produce straight lines in the uv plane, their nearest distance from the

2.4 Methods of trackfinding

origin being

$$d = 1/2R \qquad (2.59)$$

with the track radius R (Fig. 2.16).

The histogram method, if applied to this conformal mapping, will tolerate a certain distance of the tracks from the origin and still find the tracks correctly (Fig. 2.17). Overlaps will, of course, have to be treated separately, but are readily recognized because the total point count becomes too high for one track. The method will at least allow well-separated tracks to be found quickly. This method has been applied in experiment R807 at CERN (Dahl-Jensen 1979) and is based on bubble chamber trackfinding algorithms.

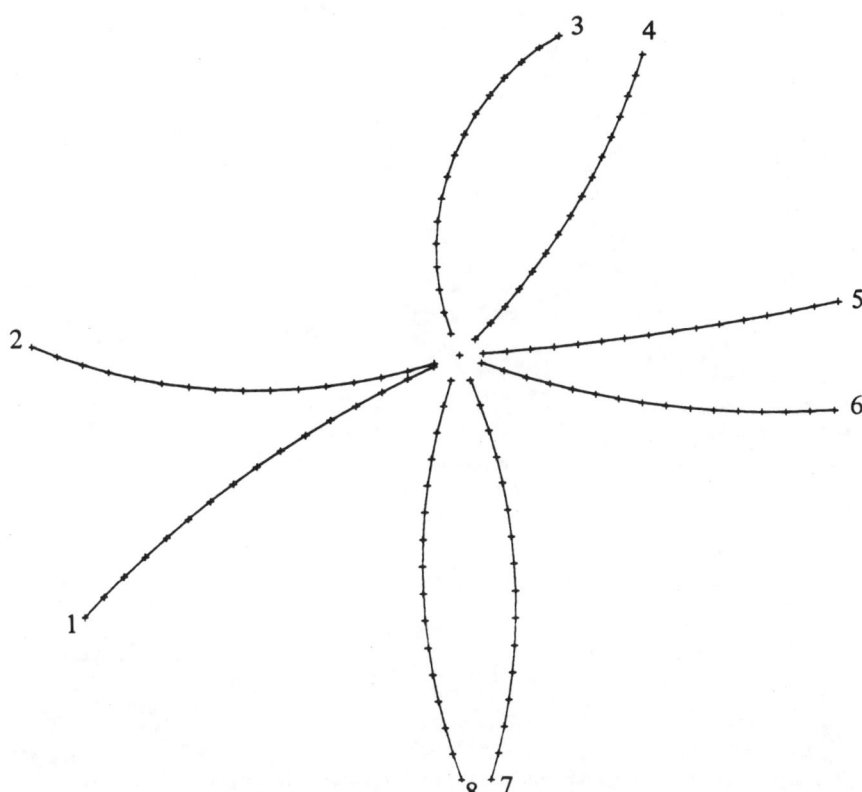

Fig. 2.15 Arcs of circles through the origin. The arcs could be caused by charged particles inside a homogeneous magnetic field, e.g. in the detector of Fig. 2.3.

182   *Pattern recognition*

One should stress that tracks which do not pass near the point used as the origin for the conformal mapping are not found in this method. This is a strength (if the track which is missed belongs to the background) and a weakness of the method (if the track belongs to the event and comes from a particle decay).

In histograms of several dimensions, recognizing the track clusters turns out to be more difficult than finding the tracks directly via a track model. This limits this method to one or two projections.

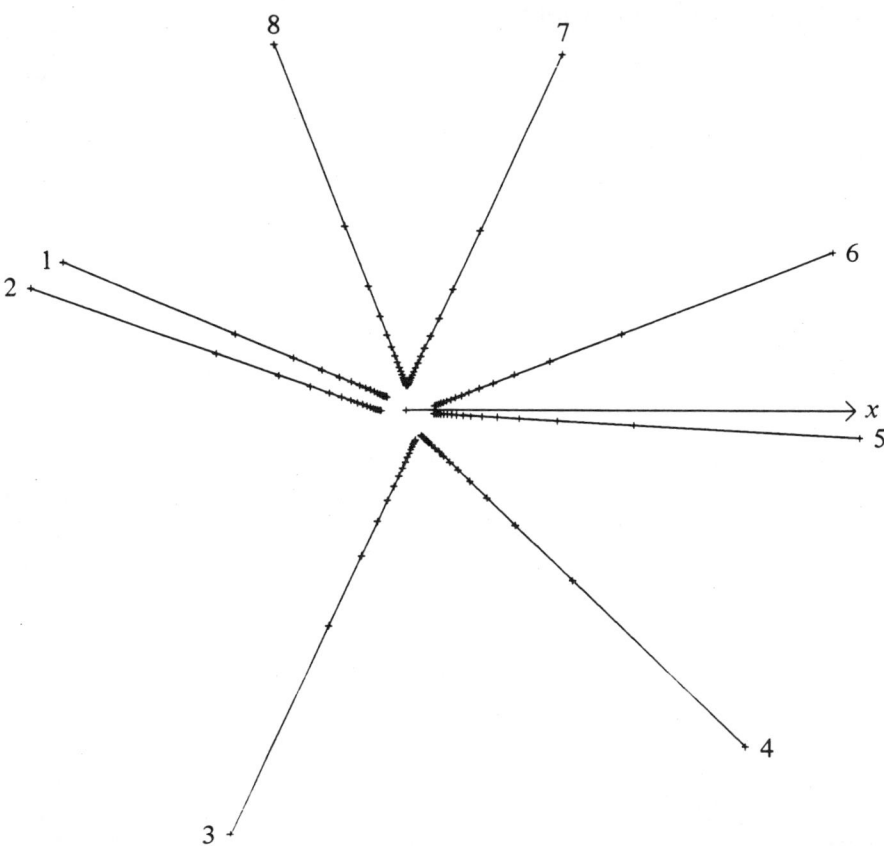

Fig. 2.16 Arcs of circles after a conformal transformation. The original arcs are given in Fig. 2.15. The nearest distance of any line to the original is proportional to the curvature of the original circle arc. This offset makes angular clusters of low momentum tracks in a histogram lower and wider than those of high momentum tracks (see Fig. 2.17).

2.4 Methods of trackfinding

Application of template matching

This method requires a dictionary of all possible classes (tracks) and can therefore be applied only in cases where their number can be kept within reasonable limits, somewhere below 10^5, in which case a binary search or a hash algorithm can yield very fast matches (see Subsections 1.5.3, 1.5.5). The method is typically applied to cases of genuine low redundancy (only a few detector elements) or artificial low redundancy (detector elements are grouped into 'cells' giving one signal each, for the purpose of low-precision trackfinding).

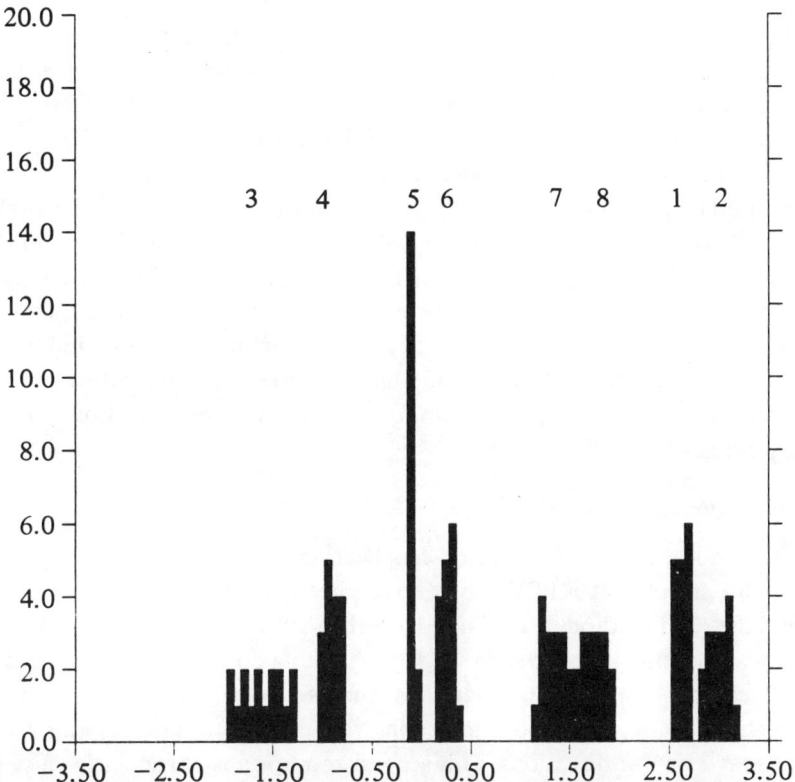

Fig. 2.17 ϕ histogram of tracks in Fig. 2.16. The angle with the (arbitrary) x axis in Fig. 2.16 is entered for all points there. The tracks form clusters which can be found rather easily. The third cluster from the right contains two tracks, as is easily recognizable from the number of entries in it. A broad cluster (if caused by one track) arises from a large distance of the straight line in Fig. 2.16 from the origin and thus from a track with high curvature according to equation (2.58).

Used in the trackfinding for the Mark-III detector at SPEAR (Becker, et al. 1984), with a total of 12832 templates, this method is reported to run three times faster than 'conventional approaches'. Each different combination of cells that can be fired by one track creates a separate template. The subdivision of the detector volume into *cells* which leads to such a low number of possible combinations means, of course, that there will be regions of confusion where tracks get very near, or cross. These ambiguous regions are then resolved in a 'classical' way, which in the case of the Mark III is a combinatorial, noniterative circle fit.

This method is particularly well suited for detectors with cylindrical drift chamber arrangements, since the division of such chambers into drift cells provides a natural basis for the algorithm. In the case of the Mark-III, it works basically as follows. During raw data conversion (*unpacking*), a cell image matrix of the detector is filled. Combinations of cells which have 'fired' are then compared with the dictionary, once all data of the event have been unpacked. The method is thus global except for the combinatorial search for tracks in regions where they overlap.

Template matching has been implemented in the fast trigger logic of the CELLO (Behrend 1981) and the TASSO detectors, both at DESY. In the first case, only 57 templates are used. For the experiment E.711 at Fermilab, a template matching algorithm was implemented on a fast vector processor. This reduced the computer time per event by a factor of 10 compared to the same algorithm run in scalar mode on the same computer. However, the scalar code had to be rewritten completely for vectorization (Georgiopoulos, *et al.* 1986).

Application of the MST

A variation of the MST method has been applied to trackfinding in the TASSO detector at DESY, where it has been shown to be more efficient than the road method (Cassel and Kowalski 1981). The basic element used in this algorithm is no longer a single point, but a point pair in two adjacent drift chamber layers. This pair has an associated pair distance and a direction. Pairs are linked into graphs when they share points and when they have similar directions. This second condition is particularly efficient in rejecting image points in drift chambers (left–right ambiguity). In this way, track segments of a certain minimum length are constructed which allows their track parameters (circle radius and centre) to be calculated. These track segments are then grouped into full tracks by using *similarity* based on the segment parameters.

For a fast search for high momentum tracks, a modified MST technique is used, where the curvature of a segment defines the edge weight. This also serves the purpose of rejecting arcs containing mirror points, since these will have much higher curvatures than the arcs made up of correct points only. The method is on the whole rather specific to drift chambers with left–right ambiguity, where it works very well.

2.5 Finding of particle showers

2.5.1 Some definitions

This chapter is entitled *Pattern Recognition*. One might think this a misleading name in the context of calorimetry: signals in calorimeters caused by elementary particle reactions can very generally be called *patterns*, but *recognition* suggests that the object to be recognized is of a known shape, predictable within narrow limits. We will describe in this section mostly *to what extent* showers in calorimeters are predictable and understood. The objective of *calorimeter analysis* is to deduce from calorimeter signals the phenomenon that is at the origin of a shower. Our objective in this section is to sketch the problems, and to show how detailed analysis proceeds in a number of typical cases, taking care to relate the local solutions to the more general concepts of pattern recognition or analysis.

Since calorimeters are still quite novel instruments, not fully understood in all aspects, our presentation will include some discussion of what calorimeters are and what characteristics they have. We recommend, for a more thorough discussion of these detectors, Fabjan and Ludlam (1982) and Ferbel (1987).

Total absorption

Calorimeters are devices that entirely absorb and hence destroy incident particles, by making them interact. (They are also referred to as *total absorption calorimeters*.) Calorimeters measure energy by annihilating it, most of it being converted into heat (*calor* is Latin for heat). Part of the energy is released in the form of a recordable signal such as scintillated light or ionization. The physical phenomena which can be observed with the help of a calorimeter are diverse in nature, but destructive measurement is common to all of them. In order to be usable, a calorimeter must allow the recording of a signal containing adequate detail of the absorption process. From this signal, the original phenomenon can then be inferred, e.g. an

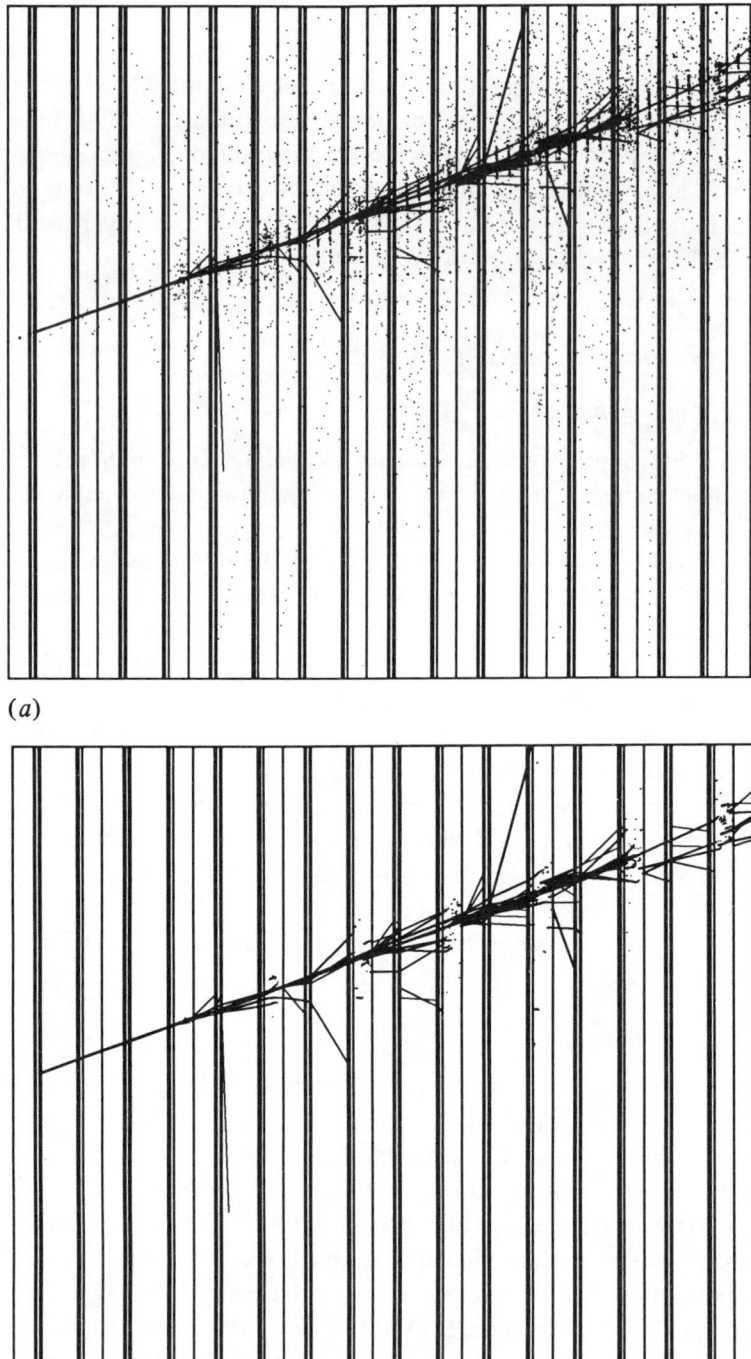

(a)

(b)

2.5 Finding of particle showers

incident particle's energy or the energy dissipation in a part of geometrical space.

As energy is deposited independently of the charge of the incident particle, calorimeters also are unique instruments for measuring the energy of neutral particles. They are further unequalled in detecting and measuring particle jets, in which mixtures of neutral and charged particles are present at small spatial separation. Charged particle tracking devices are not useful (or only marginally useful) for both neutral particles and jets. Fig. 2.18 shows the development of particle showers of several types, in the detail that only simulation can provide.

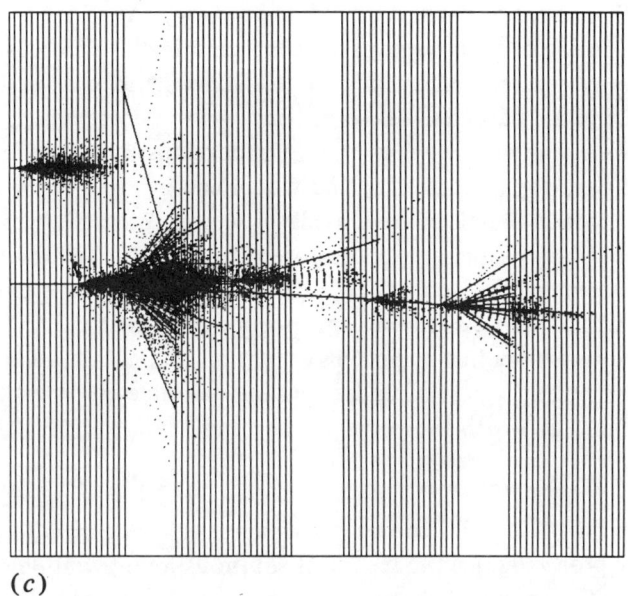

(c)

Fig. 2.18 Electromagnetic and hadronic shower simulation. (a) A 5 GeV photon enters, at a 20° angle, a shower counter as used in SLAC experiment E-137. Eight modules each consist of wire chambers with walls, a scintillator plane, air gaps, and an (aluminium) converter. All particles are shown, with neutrals as dotted lines. (b) Only charged particles in the same shower are shown. The two pictures have been generated by EGS4 (Nelson, Hirayama, and Rogers 1985) using the Unified Graphics subsystem (SLAC-TN-87-3), courtesy W.R. Nelson, SLAC. (c) Two simulated showers with 100 GeV energy are shown annihilating in a lead scintillator (50/6 mm) sandwich calorimeter. Air gaps separate the four modules, each of which is 20 radiation lengths deep. The upper shower is a single electron, the lower shower a charged π-meson. Note the multitude of local showers for the hadronic cascade, and the substantial widening by the air gaps. Photons are shown as dotted lines. The simulation was done in GEANT3, using GEISHA for the hadronic shower. Courtesy R. Brun, CERN.

The destructive measurement of particle energies in partly or fully active material is of comparatively recent origin. The spread of the method is closely connected with the advent of the availability of high-energy collisions. Calorimeters, first seriously proposed for cosmic ray physics by Murzin (1967), became popular devices in accelerator experiments as accelerator energies increased, for simple reasons of resolution. Tracking devices reach their precision limits at high energies. They allow the calculation of charged particle momenta using magnetic fields which cause a particle track to bend, as amply described elsewhere in this book. Due to the lesser bending, they give inferior relative resolution at high energies ($\Delta P/P \propto P$ for equal field and track length). Calorimeters measure the energy with smaller relative error at higher energy ($\Delta E/E \propto 1/\sqrt{E}$).

The absorption of an incident particle in a calorimeter proceeds through a multitude of interactions, strong or electromagnetic, each resulting in several secondary particles. These, in turn, will interact again, and the resulting *cascade* or *shower* will contain, as it develops in time, an increasing number of particles of decreasing individual energy. This showering has given calorimeters the alternative name of *shower counter*.

Some secondary particles, of course, are subject neither to the strong nor to the electromagnetic force. They escape calorimeters undetected (neutrinos) or with a trivial signal not proportional to their energy (muons). Some energy is further lost by nucleus excitation or breakup, and produces no visible signal. It is part of understanding calorimeters to estimate the fraction of energy which thus escapes detection.

Calorimeter properties

Depending on the physics context in which a calorimeter is to be used, *calorimeter properties*, i.e. the technical solutions found, will differ within wide limits. The recorded shower aspects will vary accordingly. In other words, calorimeters and data can have very different characteristics, caused both by the variety of possible technical solutions, and by the different physics demands.

Calorimeters are described in terms of *characteristic parameters* such as energy resolution, granularity and shower separation, optimization for certain types of particles, dynamical range, spatial coverage. All of these are determined by reaching a compromise between the physics aims of an experiment, the resources needed for construction and analysis, and the mechanical limits resulting from the use of different techniques.

Systematic 4π calorimetry, i.e. full spatial coverage for all types of single

2.5 Finding of particle showers

particles, charged or neutral, and jets, is today frequently attempted in large detectors. Often, great care is taken to minimize a calorimeter's uninstrumented areas. Such a *hermetic* detector allows the presence of escaping high-energy particles that are not absorbed, like neutrinos, to be deduced.

A further relevant parameter is the *readout and analysis level* at which a calorimeter is expected to act. If used in a fast trigger processor (e.g. Subsection 1.5.7), a signal below the microsecond level may be needed. The segmentation may then be limited by the speed that the associated electronics can handle. If used instead, as an optimal resolution signal for an off-line analysis program, possibly in combination with other parts of the detector, very different design criteria will be used.

Calorimeter construction allows many more choices than is true for a tracking device. The observed variety of techniques is proof that this type of detector is still undergoing a search for optimization. Calorimeters are largely empirical instruments and optimization is absolutely critical. They consequently offer more room for original and creative solutions than other types of detector.

Shower properties

Shower properties can fluctuate dramatically for the same energy, depending on the sequence of interactions giving rise to the showers. These fluctuations constitute a characteristic difference between tracking devices and calorimeters. Tracking chambers are usually built to interfere as little as possible with the passing track. Once a particle's parameters are given, its track can be predicted within narrow limits. Calorimeters, on the other hand, are devices based on stochastic processes. Only *average* shower properties can be predicted with some confidence and described in terms of *shower parameters*. This stochastic aspect, and the parameters describing showers will be discussed in some detail below. An important common characteristic of all showers is the statistical multiplication of particles in a shower with increasing incident energy. This multiplication is at the origin of the improved relative resolution at high energy, much like the $1/N$ law decreases the relative variance of a Poisson-distributed number N.

Calorimeter applications

Calorimeters are subdivided so that signals are obtained after integration over a limited physical volume. They, therefore, allow the energy deposition in a 'cell' of defined position with respect to the interaction point to be deduced. Thus an obvious and simple application of calorimeters is as a device to measure the spatial *energy flow* in an interaction. If the cell structure is fine enough, it may allow algorithms to associate several cells into 'clusters', and to deduce physical phenomena based on the cluster's properties.

If additional information is available (e.g. from a tracking device), clusters may also be associated to a single track and thus provide a *track energy measurement*. In conjunction with existing understanding of the showering process for different particles, a calorimeter may further be used as a *particle identifier*. Again, it must then be assumed that some independent information establishes the existence of a particle.

At high energies, the single particle aspect of calorimeters is often considered less important than its capability to *localize jets* and measure their energy. Jets are particle compounds produced by the hadronization of constituents (quarks or gluons), which manifest themselves in interactions at very short distances.

In order to understand average shower properties and their fluctuations, it is essential to look in some detail at what is known about the properties of the absorption mechanisms. This will allow us to narrow down as best we can the 'patterns' which we want to recognize. This discussion will also help us to understand why such a wide variety of methods of building and calibrating calorimeters exist, and why the data produced by these devices are analysed and used in so many different ways.

Among the relevant showering processes, two major categories can be distinguished: *electromagnetic showers* which are caused by incident electrons and photons, and *hadronic showers* which are induced by hadrons. Hadronic showers generally also contain electromagnetic shower components, but, in addition, a large variety of specifically hadronic phenomena determine their parameters. Calorimeters are, of course, not limited to strong or electromagnetic primary interactions; they are particularly valuable instruments in weak interaction physics, e.g. in neutrino physics, where they serve the double purpose of being the target for the primary particle, and the identifier for secondary leptons or hadrons.

2.5.2 Physical processes in calorimeters

2.5.2.1 Electromagnetic showers

Electromagnetic showers are those triggered by an incident e^+, e^- or γ. As a π^0, for all practical purposes, decays instantaneously into two

2.5 Finding of particle showers

photons, electromagnetic shower counters, of course, are also relevant in hadronic interactions, in which on average nearly one third of the pionic energy (which dominates other particle contents) will be found in the form of π^0s.

The reactions which predominate in electromagnetic showering, all produce further electromagnetic particles. This has two important consequences: Firstly, all the energy in an electromagnetic shower will finally be deposited as ionization loss by electrons. The longitudinal energy distribution of an electromagnetic shower is therefore often given in terms of the *Number of Equivalent Particles* ('NEP') i.e. compared with the number of (minimum ionizing) tracks present at any cross section. On the other hand, electromagnetic showers can be quite completely described by quantum electrodynamics (QED), and can be calculated using perturbation theory (Feynman diagrams). Bremsstrahlung and electron pair production are the dominant processes for high-energy electrons and photons. Their cross sections become nearly energy-independent above 1 GeV. The more relevant electromagnetic interactions in photon- or electron-induced showers are:

Bremsstrahlung (photon emission in acceleration),
Electron-positron pair creation ($\gamma \to e^+ e-$),
Compton scattering (photon-electron scattering),
Coulomb scattering (electron–nucleon scattering),
Bhabha scattering (electron–positron scattering),
Photoeffect (electron emission from γ-irradiated nuclei),
Møller scattering (electron–electron scattering),
Annihilation ($e^+ e^- \to \mu^+ \mu^-$, or $\to \gamma\gamma$).

All reactions can be suitably expressed in terms of a scaling variable, (i.e. a variable which makes the description independent of the material used), the *radiation length*, usually denoted by X_0. It is defined as the mean path (or attenuation) length of an electron due to radiation, and is given by the approximate formula

$$X_0[g\,cm^{-2}] \approx 180\,A/Z^2, \qquad (2.60)$$

where A, Z are the atomic weight and the atomic number of the material respectively. If X_0 is to be expressed as a length, it is multiplied by the density ρ: $X_0[cm] \approx 180\,A\rho/Z^2$. These formulae are approximate, but reasonably accurate for high Z. Due to the scaling property of showers with X_0, the dimensions of electromagnetic calorimeters are usually expressed in units of X_0.

Electromagnetic showers start after a small fraction of a radiation length, when track multiplication sets in. This happens somewhat faster for an incident electron than for a photon. Multiplication proceeds until the individual track energies become small, so that ionization and Compton scattering start to dominate and to absorb the low-energy part of the shower. An energy profile (dE/dx) maximum is thus reached at some depth of the shower. As the shower develops further, dE/dx will decrease, eventually exponentially. In other words, towards the end, a shower has a defined attenuation length. For approximate shower parameters see Subsection 2.5.4. Typical electromagnetic showers, from Monte Carlo simulations, are shown in Fig. 2.18.

Numerous authors have discussed the total and differential cross sections of the QED processes relevant for electromagnetic showers. Rossi (1965) has given an extensive analytical treatment of electromagnetic showers (*cascade showers*), albeit with some assumptions. His 'approximation B' resulted in shower parameters that are still a reference result today. In other early publications, similar attempts were pursued, and later more and more were combined with Monte Carlo results (e.g Hayakawa 1969; Messel and Crawford 1970). Not surprisingly, it is precisely when showers have to be simulated in detail and compared with measurements that a compilation of all available data and theories is most necessary. The slowly converging process of developing complete QED Monte Carlo programs, has resulted today in a single widely used Monte Carlo code for electromagnetic shower simulation, EGS (Nelson *et al.*, 1985), which we will discuss in some more detail in Subsection 2.5.5.

Often, global and average shower properties are sufficient or are more relevant than the detailed understanding of the underlying processes. This is particularly true in simulation, where program complication and computer time restrictions may dictate such simplicity. Global properties are discussed in Subsection 2.5.4.

2.5.2.2 Hadronic showers

Hadronic showers, much like electromagnetic showers, build up through a multiplication process, and decay when individual particles have reached low energies and are eventually stopped ('ranged out'). Unlike electromagnetic interactions, however, hadronic interactions particularly at lower energies (remember that all showers end up with particles of low energy!) can not be described with rigour, and a much wider variety of

2.5 Finding of particle showers

interaction channels is available, depending on the primary and secondary particles' mass and energy.

Hadronic showers result from a mixture of reactions, dominated by multiple particle production, including the entire hadronic spectrum as possible secondaries; π^0s may be produced abundantly in hadronic interactions, electromagnetic showering therefore is a subprocess in hadronic calorimetry. The first hadronic interaction will happen *on average* (with exponential distribution) after one nuclear interaction length, i.e. about 17 cm in iron, 18.5 cm in lead, or about 70 cm in a scintillator. This very first reaction causes by its varying π^0 content, major fluctuations in the nature of the secondaries. There will be a varying electromagnetic shower component, which will propagate in subsequent interactions and eventually in all relevant shower parameters. Fluctuations in the hadronic interactions intrinsically reduce all aspects of the performance of hadronic calorimeters compared to electromagnetic calorimeters. A detailed discussion is found in Amaldi (1981). Several examples for hadronic showers are shown in Fig. 2.18.

In contrast to electromagnetic showers which convert nearly all energy into ionization (track length), hadronic showers produce a variable fraction of *invisible energy*. Some of the available energy is used for the excitation or breakup of the absorber nuclei, and is not fully observable. Nucleus breakup absorbs the binding energy, low-energy neutrons will not always result in an observable signal. In addition, neutrinos or muons may be produced in some reactions. These particles escape the detector without depositing their energy or with only an insignificant signal.

Due to the invisible energy, a hadronic particle typically generates a signal smaller than that of an electromagnetic particle with the same energy. Without special precautions (we discuss this problem of *compensation* in Subsection 2.5.3), the effect is a signal difference of some 20%, dependent not only on the calorimeter construction, but also on the energy and nature of the particle. It is therefore necessary to use a different conversion factor for energy readings if the incident particle is hadronic.

The fluctuations between electromagnetic and hadronic components in a hadronic shower are large. Hence even a correct average compensation, made by adjusting the conversion factor, will result in a worse energy resolution for hadronic showers when compared to electromagnetic showers. This is why so much time has been spent on designing calorimeters that compensate with their hardware, and give the same response for any shower component.

2.5.3 Calorimeter parameters

Homogeneous or sampling calorimeters

Calorimeters can be built entirely of active material, i.e. as *homogeneous calorimeters*. At low particle energies, and for electromagnetic showers, this is perfectly possible, and ensures optimal resolution from the statistical point of view. As calorimeters must contain the showers as completely as possible, and large volumes have to be equated with high cost, interest focusses on heavy scintillators such as $Bi_4Ge_3O_{12}$(BGO), NaI, CsI, lead glass, or BaF_2. Their small radiation length results in an acceptable overall dimension and thus financial feasibility (Fabjan 1985; Suffert 1985).

As energies increase, containment becomes more and more difficult to achieve economically. Particularly for hadronic showers, whose lengths scale roughly with the interaction length λ (in most materials λ is much larger than X_0), *sampling calorimeters* are then the only solution. In these, inactive absorber material (iron, lead, uranium, and sometimes combinations of these) alternates with active material (scintillating material in a solid or liquid form, or ionizing gas). Typically, active and absorbing materials alternate in layers arranged normal to the preferred direction of incident particles ('sandwich' calorimeters). Many of these will be described later in this section. Comparatively new is the idea of a calorimeter using scintillating fibers embedded in absorber (Sonderegger 1987), which is still in an early phase of evolution. In sampling calorimeters, a reasonably large and constant fraction of the shower is sampled in the active layers, yielding a reasonable conversion factor between the observed signal and the incident energy. Fluctuations in the sampled fraction result in a loss in energy resolution.

Granularity

Granularity is the not-so-precisely defined parameter which describes globally the size of the elements in a calorimeter for which an integrated signal will be produced. In practice, calorimeters of very different shapes and sizes have been built, depending on the aims of the experiment and the resources available. Frequently, multiple readout with controlled attenuation allows interpolation within a physical cell, thus improving spatial resolution if a single shower can be assumed.

Before entering a more detailed discussion, some *gross distinctions* must be made. Aspects of granularity which need to be considered are:

The spatial segmentation normal to the direction of incidence of particles or jets (this determines angular resolution).

2.5 Finding of particle showers

Table 2.1 *Energy resolution for electromagnetic and hadronic calorimeters.*

Detector	Type	$\Delta E/E$ (E in GeV)	Comment
Electromagnetic calorimeters			
TASSO	Lead/liquid argon	$0.10/\sqrt{E}$	Braunschweig, *et al.* 1986
MARK J	Lead/scintillator	$0.12/\sqrt{E}$	
CHARM	Marble/scintillator	$0.20/\sqrt{E}$	Diddens, *et al.* 1980
AFS	Uranium/scintillator	$0.16/\sqrt{E}$	Akesson, *et al.* 1985
GAMMA	Lead glass	0.07	at 1 GeV; Adiels, *et al.* 1986
Crystal Ball	NaI	$0.023/E^{\frac{1}{4}}$	
CLEOII	CsI	$0.013/E$	Blucher, *et al.* 1986
Test	BGO	$0.02/E^{\frac{1}{4}}$	Dietl, *et al.* 1985; Iwahori, *et al.* 1986
Test	BGO	$0.01/\sqrt{E}$	Bakken, *et al.* 1985
Hadronic calorimeters			
MAC	Iron/prop. chamber	$0.75/\sqrt{E}$	Ford, *et al.* 1982
UA2	Iron/scintillator	$0.60/\sqrt{E}$	
CHARM	Marble/scintillator	$0.53/\sqrt{E}$	Diddens, *et al.* 1980
AFS	Uranium/scintillator	$0.36/\sqrt{E}$	Akesson, *et al.* 1985
Test	Uranium/scintillator	$0.335/\sqrt{E}$	at 3–9 GeV; Anders, *et al.* 1986

The spatial segmentation along the direction of incidence (this determines the capability to distinguish shower shape, i.e. mostly to differentiate between electromagnetic and hadronic showers); it may also allow leakage corrections leading to improved energy resolution.

The ratio of active (signal-producing) material to passive absorber in sampling devices (this directly determines the energy resolution).

The distribution of the active material along the shower (this must be optimized in function of the mix of energies and nature of incident particles).

In a given experiment, the sizes of cells, their orientation, and their relative order are optimized locally. Frequently, large fractions of the available geometrical space are covered by combined electromagnetic and hadronic calorimeters, optimizing for electromagnetic showers closer to the interaction (only relevant parameter: radiation length X_0) and for hadronic showers at larger distance (two basically relevant parameters: interaction length λ, and X_0).

Energy resolution

The energy resolution of calorimeters is usually expressed in terms of $\Delta E/E$, ΔE being the overall statistical error (root of variance) including all error sources. At high energies, statistical contributions dominate; the relative resolution therefore improves with increasing E, due to the increasing number of particles produced in the showering process. The main tendency for $\Delta E/E$ therefore is to follow the form $k/E^{\frac{1}{2}}$, where k depends on the calorimeter and readout characteristics. This law has been observed to hold up to rather high E, subject to more and more subtle calibration procedures which eliminate systematic contributions. At low energy, electromagnetic calorimeters can really become better than this law suggests (see Table 2.1). Hadronic calorimetry, however, is not a practical method below, say, 1 GeV, the reactions becoming very dependent on the particle type and fluctuations making tracking devices far superior.

Since the energy regime is mostly much higher than the particle masses, the definition of E is not usually given much attention. It should be noted that for most incident particles the best approximation is $E_{kin}(\equiv E_{tot} - M$ where M is the mass of the particle); this is what simulation programs use. If annihilation processes are likely, e.g. for incident $\bar{p}$, the situation is more complicated, and none of the mechanisms is fully understood or experimentally measured.

A general formula to give an approximate resolution is $\Delta E/E = \sqrt{(A^2 + R^2)}$, where A is related to the photon statistics, i.e. the number of photons arriving at the photomultiplier (this contribution is inevitable in most calorimeters), and R deals with the sampling fluctuations in the active material. The fluctuations in sampling calorimeters are now understood to be of prime importance, a single sampling fraction is a very poor approximation for the large variety of physics processes involved (Brückmann, Behrens, and Anders 1987). In Table 2.1, some characteristic resolution functions for single particles are compiled, all of them obtained from experimental data.

Incident $\bar{p}$s have a high probability of annihilating when interacting, at low energy they range out and invariably annihilate. The annihilation energy of $\bar{p}$ and the nucleon is then converted into mostly kinetic energy. Despite the nonannihilation cross section, a $\bar{p}$ should therefore be distinguishable with a good probability at low energy from K or p, by calorimetric methods. The method, however, has only been tested in principle (Fabjan, private communication 1986), and has not seriously been used.

Containment

Containment is a very important boundary condition for any calorimeter. If showers are even occasionally partly converted in the absence of active material, the recorded energies will have a distribution with a tail towards low energy values. Depending on the physics intended, this may result in missing triggers or in severe biases, or simply be detrimental to the energy resolution. The effect of leakage on resolution has been estimated by several authors. Blucher *et al.* (1986) give error terms for a CsI crystal calorimeter. They note a side leakage term scaling with $E^{\frac{1}{3}}$, and a longitudinal leakage term scaling with $1/E^{\frac{1}{2}}$ (both terms refer to $\Delta E/E$). The terms have been derived using the Monte Carlo program EGS for incident photons/positrons, in the energy range 100–5000 MeV.

Generally, calorimeters are built to avoid leakage. Cashmore, *et al.* (1985) and Grassmann and Moser (1985), on the other hand, give a simple and unoptimized algorithm, based on Monte Carlo calculations, for how rear leakage can be corrected assuming that information is available from the last part of the calorimeter (by longitudinal structure in the cells). It is therefore conceivable to construct compacter and cheaper calorimeters, that accept the degradation in resolution remaining after such corrections.

Compensation

Attempts were made early (Fabjan, *et al.* 1977) and successfully (Akesson, *et al.* 1985; Anders *et al.* 1986) to compensate for the invisible energy by choosing absorber and active materials with an electron/hadron (usually called e/π) signal factor as close to 1 as possible. The basic idea proposed originally was to use uranium as the absorber material which would contribute an additional, i.e. compensating signal due to nuclear fission caused by nuclear excitation. Indeed, calorimeters were built which had an e/π factor very close to 1 over a wide energy range, and showed experimentally much improved resolution for hadronic showers. Understanding the compensation mechanism, however, must go beyond this simple explanation (e.g. Workshop on Compensated Calorimetry, Workshop 1985). This relative importance of the components of invisible energy and their compensation has only very recently been studied, both in experiment (Leroy, Sirois, and Wigmans 1986) and in detailed simulation (Wigmans 1986). The research is still far from completed, an up-to-date overview is given by Brückmann, *et al.* (1987).

Several somewhat surprising effects have been found in the very important attempt to pin down the precise role of low-energy reactions in compensating for invisible energy in hadronic interactions. Among the findings is the recognition that it is the shower yield in low-energy neutrons and the active material's capability to convert these into a signal (neutron slow-down, deexcitation photon detection), that determines resolution and the e/π signal ratio. Most important in this respect is the free proton content of the active material, and the relative thickness of active and passive materials. Nuclear fission results in only one contribution to low-energy neutrons. Galaktinov, *et al.* (1986) have shown this experimentally, and added the caution that low-energy neutrons are themselves subject to large fluctuations, hence their detection may result in average compensation without improving resolution. All the same, it is still with uranium that compensating calorimeters with optimal resolution can be built in a compact form and hence most economically.

It may be expected that future experiments which build on present understanding, will clarify these findings, and will push the energy resolution in hadronic calorimeters to its theoretical limits, possibly with fewer technical resources involved (uranium is a hazardous material to machine, and its world-wide supply is limited).

Backscattering

A small fraction of energy in any hadronic calorimeter is backscattered from the initial calorimeter surface. This so-called *albedo* effect (from the astronomical 'whiteness' of planets by backscattering light) has its origin in the nuclear breakup products, which show approximately isotropic distribution. The fraction of backscattered energy is strongly dependent on the incident energy. For hadronic showers which start early in the calorimeter, the number of backscattered tracks is large, but their integrated energy is not an important contribution. It is of the order of 5% at 0.8 GeV incident energy, 3% at 1 GeV, and 1% at 10 GeV. At higher energies, the amount of backscattered energy is reported to become constant at around 150–200 MeV, (Dorenbosch, *et al.* 1987; Ellsworth, *et al.* 1982).

Calibration and monitoring

One aspect of prime importance, that does determine the effective resolution of calorimeters in a critical way is the understanding, at any moment in time, of the relation between the observed signal and the energy deposition which is the signal's origin. A calorimeter responds to showers by producing light in scintillating material, or ionization in gaseous detectors. The conversion of either into small electric signals goes through very different stages, whose detailed *calibration at any point in time* constitutes an important aspect of calorimetry. A high-resolution

2.5 Finding of particle showers

calorimeter requires substantial care to be taken in understanding every detail of the process of converting and transporting signals through the various stages. Light goes through scintillators, wavelength shifters, light guides, etc until converted into a signal in a photomultiplier or diode. Electron charges drift to a wire and are collected directly or induce a signal. Electric currents are amplified and stored until triggers have decided that digitization should occur. And the conversion of analogue signals to digital numbers itself relies on components whose performance must be understood and monitored. As calibration typically introduces new components, their understanding will also become relevant.

Calorimeters are entirely *empirical* instruments. Whatever the *monitoring* procedures are, they are *calibrated* empirically in a test beam whose parameters are known, or ideally by suitable events taken during operation. Useful calibration events are provided in e^+e^- colliders in the form of Bhabha scattering ($e^+e^- \to e^+e^-$), which is produced abundantly and is easy to recognize. In hadron detectors, cosmic muons are frequently used for the limited monitoring a minimum ionizing particle can provide. All supplementary calibration and monitoring, e.g. with specially installed light or electric pulsers usually serves the purpose of interpolating, in energy or time, between test beam results.

Excellent published examples exist for the variety of problems encountered in calorimeter calibration. Some of these are as follows:

Dorenbosch, *et al.* (1987) describe in detail the calibration problems encountered with the fine-grained marble/scintillator calorimeter in the CHARM neutrino experiment. The calibration relies rather exclusively on e^- and π^- test beams at energies from 5 to 140 GeV c^{-1}.

Cochet, *et al.* (1986) gives the procedures used in the barrel electromagnetic calorimeter of the UA1 experiment at the CERN SPS $\bar{p}p$ collider. In calibration, sparse test beam measurements are combined with very extensive and repeated masurements using a penetrating ^{60}Co source. Due to the large surface and multiple readout of single calorimeter elements ('gondolas'), such a fine irradiation was judged to be the only realistic way for a fine-grain calibration of all elements.

Drescher, *et al.* (1986) refer to the ARGUS calorimeter at DORIS II, a lead-scintillator sandwich type calorimeter with wavelength shifter readout. A central laser is used as a constant light source for monitoring, and cosmic muons are also recorded. The correction procedures described are most typical for calorimeters of this type.

2.5.4 Shower parameters

In the following, the current state of knowledge concerning the *average* parameters of showers will be presented. It should be understood that they describe *individual* showers rather poorly; in particular, they are of limited usefulness for simulating hadronic showers. Few efforts have been made to describe the important shower *fluctuations* in more than qualitative terms. For the lateral shower shape, even average formulae are not available. If showers are recognized and used in some specific physics context, the relevant parameters are, of course, those that describe the physical track, the jet, and the energy density in part of the geometrical space. Shower parameters as given in this section may be useful for interpolating between measured points, or for some aspects of simulation.

2.5.4.1 Longitudinal shower shape

Electromagnetic showers

Many authors have based an average electromagnetic shower description on analytical calculations published by Rossi (1965). His so-called approximation B (constant energy loss ε per radiation length X_0, valid at all energies of the high-energy approximation for radiation and pair production) results in formulae for the shower maximum (the depth at which dE/dx in the shower reaches a maximum), the shower energy median (the depth at which half the shower's energy has been dissipated), and the shower attenuation (the exponential damping slope in the shower tail). For incident electrons of energy E (incident photons result in a slightly stretched shower), the parameters are given as

Shower maximum at: $X_0[\ln(E/\varepsilon) - 1]$
Shower energy median at: $X_0[\ln(E/\varepsilon) + 0.4]$ (2.61)

In these formulae, ε is the *critical energy*, which is defined as the energy loss by collisions per X_0. The approximate crossover point, at which an electron loses energy equally by bremsstrahlung and by ionization is at $E = \varepsilon$. An approximate value for ε (good for high Z materials) is given by $\varepsilon = 550/Z$ [MeV], numerical values are given by Iwata (1980), e.g. 7.2 MeV for lead, 20.5 for iron, 29.8 for liquid argon, 87 for plastic scintillator. Iwata has also discussed the attenuation of the shower, and proposed, based on measurements,

Shower attenuation length: $3X_0$

2.5 Finding of particle showers

Fabjan, (1985), Amaldi (1981), and Iwata (1980) have given more details. Abshire, et al. (1979), have published a tabulated shower shape, and underlined the sensitivity to the absorber-dependent energy cutoff parameter.

A more complete shower description has been suggested by Longo and Sestili (1975), and has been found to be rather universally applicable. It parameterizes the longitudinal energy density by the ansatz

$$dE/dx = k_{\text{norm}} t^a e^{-bt} \tag{2.62}$$

in which t is the shower depth expressed in radiation lengths ($t = x/X_0$), and a, b are parameters. They are meant to be independent of the calorimeter material (which enters, however, the depth variable t) and b has been shown to vary little with the incident particle's energy E. The parameter a is related to the shower peak by $(dE/dx)_{\max} = a/b$, and hence varies logarithmically with E, if b is constant and the Rossi approximation is accepted. Normalization to unity is obtained by dividing the density function by its integral from zero to infinity, given by the quotient $b^{a+1}/\Gamma(a+1)$, where Γ is Euler's gamma function defined by

$$\Gamma(a+1) = \int_0^\infty t^a e^{-t} dt$$

Normalizing to the incident energy is then straightforward, i.e.

$$k_{\text{norm}} = E b^{a+1}/\Gamma(a+1) \tag{2.63}$$

The often parametrized normalization to a *total track length* (e.g. Rossi 1965) is not as useful. It builds on the notion of 'visible' tracks, i.e. on a parameter describing a minimal energy above which a track becomes visible by ionization. Depending on the active medium in a given calorimeter, this cut-off parameter may be very different. As calorimeters are meant to measure energy, we prefer a direct conversion into MeV. To express dE/dx in the NEP, however, can be useful when calibration procedures include the use of minimum ionizing particles, as provided by cosmic ray muons. It is also in terms of NEP that the statistical properties of calorimeters (the law $\Delta E/E \propto E^{\frac{1}{2}}$) is intuitively most easily understood.

Longo and Sestili (1975) give numerical values for a and b, obtained by a Monte Carlo program in lead glass for $0.1 \leqslant E \leqslant 5$ GeV. Their values are reasonably parametrized as

$$a = 1.985 + 0.430 \ln E, \quad b = 0.467 - 0.0211 \ln E \quad (E \text{ in GeV})$$

Measurements in lead and iron calorimeters sandwiched with scintillator, using electrons from 5–92 GeV c^{-1}, have resulted in

$$a = 2.284 + 0.7136 \ln E, \quad b = 0.5607 + 0.0093 \ln E \quad (E \text{ in GeV})$$

This result (Bock, Hansl-Kozanecka, and Shah 1981) gives faster tail attenuation than Longo and Sestili (1975), but the two agree in a/b (the position of maximum) at the common point $E = 5$ GeV. Compared with Rossi's approximation, both results correspond to a shorter attenuation length, and the shower maximum for $E = 5$ GeV would translate into a critical energy ε of 4–5 MeV. More recent or generally accepted values do not seem available, despite the wealth of data recorded and published in recent years.

Hadronic showers

With a much wider range of physics phenomena involved, and with much larger fluctuations resulting, the description of hadronic showers either in approximate form, or in detail, is not nearly as advanced as for electromagnetic showers. Even the basic cross sections are not fully known down to the small energies relevant for shower development. Further, there is no simple scaling variable, both the radiation length X_0 and the nuclear absorption length λ (the mean free path for inelastic interactions) play a decisive role. Fabjan (1985) gives a few simple formulate for the average shower shape

Shower maximum at: $\quad \lambda(0.7 + 0.2 \ln E) \quad (E$ in GeV$)$
Shower attenuation length: $\quad \lambda^8 \sqrt{E} \quad (E$ in GeV$)$ (2.64)
Full containment lengths $\quad$ maximum + 2.5 attenuation lengths

More attempts at a simple description are collected in Iwata (1980). One should also retain the result of extensive measurements in iron by Holder, *et al.* (1978), resulting in

Shower median: $\quad \lambda(0.82 + 0.23 \ln E) \quad (E$ in GeV$)$

A more complete ansatz has led to the parameterization (Boch, *et al.* 1981):

$$dE/dx = k_{\text{norm}}[w t^a e^{-bt} + (1-w) u^c e^{-du}] \quad (2.65)$$

where t is the shower depth, measured from the shower origin, in radiation lengths (x/X_0), u is the same depth expressed in interaction lengths (x/λ), k_{norm} is a normalization constant, w and $1-w$ are the relative weights of the two curves, and a, b, c, and d are shape parameters. They have been fitted,

2.5 Finding of particle showers

together with w, using data taken in a variety of materials and with pion beams at momenta from 10 to 400 GeV c^{-1}. The quoted results are

$a = -0.384 + 0.318 \ln E$ (E in GeV)
$b = 0.220$
$c = a$
$d = 0.910 - 0.024 \ln E$
$w = 0.463$

Normalization has to be done separately for the two terms, using two partial gamma functions as in Equation (2.63).

Within the limitations arising from the data used, this description seems applicable to quite different calorimeters. The parameter d is closely related to the hadronic shower attenuation length, and in agreement with numbers given in Iwata (1979) ($\lambda_{att} \approx \lambda$ at $E = 1$ GeV, and increasing with E). As for electromagnetic showers, test results published during the last few years are so abundant that a substantial refinement of this approximation seems possible with a limited effort.

2.5.4.2 Lateral shower shape

Electromagnetic showers

Rossi's approximation B explicitly treats showers as linear objects without Coulomb scattering. Hence it does not make any predictions for their lateral extension, which is mainly caused by the multiple Coulomb scattering of electrons. General considerations (Amaldi 1981) lead to a transverse radius of the shower equal to the Molière radius R_m, which is defined as

$$R_m = X_0 E_s / \varepsilon \tag{2.66}$$

with $E_s = 21$ MeV (from multiple scattering theory), and ε is again the critical energy as in Equation (2.61). The double Molière radius does, indeed, contain the shower rather fully. In practice, however, the assumption of scaling with R_m is a poor one. The lateral extension is determined by competing processes, it depends on depth (showers are cigar-like with a broadening developing slowly), and has a clear energy and material dependence. Further, lateral distributions are very non-Gaussian with at least two components clearly visible, a collimated central (core) and a broad peripheral (tail) component.

An approximation with two exponentials (*core* and *tail*) has been tried

(Iwata 1980), where the relative amplitude and at least one attenuation length are functions of depth. Abshire, et al. (1979) have parametrized test results in lead/scintillator by superimposing two Gaussians. They use electrons at $3\,\text{GeV}c^{-1}$ in a detector with a longitudinal segmentation of length $2.6\,X_0$, $4.1\,X_0$, and $6.9\,X_0$. From their data they extract the ratio of the integrated area of the central shower part over the same for the tail, and the widths of these shower parts. For the first (last) segment they obtain the ratio 100 (5). The central part's width is given by $\sigma_{\text{core}} = 0.5X_0(1.0X_0)$, σ_{tail} is of the order of $5X_0$ (in both the first and the last segment). Bugge (1986) has also fitted data with two exponentials, using electron showers at $E = 4\,\text{GeV}$ integrated over $4.7\,X_0$. If we compare the two exponential curves in terms of their attenuation factor (slope) and their relative contribution (integral), then the core part has an attenuation nine times higher than the tail part, and contains over 80% of the energy.

We should note that lateral results are strongly dependent on the longitudinal homogeneity of a calorimeter; any gaps will result in a widening of the lateral profile (see Fig. 2.18 for an illustration).

Hadronic showers

The lateral development of hadronic showers is dominated by the average transverse momentum in hadronic interactions, which is of the order of $300\,\text{MeV}c^{-1}$. The variety of interactions and their dependence on material parameters has, however, not permitted a global description of the lateral shower shape as function of depth. Some measurements obtained in test beams with calorimeter set-ups have been published in Holder, et al. (1978) and Diddens, et al. (1980). Holder, et al. (1978) show that a simple Gaussian or exponential will not approximate to measured lateral shower shapes. The hadronic shower width increases linearly with depth, and at least two Gaussian-like curves have to be superimposed to approximate the measured integrated shower profile. If we characterize the curves by their Full Width at Half Maximum ('FWHM', the Cauchy width), then FWHM $= 0.28\lambda$ and FWHM $\geqslant \lambda$ are characteristic for the two components. Further, the large difference in width between full containment (defined by less than one minimum ionizing particle remaining) and 95% containment is also measured. The depth dependence of the 95% cone can be expressed by its radius

$$R_{95\%} = \lambda(0.29 + 0.26\,u) \quad \text{at } E = 50\,\text{GeV} \tag{2.67}$$

2.5 Finding of particle showers

and by

$$R_{95\%} = \lambda(0.29 + 0.17\,u) \quad \text{at } E = 140\,\text{GeV}$$

Here u is the depth in units of λ. The increase with rising energy of the relative contribution from the core part is apparent. Diddens, et al. (1980), and Jonker, et al. (1982) show the significant difference in lateral spread between electromagnetic and hadronic showers. In their experiments this difference has been used as prime information for π/e^- separation (see Fig. 2.18, and the case studies in Subsection 2.5.6).

Muraki, et al. (1985) have reported lateral shower shapes obtained at high energy (300 GeV), when exposing stacks of X-ray film interspersed with iron and lead as absorbers, to a proton beam. They propose a multiparameter formula to fit their data, which underlines the complexity of the parametrization approach. Despite such rather clear evidence, the approximations used in practical applications often do not go beyond the simple Gaussian, e.g. Akesson, et al. (1985).

2.5.5 Shower simulation

We have mentioned above that shower simulation by Monte Carlo methods, in particular the simulation of hadronic showers, has been a formidable tool in designing calorimeters, and that the completeness of some of the resulting programs is at a very remarkable level. We refer here to programs that embody the complete understanding of the showering process, and have been tuned until all relevant experimental results can be reproduced. Such programs are inherently ideal guzzlers of computer time, and many efforts have been made in the past to approximate locally what the full programs generate in detail. In their crudest form, the approximations do not exceed the formulae given in Subsection 2.5.4. Indeed, many gross detector studies can usefully be done at the level of average showers.

Probably the most widely used *full shower simulation* codes are EGS (Nelson, et al. 1985) for electromagnetic, and GEISHA (Fesefeldt 1985) for hadronic showers. For the inclusion of the subtleties of compensation mechanisms at low energy (see Subsection 2.5.2) the HETC (Armstrong, et al. 1972; Armstrong et al. 1984) and HERMES/DYMO codes (Brückmann, et al. 1987) represent the present state of the art.

These codes are kept in many of the computer centre program libraries of the high-energy physics community, though not always in identical versions. Their use is relatively straightforward, but a high level of expertise is required for parameter tuning and in interpreting results. The use of multiple sensitive cutoff parameters in the infrared (at low energies), requires application-dependent decisions. Also, continuous improvements are being made to these programs, and their evaluation and possibly their introduction by nonexpert users is not always trivial. The completeness of all cross sections, particularly at critical low energies, is one area of ongoing work, which the authors of these programs pursue in a painstaking way the merits of which cannot be overestimated.

Another area where specialized experience and expertise are required is detector description, both in structure and content. Both EGS and GEISHA have been embedded in many local codes that introduce detector details, e.g. in GEANT (Brun, et al. 1986). Calorimeters come in ever newer shapes and often contain materials whose behaviour is not well understood, i.e. material compounds (molecules or mixtures of materials). Scintillators of various composition or BGO are prominent (and understood) examples. Fesefeldt (1985) discuss ways of calculating the relevant numbers for such compounds.

Several nonauthor comparisons of GEISHA and/or EGS results with existing data have been made, and some of them are found in the literature. Cashmore, et al. (1985) have gone from hadronic shower comparisons over a wide range of energies (2–300 GeV) to the use of simulated data for optimizing strategies for correcting for leakage. Adiels, et al. (1986), have compared the performance of electromagnetic calorimeters at energies below 1 GeV to EGS predictions. This work was done in the context of an experiment at the LEAR at CERN, detecting π° and η; the comparison was satisfactory.

The 'standard' programs are not accepted without some healthy competition, of course. There are applications in electromagnetic shower simulation which can be solved with the same precision as EGS, but more efficiently in computer time (Brun, et al. 1985). FLUKA (Aarnio, 1986) and NEUKA (Kowalski, Moehring, and Tymieniecka 1987) are other hadronic cascade programs that should be mentioned. Several other codes have reached publication stage and are locally popular (e.g. CASCADE, Grant 1975). It is probably fair to predict that most future improvements are likely to happen *in the framework* of the large existing simulation programs, simply due to the fact that the major investments in acquiring experience with the applications (and limits) of a system must be maintained, and makes it hard for a radically new approach to succeed.

At very high energies, i.e. for the multi-TeV accelerators that are now

2.5 Finding of particle showers

only a vision of the future, these codes may not be applicable, because they would be too detailed and prohibitive in computer time. Currently, a number of programs exists for very high energies, written in connection with studies for future accelerators. From various comparisons one can conclude that these programs, used all over the world in accelerator development groups, are in agreement in the relevant results, but they each have their own specific advantages.

In many applications, criteria other than the completeness of treatment may dominate in judging shower simulation programs, such as ease of use for detector studies, good interfacing with existing event generators for certain physics classes, or efficiency, particularly on specific computers like vector processors. These aspects have been considered perhaps most generally in the all-embracing program GEANT (Brun, et al. 1986), which includes as options more or less all confirmed shower generators, and surrounds them with a very general description of detector details, tracking facilities for charged track detectors, and graphical output options.

When the efficiency of shower simulation is of more concern than the shower detail, general-purpose programs are not what comes to mind. *Empirical shower shape parameterizations* have been developed which are considered satisfactory in many instances where the detailed understanding of the calorimeter is not demanded. Parameterizations as given in Subsection 2.5.4 give an overall shower shape and distribution, but largely fail to provide the fluctuations of showers (Bock, et al. 1981). They are applicable when a global detector optimization is necessary, the effect of dead spaces has to be studied, or grossly different physics choices are to be compared. In such applications, parametric approximations permit computer time gains by orders of magnitude.

An equally important saving in computer time has been achieved by *pregenerating many individual showers* in all details, and storing the details on a file. Random access to the showers on the file will then result, after some tuning, in satisfactory to excellent simulation, limited mostly by the number of showers pregenerated. The method requires similar starting conditions for many showers. It therefore has been used most successfully, and with major computer time savings, for the shower details after a certain cutoff energy has been reached in individual track energies. Details were published in Longo and Luminari (1985). Also the combination of single-particle simulation (e.g. for high-energy particles) with parametric approximations (at lower energies) has been used. This problem is discussed in some detail in Fesefeldt (1985).

2.5.6 Calorimeter algorithms: case studies

The preceding discussions should have established several facts: Calorimeters rely on *stochastic processes* much more than other types of detectors. Their precision limits are defined less by the luxury of instrumentation than by the fluctuations of the underlying physical processes. Calorimeters are built and instrumented with a *specific physics problem* in mind. A wide and ingenious variety of solutions have successfully been implemented. Calorimeters are *empirical* devices whose understanding depends largely on test measurements. Calibration procedures with test beams or events of known particle composition play a dominant role. Simulation by Monte Carlo procedures is tuned to reproduce such calibration data correctly, and may then be used to optimize and interpolate.

A consequence of these facts seems to be that a general method for reconstructing particle or jet energies, for identifying particles, and for separating showers of different origin, on the basis of calorimetric data, can not be given. Our approach, in this subsection, will be to discuss calorimeter applications with different characteristics in some detail, suggesting that they are representative for many others. Undoubtedly, calorimetry is a field where ingenious and creative approaches are possible and even necessary, and surprising future optimizations are likely to be found, that are not covered by this short review.

The following series of case studies is taken largely from calorimeters which have successfully solved the intended physics problem. Some case studies are also included that have not yet produced published results, but which are under serious development. Our description of the experimental details will necessarily have to be condensed, but most cases are properly documented in the literature for anyone who is interested in more detail. The categories of applications are the following:

Spatial energy flow measurements (UA experiments)

This task is the one most directly performed by calorimeters in the absence of any external information. The physical space of interest is subdivided into cells of the desired granularity, with the intercell gaps minimized. Every cell measures an energy value. By suitably weighted sums over energies measured in the cells, one can form total event energy, total transverse energy, asymmetries like left–right or up–down, or escaping (unmeasurable) energy. Inhomogeneities in the energy flow (i.e. local clustering not explained by the physics model) have also been shown using simple cell energy distributions.

π^0/γ discrimination (TASSO)

Calorimeters attempting to discriminate clearly between the shower left by a single photon and showers originating from two γs from the electromagnetic decay of a π^0, typically need to put the accent on position resolution. They are also in need of vetoing against the presence of a charged particle, i.e. of calorimeter-external information from scintillators or tracking devices. This discrimination is comparatively easy at low energies (< 1–2 GeV), i.e. for large-angle π^0 decay, but at higher energies, it is necessary to use an optimally discriminating 'feature space' (see Sections 2.1.4 and 4.2), typically found by using training samples of both event types.

Electron/charged hadron identification (Mark III, CHARM, UA1)

This application of calorimeters requires an independent signal from a tracking chamber or equivalent device which establishes the existence of an isolated charged track. Hadronic showers have large enough fluctuations to simulate, with some finite probability, electromagnetic shower shapes at all energies. As a minimal background of this type, one may visualize the charge exchange reaction $\pi^- + p \rightarrow \pi^0 + n$, when this occurs near the beginning of the calorimeter and with small energy transfer to the neutron. The hadronic π^- shower will then indeed be a purely electromagnetic shower caused by the π^0. The tuning of charged e/π discrimination will, therefore, be a delicate one, and will never produce correct decisions for all individual events.

Jet finding (UA experiments, axial field spectrometer (AFS), JADE)

Jets are quark- or gluon-induced compounds of mostly hadronic particles, both neutral and charged, that follow near-by trajectories. Their origins are the partons which emanate from an interaction and which 'hadronize' at medium to long distances from the interaction point. Although no firm theory exists as yet for the *hadronization* of partons, phenomenological descriptions exist, and the existence of jets and their most relevant properties have been clearly established and shown to be in agreement with the theory describing the *production* mechanisms of partons. Jets were first observed by e^+e^- experiments at DESY, and described by global variables (e.g. Brandt and Dahmen 1979; Wu 1984). Jets originating from hadronic collisions were found at the CERN SPS $\bar{p}p$ collider and at the CERN ISR. For detailed results, refer to Banner, *et al.* (1982), Arnison, *et al.* (1983b), Akesson, *et al.* (1984), and Arnison, *et al.* (1986).

Calorimeters are the most prominent devices for detecting jets. Jet algorithms typically define windows in suitable variables and use a maximization procedure to find energy clusters inside such windows. In other words, the problem is that of object classification in the sense of Subsection 2.1.4, with the metric adapted to the measurement variables a calorimeter can provide. Training samples (see Subsection 2.1.3) and physics modelling have to provide the parameters that are used to assign a 'jet likelihood' to clusters so identified, and to guard against random event fluctuations.

We will now discuss examples in more detail.

2.5.6.1 Global energy flow in UA2

The UA2 Experiment at CERN was one of two using the CERN SPS $\bar{p}p$ collider for general studies on $\bar{p}p$ collisions at the highest available energies. The central calorimeter is described in detail in Beer, *et al.* (1984), and has already been introduced in Subsection 1.5.7. It consists of a highly and regularly segmented electromagnetic and hadronic calorimeter of the sampling type, over the angular region $40 < \theta < 140°$ and full azimuth (ϕ). For a definition of the coordinates frequently used, see Fig. 2.19. The

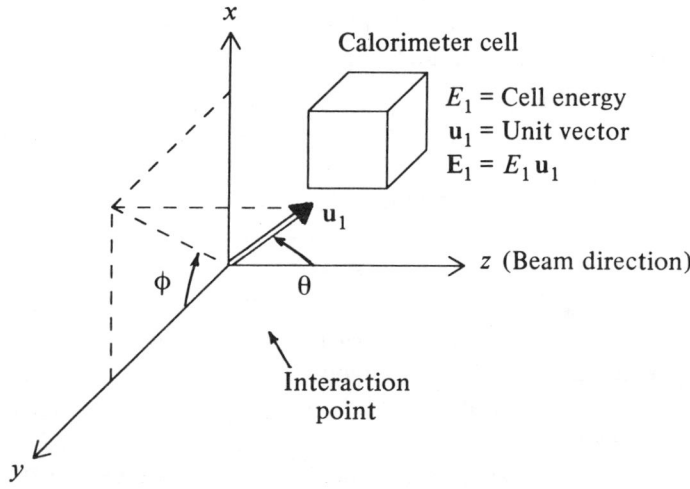

Fig. 2.19 The coordinate system and energy vector definition as used in calorimetry.

2.5 Finding of particle showers

geometry is spherical, and all elements cover a constant polar and azimuthal slice, $\Delta\theta = 10°$ and $\Delta\phi = 15°$. Cells are stacked as truncated pyramids arranged to point towards the centre of the interaction. The first (electromagnetic) compartment starts at a radial distance of 0.60 m (see Fig. 1.43). It is a stack of 26 lead absorber plates, each 3.5 mm thick, interspersed with 4 mm scintillator plates. Light collection is over the entire length via two wavelength shifting bars on opposite sides. The second (hadronic) compartment is subdivided longitudinally into two readout sections, and consists of 18 + 22 plates of stainless steel, each 15 mm thick, interspersed with scintillator 5 mm plates. The readout of the two sections is again through two wavelength shifter bars.

The energy flow results were amongst the earliest from the experiment and were reported by Banner, *et al.* (1982). The algorithm consists of the following steps (for definition of the variables, see Fig. 2.19):

Obtain the energy, E_i for every cell i.

Sum the energies from the three depth compartments, i.e. for every group of three with the same θ_i, ϕ_i coordinate pair for its centre.

Disregard sums made from cells which are below a cut off value (of 150 MeV).

Multiply every sum of three with a factor $|\sin\theta_i|$ to obtain the transverse energy.

Sum over all cells to obtain the total transverse energy of an event.

The analysis of energy flow in this experiment also included a clustering algorithm similar in concept to the one described in Subsection 2.5.6.8. This provided a convincing demonstration that, for events with high total transverse energy, a large fraction of the energy tends to be concentrated in small angular areas, thus establishing the dominance of two-jet-like events.

2.5.6.2 Missing energy in UA1

If a calorimeter can be constructed to be hermetic, i.e. such that no major insensitive areas exist over the entire acceptance volume of tracks, it provides complete energy flow information on each event. If of sufficient accuracy, and coupled with fully hermetic charged track detection (for tagging muons), such a detector allows firm statements about the energy escaping undetected in form of neutrinos.

A fine example of this technique, limited to the transverse plane, has been provided by the experiment UA1. To demonstrate the existence of the

Pattern recognition

intermediate vector boson $W^{\pm}$, the experiment used not only the active signal of muons or electrons, with their respective identifications, but also the measurement of the missing transverse energy corresponding to the conjectured neutrino (Arnison, *et al.* 1983a).

The electromagnetic shower counters used in the large-angle region of UA1 are crescent-shaped 'gondolas' of lead scintillator type (Fig. 2.20). They cover 180° in ϕ and 23 cm along the beam, at an inner radius of 1.3 m

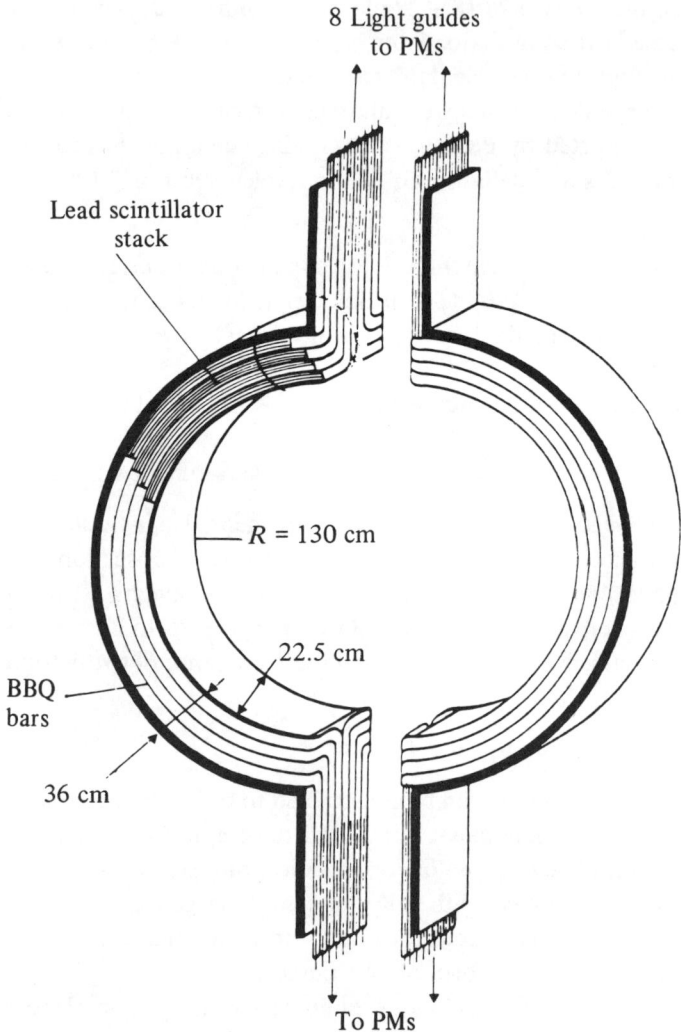

Fig. 2.20 A single electromagnetic calorimeter module ('gondola') as used in the barrel part of the UA1 experiment. The total number of modules is 48.

(Cochet, et al. 1986). The coordinates used are again those shown in Fig. 2.19. Light transport is over wavelength shifter plates (BBQ, Benzimidazo-Benziso Quinoline-7-one) covering the full ϕ range of the modules on both sides. At both ends the BBQ bars are connected to light guides leading to photomultipliers. Four separate wavelength shifters provide depth segmentation (3.3 X_0, 6.5X_0, 10.0X_0, and 6.6X_0).

For a single electron, this calorimeter offers substantial redundancy in the recorded signals. The electromagnetic shower counter is complemented by a hadronic calorimeter covering the entire available space, consisting of an iron/scintillator sandwich with some longitudinal segmentation (instrumented magnet yoke). It is described in detail in Corden, et al. (1985). The two devices are not geometrically matched, either in the central or in the forward region.

The algorithm for obtaining the missing transverse energy vector was similar to the one used in UA2 for the calculation of total transverse energy, but adapted to the calorimeter cell properties of UA1 which are inhomogeneous in θ, ϕ for different calorimeters, and do not allow simple summing over depth:

Obtain the energy E_i for every cell i.
Disregard cells which are below a cut off value.
Form a transverse energy 2-vector by multiplying every cell energy with the projection angles, i.e.

$$E_x = E|\sin\theta|\cos\phi$$
$$E_y = E|\sin\theta|\sin\phi$$

Sum the vectors over all cells to obtain the total transverse energy vector of an event.

Additional corrections were applied for identified muons, only measured in tracking devices. Errors were such that for $E_T(v) > 10\,\text{GeV}$, the signal could be considered significant, provided the missing transverse energy vector did not point towards the known uninstrumented parts of the device.

2.5.6.3 π^0 selection in TASSO

The TASSO detector at the PETRA e^+e^- storage ring of DESY has measured the inclusive π^0 cross section at various energies. One critical task in this analysis was the separation, in the barrel electromagnetic calorimeter, of the π^0 signal from genuine single γs.

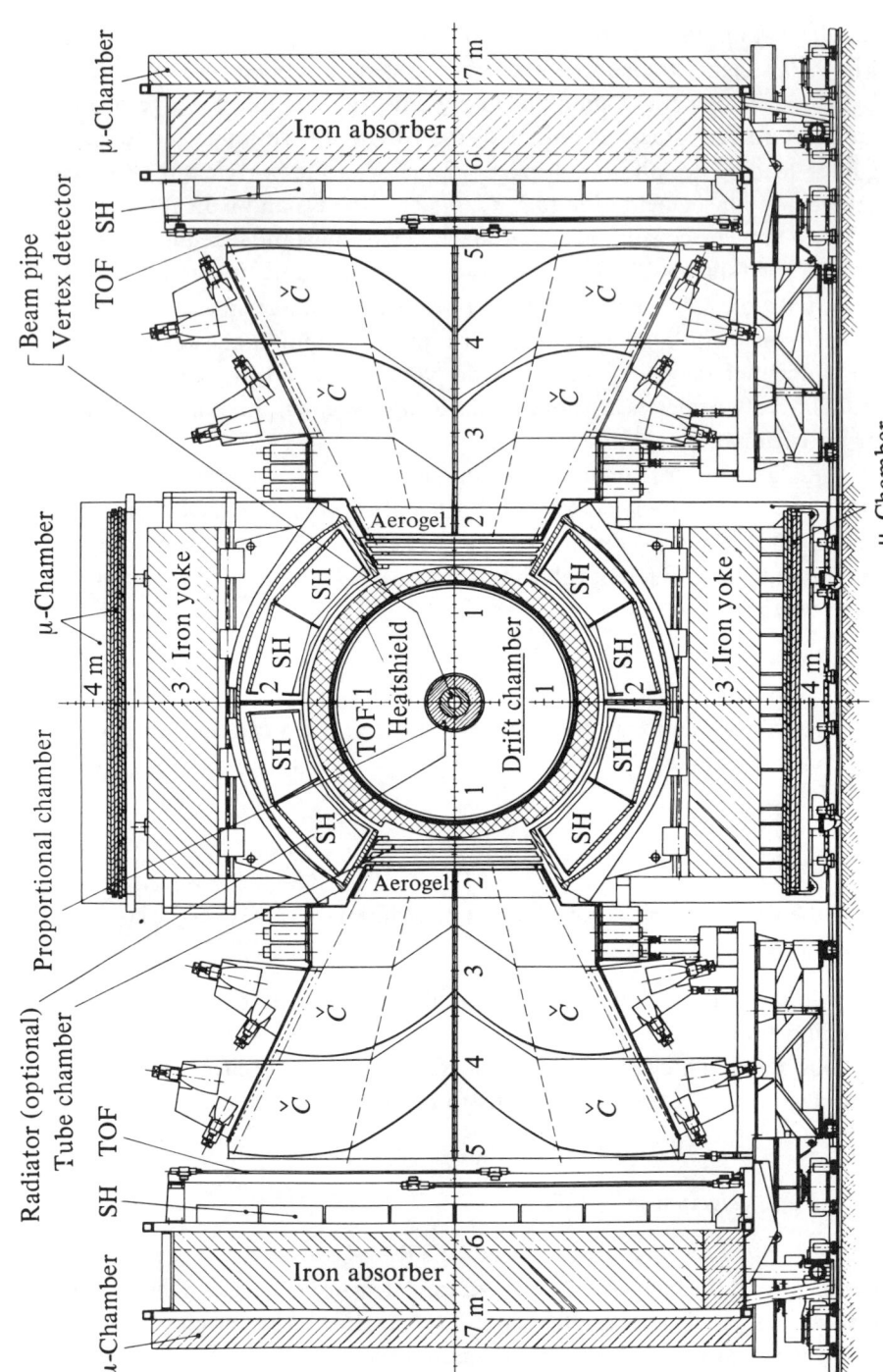

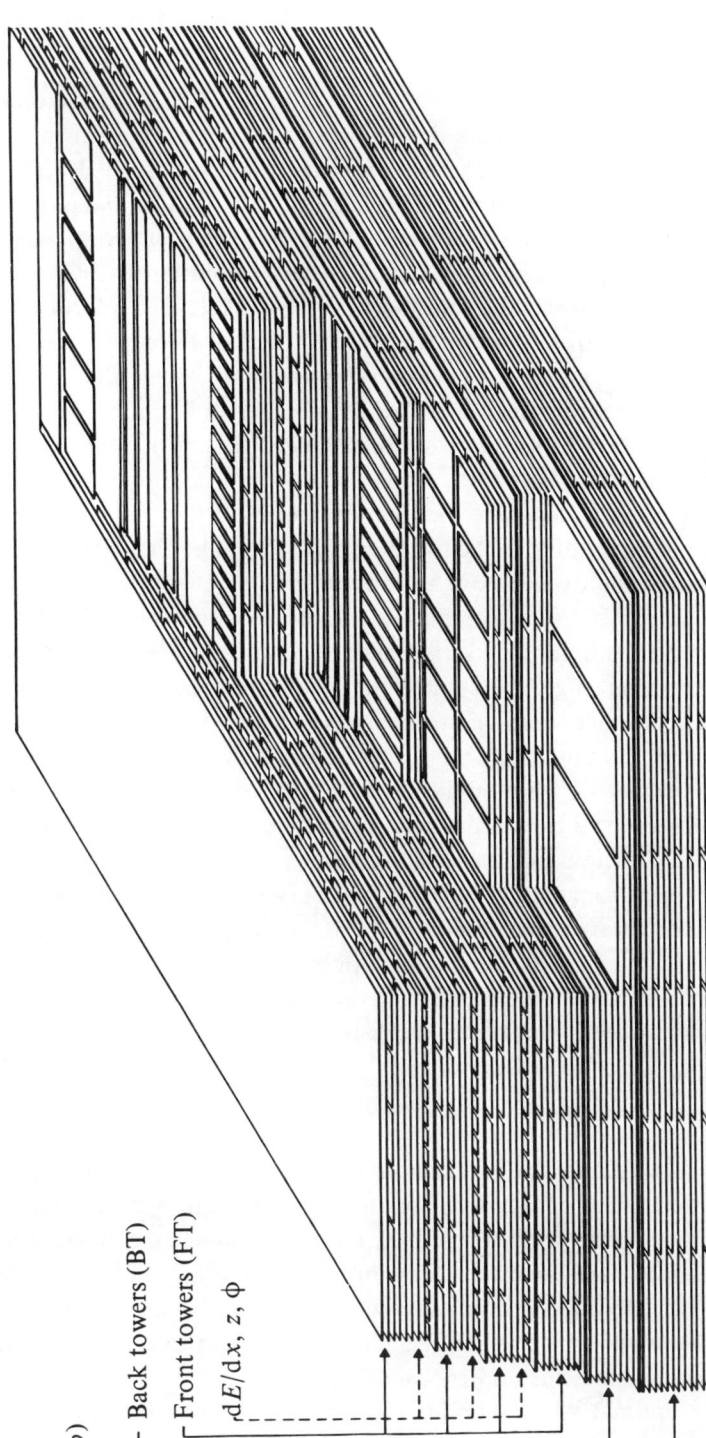

Fig. 2.21 The TASSO detector. (*a*) In this overview, the electromagnetic shower counters are labelled SH. (*b*) A detailed view showing the construction of a single tower, with different interspersed readout layers.

The barrel electromagnetic calorimeter in the TASSO detector has a partial azimuthal coverage, and is of the sandwich type. The absorber plates are 2 mm thick lead plates, between which is a 5 mm gap filled with liquid argon as the active material. The calorimeter modules (labeled SH in Fig. 2.21) contain a stack of 35 lead plates. The readout is organized into towers with depth segmentation for energy resolution (it has achieved $\sigma_E/E \approx 0.1/E^{\frac{1}{2}}$). Position resolution is the most relevant information for π^0/γ separation. It is obtained by having a total of seven planes equipped with readout strips in 2 cm width, aligned along either the z or the ϕ direction (the coordinates are explained in Fig. 2.19). The modules start at a radial distance of 1.79 m. For a more detailed description, see Braunschweig, et al. (1986).

At low π^0 energies, a reconstruction can be done using the separate clusters corresponding to the two decay photons. The clustering algorithm merges adjacent towers above a threshold energy into clusters, if they have a common edge. Clusters compatible with depositions from a charged track observed in the inner drift chambers are eliminated (Zeuner 1984).

The opening angle for symmetric decay (minimal angle) is given by $\alpha_{min} \approx 2m/E$ (where m and E are the mass and energy of the π^0 respectively), and most decay angles are not much larger than α_{min}. The recognition of clusters strongly depends on the lateral size of the towers; clusters from γs in the TASSO barrel electromagnetic calorimeter tend to merge when their angular separation falls below 80 mrad.

At energies from about 4 GeV up, the analysis therefore has to rely on the position detectors (strip planes) to decide if merged double shower exists. The decision between a single or a double shower is carefully optimized, and finally is based on two test statistics called 'dispersion' D (Tysarczyk, Mättig, and Lohrmann (1985). The dispersion is defined in two directions by

$$D = [\sum E_i(x_i - \langle x \rangle)^2 / \sum E_i]^{\frac{1}{2}}$$

where x_i and E_i are the position and energy measured in a strip i of the cluster, $\langle x \rangle$ is the energy-weighted average position of the cluster, and x denotes one of the position variables, z or ϕ (ϕ is taken here to be a length in the direction of the azimuth). More correctly, one should talk about $r\phi$, as it is customary to give the dimension of length to the variable along ϕ. In the TASSO example, $r = 1.79$ m. The summing extends over all strips of one orientation in the cluster, hence a D_z and a D_ϕ are found separately. The retained optimal two-dimensional decision (see Section 4.2) is used as a condition for a double shower

$$D_z \geqslant 2.49 \text{ cm} \quad \text{and} \quad D_\phi \geqslant 2.13 \text{ cm},$$

independent of cluster energy. Losses of π^0 (signal) and contamination from single γ are 20–40% and 10–15% respectively, depending on the energy. They have been determined, along with corrections, in Monte Carlo studies based on the EGS code, and including, in particular, data from the measurement of hadronic reactions and of $e^+e^- \to \gamma\gamma$.

2.5.6.4 e/π separation in the Mark III electromagnetic shower counter

This calorimeter at the SPEAR e^+e^- storage ring at SLAC covers the entire solid angle, with the exception of a cone of half-angle 11° around the beam axes. Its central (barrel) part is cylindrical, the end caps are flat, and both follow a very similar design with 24 layers of rectangular proportional tubes alternating with 23 sheets of a lead-antimony alloy ($0.5X_0$ each). All tubes are parallel along the beam for the barrel, and vertical in the end cap. Figure 2.22 illustrates the Mark III calorimeter. It has been described by Toki, et al. (1984).

Shower coordinates along the wire are obtained by charge division. The calorimeter is very highly segmented, resulting in excellent position resolution, ± 7 mrad/± 7 mm perpendicular to the tube, about $\pm 1\%$ wire length (± 44 mm) along the wire. The electromagnetic energy resolution is comparatively modest ($\Delta E/E \approx 0.17/E^{\frac{1}{2}}$), due to the small sampling fraction in the proportional tubes. The energy variable used for photons produced in connection with unstable particles in the later analysis is $E^{\frac{1}{2}}$, which is preferred to E since it has a more Gaussian-like behaviour. Efficiencies are excellent even at photon energies below 1000 MeV, mostly due to the small amount of matter ($0.4X_0$) in front of the first layer of proportional tubes.

Recognition of showers starts in the high-quality projection (e.g. ϕ for the barrel) alone, with clusters defined as at least two cells with hits separated by no more than two cells. Subsequently, the lower-quality projection (e.g. along the beam) is used to check for consistency or for splitting signals. Finally, hits are examined in depth (e.g. radial variable r), and may be split if early hits show a clear two-prong structure (indicative of two close photons from fast π^0 decaying symmetrically). The relevant position information is extracted by fitting a straight line to the detailed hits. The photon entry point is defined to be that of the fitted line at calorimeter entry surface (Richman 1986).

218 *Pattern recognition*

For π/e separation, an algorithm was designed with great care, combining the shower detector information with a time-of-flight (TOF) counter probability for e or π, and with tracking (momentum) information. Eight variables were chosen as the most efficient test statistics to separate π from e, after systematically comparing more than 50 variables. The decision is taken by successive cuts in a *decision tree*, using one of the variables at each node. In short, the test statistics retained are these: total shower energy, longitudinal shower barycentre, shower width, layer-wise energy correlation (using an experimentally determined dispersion matrix as described

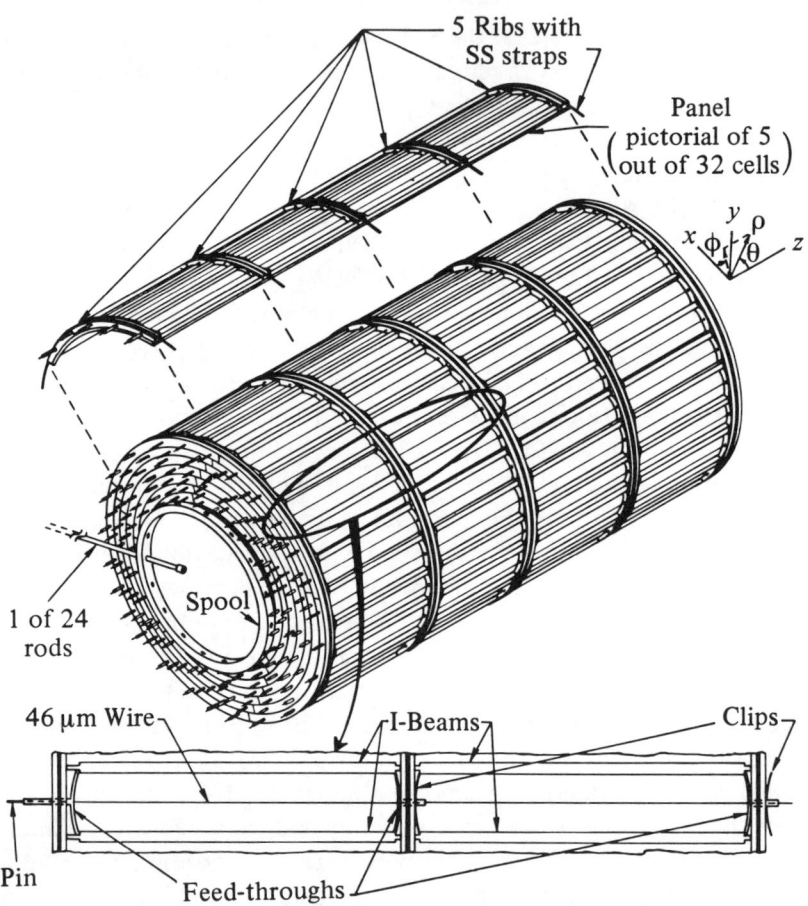

Fig. 2.22 General view of the barrel electromagnetic shower counter of the Mark III detector.

2.5 Finding of particle showers

in Subsection 2.18, Equation (2.22)), energy/momentum ratio, energy in the early layers (two groups), and TOF probability. The precise tree and the cuts used are energy-dependent; altogether seven different decision trees were defined. Separate tuning was performed for most efficient electron and pion selection. The final performance of both is approximately the same: 20% loss of signal, 4% of background as contamination.

In optimizing the decision tree, a method called *recursive partitioning* (Friedman 1977; Breiman, *et al.* 1984) was used. This is based on the notions of sample *purity* and *cost* of misclassification. These two terms describe variables equivalent to *contamination* and *loss*, which we use in our discussion of Section 4.2. A training sample of events whose classification is known, is subjected to a large number of decisions (start with a very large tree, with few events in each decision class). An algorithm then removes more and more decision nodes, attempting to minimize the cost and maintain maximum purity (Coffman 1987).

2.5.6.5 e/π separation in CHARM

Calorimeters in neutrino experiments have a different role from those used in other experiments. They are not *complementary* to other detectors, but have to be sufficiently instrumented to be simultaneously the *target* for the incident neutrino, and the *detector* for all interaction products. The well-analysed CHARM detector, and its successor CHARM II (which is in early operation at time of writing) are target calorimeters that have been built with these specific constraints in mind. They are of the sandwich type, with multiple active components embedded between layers of absorber.

The CHARM vertex calorimeter (Diddens, *et al.* 1980), made primarily to study neutral current interactions, consists of an assembly of 78 identical subunits, each of about 20 cm depth and $3 \times 3\,\text{m}^2$ surface, and comprising an absorber and three active layers, a marble plate of 8 cm thickness, an array of scintillation counters, an array of proportional drift tubes, and an array of limited streamer tubes (Fig. 2.23). The scintillators provide triggering and energy resolution, the proportional and limited streamer tubes together give precise tracking information (e.g. on muons) and shower positions/directions. The shower directional resolution is energy-dependent and reaches ± 6 (± 20) mrad for electromagnetic (hadronic) showers above 20 GeV. Marble was chosen as the absorber

220 *Pattern recognition*

because it approximately equalizes the average shower lengths of electromagnetic and hadronic showers (due to the small ratio $\lambda/X_0 = 4$).

When using these detectors for electron identification (against a background of both hadrons and photons), use is made of the different quality of the signal in these detectors. Electron showers will typically not deposit more than the equivalent of a minimum ionizing particle in the first scintillator plane, assuming it is not preceded by absorber; π^0- or γ-initiated showers will typically start with two or more charged tracks

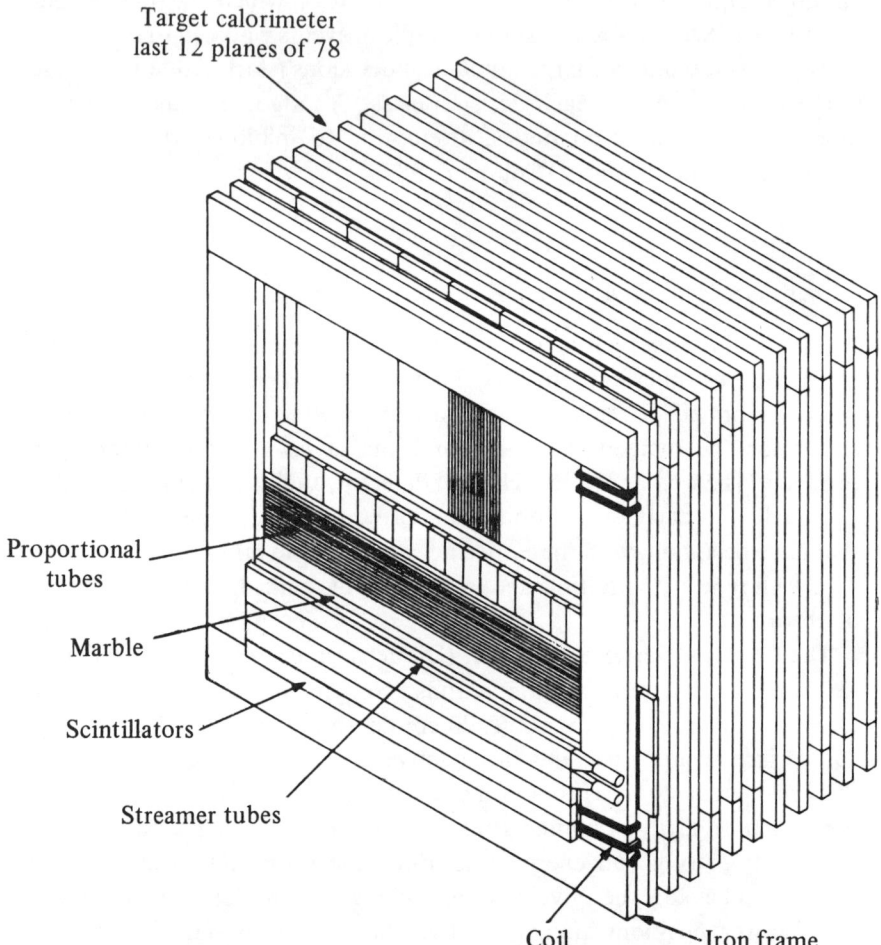

Fig. 2.23 The vertex calorimeter in the CHARM detector; only twelve of a total of 78 planes are shown.

2.5 Finding of particle showers

(Dorenbosch, et al. 1987). For discrimination against hadrons, the lateral shape is used which for electromagnetic showers is narrower and, in particular, has much less tail than for hadronic showers. In the specific set-up, however, it is not possible to make a distinction from the longitudinal shape.

Lateral shape measurements are expressed as a Cauchy width Γ for the lateral energy flow, as obtained from the scintillator. Despite the coarse scintillator information, the long tails of the Cauchy distribution allow Γ, the full width at half maximum, to be deduced. The Cauchy distribution is given by the probability density function $P(x) = (1/\pi)/(1 + x^2)$, which has a full width at half maximum of 2. When generalized, this becomes the Lorentz distribution with a parametrized width, expressed as $P(x) = \text{const}/[(x - x_0)^2 + \Gamma^2/4]$.

The RMS width σ as obtained from the proportional tubes is also used in discriminating between electrons and hadrons. Simultaneous constraints on all quantities result in a hadron rejection factor of more than 100, with an efficiency of 60% for electrons.

CHARM II is the successor experiment to CHARM, with improved instrumentation (Busi, et al. 1983). It relies on finer depth segmentation, using glass as the absorber, and more complete coverage with streamer tubes including cathode readout by (2 cm) strips. In this 'imaging' detector which allows very fine-grained observations, a very systematic approach to shower recognition is intended. A minimum spanning tree (see Subsection 2.1.9) algorithm is used, based exclusively on the streamer tube information. The metric is the separation of hits in geometrical space. The resulting unambiguous tree is cut into clusters by eliminating edges above a threshold. For each cluster two characteristic parameters (test statistics) are then calculated: the total path length L, summing over all edges in the cluster, and the average path length $\lambda = L/n$, where n is the number of edges. Both L and λ are energy-dependent, and show a very different statistical behaviour for π and e^-, as shown in Fig. 2.24, with data taken from training samples.

The algorithm's (Grancagnolo, private communication) goal is to lose a minimum of e^-. The objective, in particular, is to define losses with optimal precision. To achieve this, the test has to tolerate a sizeable contamination. In the terminology of Section 4.2, the test favours small *significance* over large *power*. Since L and λ are correlated, an error ellipse in the $L\lambda$ plane defines equal probability of losing events. The error ellipses corresponding to 2.5σ (i.e. 1.2% signal loss) at all energies have an envelope which defines a cutoff line to distinguish between pions and electrons; it has been drawn in Fig. 2.24.

2.5.6.6 Identification of e^- in UA1

The clear identification of electrons was a key issue in the UA1 experiment in tracking down the intermediate bosons $W^\pm$ and Z^0. The calorimeter properties in the experiment have been described in Subsection 2.5.6.2. For the small number of events that comprised the W/Z samples, individual signals in the electromagnetic calorimeter's photomultipliers were compared for consistency with the hypothesis of a single showering electron. All knowledge about electromagnetic showers and all available information concerning the attenuation of light in scintillator and

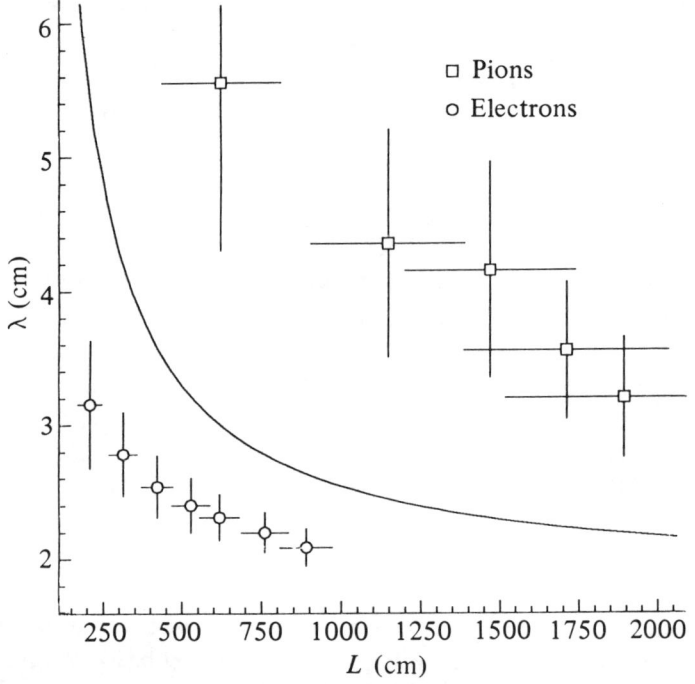

Fig. 2.24 Intended e/π discrimination in CHARM II. Clusters are formed from streamer chamber data, using a MST definition. For each cluster's tree, the total path length L and the average length of edge λ are found. The points shown are obtained from discrete energy samples (between 3 and 40 GeV) of electrons and pions; they lie in very different domains of the $L\lambda$ plane. The dividing line drawn is a parametrization of the envelope corresponding to the 2.5σ distance from the electron points.

2.5 Finding of particle showers

wavelength shifter were used. The fully analysed track in the central tracker was a necessary complement, as was the missing signal in the hadronic calorimeter behind the electromagnetic shower (Arnison, *et al.* 1983a).

For later studies involving larger samples and requiring electron identification on a less individualistic basis, longitudinal segmentation was used to calculate a test statistic called χ_r^2. Shower shapes were derived from test beam results on gondola modules, in π^- and e^- beams from 5 to $20 \text{ GeV } c^{-1}$ and at various angles of incidence. They were then parametrized using the model for electromagnetic showers given in Subsection 2.5.4.1, refitting the coefficientis a and b. Both the model and the data were expressed in terms of ratios $R_i = E_i/E_0$, where E_i is the energy recorded in depth segment i (i goes from 1 to 4), and E_0 is assumed to be known independently (from the beam momentum or a tracking device). The test statistic χ_r^2 was then defined as $\sum[(R_i - P_i)/\sigma_i]^2$, where R and P are the energy fractions derived from observations and the model, and σ^2 is the variance derived from test data. In addition to using the ratios from the four layers, all possible layer sums (of adjacent layers) were entered into the summation, nine quantities altogether. Although these sums are not treated as correlated quantities, as they should, they have the advantage of smaller variances, and increase the sensitivity of the test substantially (Eisenhandler, *et al.* 1984). The net result of the condition $\chi_r^2 < 9.0$ is a π rejection level of 99.6% (4×10^{-3} background) with 70% of the e^- passing (0.30 loss), for track momenta of 10 GeV.

2.5.6.7 Future DELPHI High-density Projection Chamber (HPC) electromagnetic shower recognition algorithms

DELPHI is one of the four experiments planned for the LEP e^+e^- collider ring at CERN, due to start operation in 1989/90. Its central (barrel) electromagnetic calorimeter is known as the High-density Projection Chamber (HPC). It is a sampling calorimeter based on lead converter plates inter-leaved with gas sampling volumes (Fig. 2.25). The ionization electrons drift in a direction parallel to the beam, to a single proportional wire plane in TPC style (see Subsection 2.1.4). Signals are picked up exclusively by pads of variable size ($4 \times 4.5 \text{ cm}^2$ minimum, larger as calorimeter depth increases). The least count in direction of electron drift correponds to 3 mm.

This chamber is potentially of extraordinary granularity, limited only by space for signal cables and financial considerations. It is intended to use optimal statistical methods for the recognition of showers and the

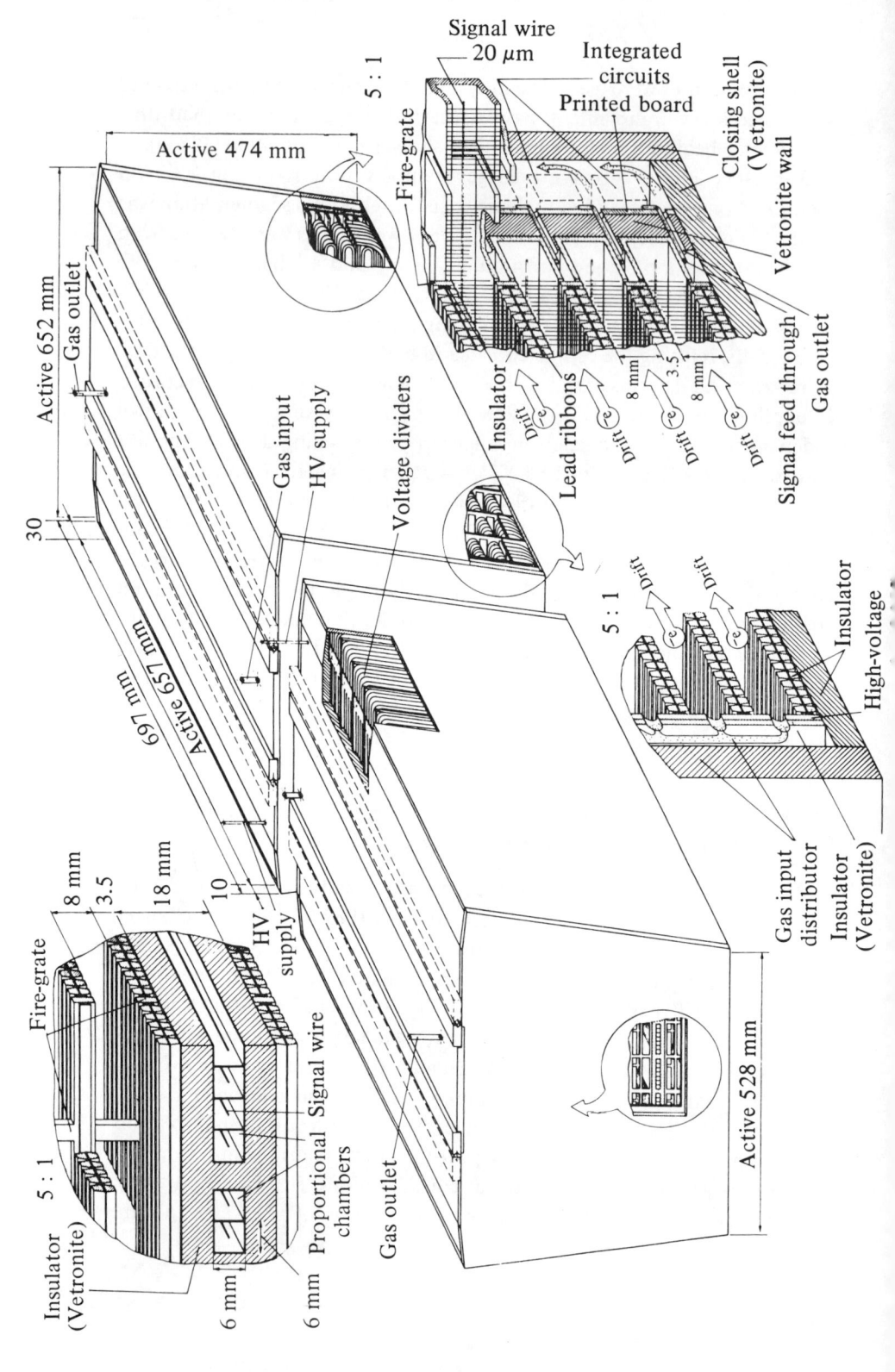

2.5 Finding of particle showers

distinction between showers of different origins. It might seem conceivable to use the fine spatial resolution for finding showers at a level of detail unequalled in other calorimeters.

In practice, whilst very general statistical methods may be used and thus the mathematically soundest approach can be selected, the performance of such fine-grain readout does not seem to be very clearly superior to numbers obtained elsewhere. The precise algorithms are presently under study. Training samples from test beam exposures and Monte Carlo data are being used to find the best variables which allow the discrimination between

A single electromagnetic shower and several superimposed showers,
An electron shower and a photon shower in coincidence with a charged particle (noninteracting),
A single electromagnetic shower and a hadronic shower
Any of these and other classes of showers.

A general method has been proposed (Calvi 1987), which relies on a standard method of statistics called 'canonical discrimination', and which is equivalent to the linear feature extraction or principal component analysis of Subsection 2.1.8. The method searches for a set of linear transformations applied to the variables of the showers in the training sample, such that the transformations are orthogonal with respect to each other, and, in turn, maximize the correlation with the (known) groups of showers. It is expected to be able to compare only two clusters of events (two types of showers) at a time. Maximal separation in such a simple case is relatively easy to define, and the method will lead to different 'canonical variables' for each type of comparison.

Another proposal sets out to use the method of recursive partitioning, that has already been described with Mark III (see Subsection 2.5.6.4). The choice of starting variables will be difficult (the method only consider linear combinations of variables chosen by the physicist). Another general statistical method is being used to find the shower centres. Integrating in depth, a two-dimensional peak finder is being used which is based on the well-known simplex algorithm used in minimizing problems (Blobel 1984a).

Fig. 2.25 The DELPHI HPC. The construction using lead ribbons between drift volumes, and the sense wire arrangement in the TPC style are apparent. The arrangement of readout pads is not shown in this drawing.

2.5.6.8 Jet finding in calorimeters

Jets are conglomerates of particles, charged and neutral, originating from the 'hadronization' of quarks or gluons ('partons'). When the energy of the underlying parton is sufficiently high, the particles travel close together in space, and the more energetic hadrons in a jet at least can be associated and serve as an approximate measurement for the original parton's energy vector.

The energy regimes of the CERN ISR and SPS (Super Proton Synchroton ($\bar{p}p$ colliders made them the first hadron colliders to give access to the observation of hadronic jets and a study of their properties. In the hadronic environment, hard collisions involve only part of the colliding particles, and parton debris obscure the hard collision products. General jet finders based on local clustering were first developed for these experiments. In the cleaner e^+e^- collisions, where jets were originally observed, global test statistics were used to describe hadronic events, and general jet finding algorithms have only recently been introduced.

We describe here four algorithms for jet finding, with different characteristics; a cluster finder relying on good matching of cells, a cluster finder using a rapidity/azimuth metric, another similar one relying on a metric of effective mass, and a windowing algorithm.

We will ignore here *global* jet variables, test statistics used extensively on e^+e^- machines, where only jet-related particles exist without the diquark remanants. Global jet variables are neither calorimeter-specific nor can they be classified under pattern recognition. These test statistics have proved useful in comparing observed event samples and physical models at e^+e^- colliders. In the presence of a hadronic background reaction, however, they would fail to reveal jet properties reliably. See Wu (1984) for a detailed discussion.

Jets by clustering adjacent cells

The UA2 central calorimeter is described in sufficient detail in Subsection 2.5.6.1. Jets are identified as calorimeter *cluster*. The clustering algorithm proceeds as follows: Cell energies are summed as in Subsection 2.5.6.1, over the three longitudinal compartments. All cell sums which share a common side (in the fine-grain mesh of θ/ϕ) and above a threshold energy of 400 MeV are joined into a cluster. If, inside a cluster, two or more local maxima are found which are separated by a valley deeper than 5 GeV, they are split.

2.5 Finding of particle showers

Clusters thus obtained typically contain three cells for a transverse energy of $E_T = 2\,\text{GeV}$, or ten cells for $E_T = 40\,\text{GeV}$. In order to eliminate the effects of gluon bremsstrahlung, lower-energy jets that are separated in the axis direction by a space angle less than $78°$ ($\cos\omega > 0.2$) from a jet of $E_T \geqslant 10\,\text{GeV}$, are merged with the latter (Appel, *et al.* 1986). This procedure is applicable in UA2 only because the analysis was limited to jets with axes in the central region ($50 > \theta > 130°$).

Jets by clustering, in two different metrics

(*a*) The calorimeter cells of UA1 have been described in Subsection 2.5.6.2. Jet finding in UA1 uses individual cell readings (Arnison, *et al.* 1983b). Each cell (electromagnetic or hadronic) with a signal above some threshold enters a jet algorithm with the measured energy. The energy is converted into an energy vector using as the direction vector the cell centre, a point derived from pulse height sharing, or position detectors where this is applicable. Cells are now considered in order of decreasing $E_T (= E|\sin\theta|)$. In a first cycle, only cells with E_T above a given threshold are considered. A closeness criterion is used for clustering which uses as the metric the variable

$$\Delta R = (\Delta\phi^2 + \Delta\eta^2)^{\frac{1}{2}} \tag{2.68}$$

where $\Delta\phi$ and $\Delta\eta$ are differences between the angular coordinates of cell centres: η is the 'pseudorapidity' variable, a θ-equivalent, differences in which are Lorentz-invariant; it is defined by

$$\eta = -\log\tan(\theta/2). \tag{2.69}$$

The highest E_T cell now initiates the first jet. To this 'initiator' cell are added all cells close in ΔR in order of decreasing E_T as long as $\Delta R < 1.0$. 'Added' must be interpreted to mean that at each merger, the jet axis, initially defined by the energy vector of the initiator cell, is updated by vectorially adding the energy vector of the new cell. If a cell cannot be merged by this criterion, it initiates a new jet. In a second cycle, all low-energy cells are associated to the nearest jet, as long as $\Delta R < 1.0$.

(*b*) A clustering algorithm used in e^+e^- collisions has been published by Bartel *et al.* (1986). It uses as the metric a parameter which is closely related to the effective mass between particles; for all combinations of particles i and j (both charged tracks and observed photons are considered), the closeness parameter

$$y_{ij} = M_{ij}/E_{\text{vis}} \tag{2.70}$$

228 *Pattern recognition*

where

$$M_{ij} = [2E_i E_j(1 - \cos \delta_{ij})]^{\frac{1}{2}}$$

is computed. The smallest y_{ij} will cause particles i and j to be combined into a single 'pseudoparticle'. The procedure is then repeated, until the smallest possible y_{ij} exceeds a threshold. In the above, E is the particle energy, δ is an angle between tracks, and E_{vis} is the sum of the energies of all particles. The variable y is Lorentz-invariant (like $\Delta\eta$ and $\Delta\phi$ above). The dependence of jet counting on the cutoff in y has been studied extensively. It should be noted that a very similar algorithm has also been used by the TASSO collaboration, using charged particles only (Burrows, private communication).

Jets by selection of a window

The AFS operated at the CERN Intersecting Storage Rings until this was closed down in 1984. The calorimeter (Botner, *et al.* 1987; Akesson, *et al.* 1985) was a uranium and copper sampling calorimeter which used scintillator as the active detector. It had full azimuthal coverage in the central region, with pseudorapidity (Equation (2.69)) coverage up to

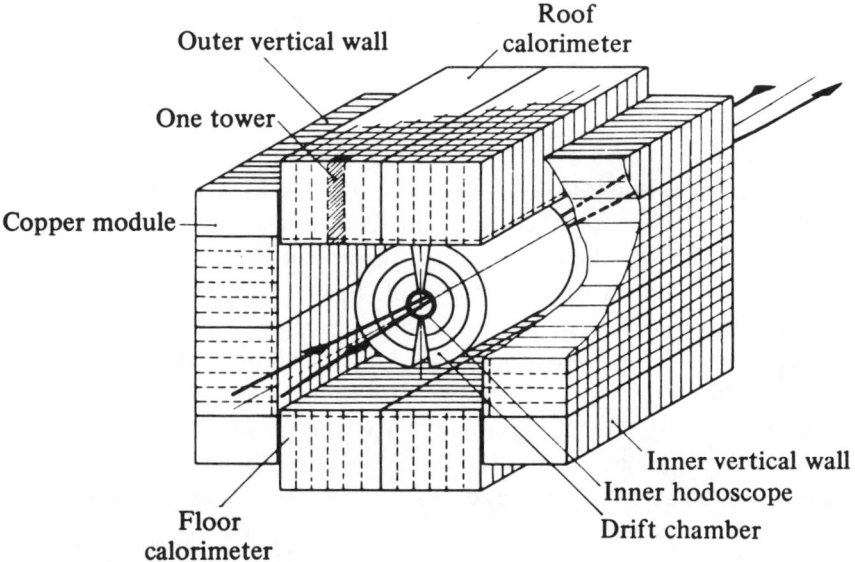

Fig. 2.26 The AFS calorimeter, used at the CERN ISR.

2.5 Finding of particle showers

$|\eta| = 1.2$. The walls were 118 cm from the interaction region, and were segmented into (square) towers of 20 cm side length. There were two segments, the first ('electromagnetic') one extending to $6X_0$ (Fig. 2.26).

The jet algorithm in the AFS (Akesson, et al. 1986) used a fixed rectangular window in η, ϕ space, of typical dimensions $\Delta\eta = 0.9, \Delta\phi = 45°$. Different window sizes were tried to verify the robustness of the results. The window was moved in the $\eta\phi$ plane until a maximum E_T was reached. After the elimination of all cells in the window of maximum E_T, the procedure was repeated on all the remaining cells, until eventually the maximum window E_T fell below a minimum (4 GeV).

2.5.6.9 Small correction and patchup algorithms

Correction for inadequate compensation

Abramowicz, et al. (1981) have shown that in a hadronic calorimeter with an e/π signal ratio very far from 1, a correction algorithm can be applied which partly recovers the loss in resolution. Their iron/scintillator sandwich had a very fine-grained readout, and they found, based on the analysis of recorded test beam data, that the resolution can be substantially improved by an empirical correction function. They determined a weight which depended on the ratio (highest signal in a layer)/(average signal of all layers), which was applied to all layers above a threshold. The weighting is found to result in a 30% improvement in the hadronic resolution, for beams of 140 GeV c^{-1}. The reasoning is, of course, based on fluctuations in the π^0 content of showers. A similar method has been reported by Braunschweig, et al. (1988), for a copper-lead-liquid argon calorimeter.

Corrections for leakage

In an extensive Monte Carlo study using GEISHA and EGS for showers from π^- beams in the range 2–300 GeV, Cashmore, et al. (1985) have shown that good longitudinal segmentation of a calorimeter can be used for leakage corrections. Where leakage occurs the longitudinal energy profile is shifted downstream, and rejection of showers with a high-energy fraction in the last segment results in improved resolution.

A similar Monte Carlo study for electromagnetic showers in a high-precision electromagnetic calorimeter (Grassmann and Moser 1985) again indicated the potential use of longitudinal segmentation in calorimeters. They attempted to find a correction function for leakage, and the study shows that in the low-fluctuation environment of an electromagnetic shower the technique might even induce experimenters to build calorimeters with less than full containment intentionally to save resources and reduce geometrical space.

Making up for projective readout

The TPC calorimeters at SLAC have been described in Buchanan (1982). The electromagnetic calorimeters in the TPC facility cover more than 90% of the full geometrical space. They are fine-grained electromagnetic gas calorimeters with projective readout throughout. The barrel part uses lead-aluminium-fibreglass laminates as the absorber ($10X_0$ deep), and a total of 40 gaps of 6 mm containing a Geiger chamber as the active medium, with sense wires along the beam and two cathode strip assemblies at $\pm 60°$. The entire arrangement is in rectangular cells, which are assembled in a hexagonal form. The end cap (pole tip) calorimeter is similarly made of 54 lead-steel-epoxy laminates interspersed with wires strung at 1.1 cm distance and operating in the proportional mode. The total depth of $17.5X_0$ is longitudinally subdivided into three equal parts.

In this projective geometry, a clever *triple product algorithm* is used (Hauptmann 1979), which is of a different class from that typically used in projections, in that it is global in approach (the author calls this a 'class II algorithm'). As the three projections were built with perfect matching they map out equal-size (triangular) cells, each of which is seen under three angles (projections called p, q and r). Each projection integrates over a group of cells. An integral over depth is also possible in all projections. The regularity of the readout thus results in a full coverage of the calorimeter face by cells of equal surface. If we call the projected readings of a single cell respectively E_p, E_q, and E_r, then we obtain a meaningful test statistic defined by

$$t = (E_p E_q E_r)^{\frac{1}{3}} \tag{2.71}$$

for each triangle, a quantity that has the dimension of energy (MeV).

This quantity associated to each cell, brings the projective readings back to the level of surface two-dimensional readings, and allows an algorithm to be used that is similar to that used for many other calorimeters: all cells with t below a specified threshold are excluded. Each triangular cell has an associated energy sum t_{sum}. Local maxima can be defined as cells for which t_{sum} is larger than for all surrounding cells. For all local maxima, and in order of decreasing t_{sum}, a positional fit to the best x-y impact point is then performed using (a variable number of) adjacent channels.

2.6 Identifying particles in ring imaging Cherenkov counters

The technique of Ring Imaging Cherenkov (RICH) counters is still a very recent one. A number of working devices already exist, others are in design or under construction. Therefore, there is at present only limited experience concerning the use of RICH counters in fully analysed experiments. Consequently, we cannot give in this chapter more than indications as to where it seems likely that this type of detector will find applications in the future. Also, in existing applications, RICH information is *not* used for recognizing the existence of particles, but only for adding information to an otherwise established measurement. We nevertheless consider the device sufficiently promising to warrant inclusion despite the rapid initial evolution it is undergoing.

2.6.1 The RICH technique

The RICH technique has been pioneered by Séguinot and Ypsilantis (1977). The idea is a simple one in principle, leading, however, to devices of substantial technical refinement. Somewhat like calorimeters, RICH devices are being employed in quite different contexts, and also allow quite different implementations.

Charged particles which enter a medium with a speed greater than that of light in this medium, undergo a deceleration due to polarization effects. Polarized molecules of the medium return to their ground state emitting the characteristic Cherenkov radiation *at a single emission angle* with the incident particle. The relation between the velocity of the particle v, the velocity of light in this medium v_0, the medium's refractive index n, and the Cherenkov emission angle δ, is given by

$$\left. \begin{array}{l} v_0 = c/n \\ \cos \delta = v_0/v = c/(vn) = 1/(\beta n). \end{array} \right\} \quad (2.72)$$

where c is the speed of light in vacuum.

If $v > v_0$, i.e. if the velocity β is above the 'Cherenkov threshold', measurement of δ becomes possible and corresponds to measuring β (or v). Combined with a momentum measurement it provides a way to identify particle masses. The generated photons all propagate at angle δ with respect to the original particle direction, but with random 'azimuth' around the incident particle, hence for a given point of origin all photons are found on the surface of a cone with a half opening angle δ. RICH devices attempt optimal measurement of δ and therefore of β, by directing the photons into a *photosensitive chamber*. This allows the reconstruction of the cone, if the origin of the photons is also known. Several problems have to be solved or understood.

Photons are emitted with varying wavelengths, and the photosensitive chambers as well as the light-transmitting surfaces, must be optimized to remain sensitive to the largest possible wavelength spectrum. This is typically in the ultraviolet region (wavelengths 150–250 nm, or 8.3–5.0 eV of energy). Light output and transmission are major problems. Different wavelengths also introduce 'chromatic aberrations': Radiating and transmitting materials have refractive index which is dependent on the wavelength. A finite wavelength width therefore results in a smearing of the signal, and a resolution limit in measuring β. Further sources of statistical errors result from the multiple Coulomb scattering of the radiating particle in the radiator, from its curvature if a magnetic field is present, from the finite resolution of the photosensitive chamber, and from aberrations of the focussing mirror, if one is used.

The energy loss due to ionization of the charged particle also traversing the chamber, is orders of magnitude higher than the energy lost due to Cherenkov radiation. Typically, high quantum efficiencies are needed in order to be able to reconstruct a ring image from the few photons impinging on the chamber. The simultaneous existence of a direct track signal in the same photosensitive chamber is an additional perturbence.

Cherenkov radiation occurs over a certain depth of a radiator, chosen for optimal identification. If this depth is more than a negligible layer, the origin of each photon will not be known, and a δ measurement will not be directly possible. Liquid or solid radiators are typically contained in layers thin enough to define the cone tip with a good approximation, and a cone development zone allows photons to move far enough from the particle axis for a useful measurement. Gaseous radiators, on the other hand, need substantial depth, and large-surface spherical mirrors will be needed to focus all parallel photons onto a single point, thus generating a unique circular image for all particles emitted under the angle δ over a full depth

2.6 Identifying particles in RICH counters

(see Ypsilantis (1981) for details). An illustration is given in Fig. 2.27. If the detection layer and the mirror surface are both concentric with normal incidence of the radiating track, the circular image will be centred about the track's impact point in the detection layer. If the magnetic field or the design make the track enter off-axis (e.g. in the 'cupola geometry' chosen for the DELPHI experiment at LEP, see Eek, et al. (1984)), then the ring may be found at some distance from the track's impact point.

Photosensitive chambers may be proportional or drift chambers, depending on the required signal speed and precision. A small ($< 10^{-3}$) admixture of triethylamine (TEA) or tetra(dimethylamine)ethylene (TMAE) to ordinary chamber gas mixtures will provide the photosensitive effect, but needs careful monitoring of gas cleanliness, concentration, and temperature. Barrelet, et al. (1982) discuss many of the problems associated with using photosensitive drift chambers in the RICH context. Arnold, et al. (1986) show some of the technical limits that have to be explored in order to operate a large RICH detector successfully.

Modern large-scale experiments have major RICH devices as part of their detector, in several cases using different types of radiator simulta-

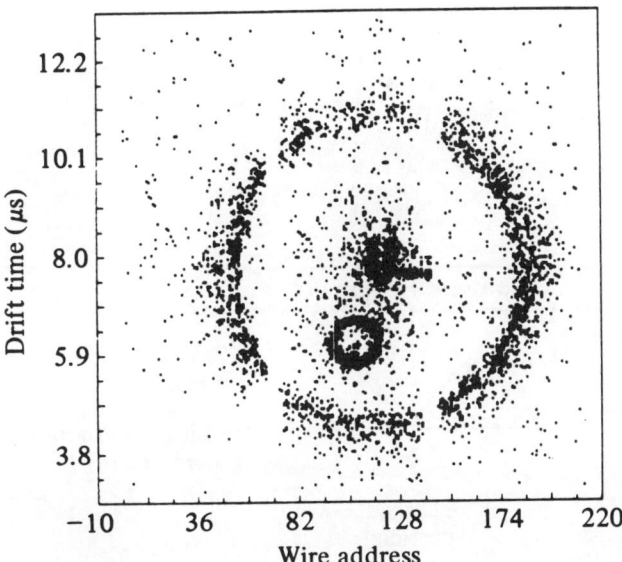

Fig. 2.27 A RICH image, with a centred (unfocussed or 'proximity focussed') ring around the track impact point, and an indirect (mirror focussed) ring slightly off centre. Superposition of multiple tracks with normal incidence.

neously. Fig. 2.28 shows the principle of using a single photosensitive chamber to provide RICH images from two different radiators, one a gas radiator using spherical mirrors for imaging, the other a liquid radiator producing a direct image (i.e. unfocussed or 'proximity focussed').

It should be noted that most tests and experiments so far have been performed over small solid angles. The assumption of normal incidence of the radiating track therefore was approximately true. Using unfocussed images with a large-angle liquid radiator, as is being proposed in some experiments, leads to photon images that are elliptic rings or that resemble conic sections, e.g. a hyperbola. No experience exists as yet in reconstructing such images.

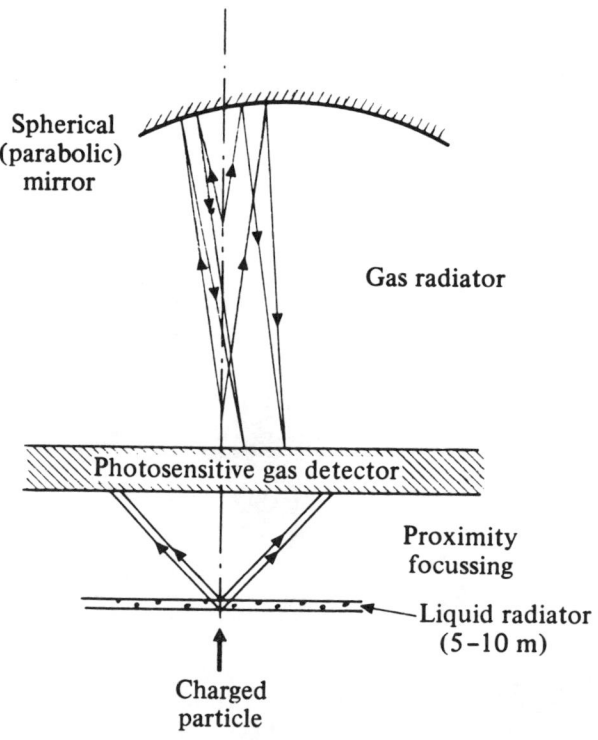

Fig. 2.28 The principle of a RICH counter with two radiators and a single photosensitive chamber.

2.6 Identifying particles in RICH counters

2.6.2 RICH applications

As in the previous section on calorimeter analysis, we will conclude the discussion of RICH devices by discussing a few practical applications. Some of them are proven, others are part of future projects. For more detailed reading, the source of information is indicated.

2.6.2.1 Electron identification in a test set-up

A 60 cm long gas radiator (C_2F_6) was used as a life-size test set-up in the CERN $\bar{p}p$ collider, in one of the 12 azimuthal sectors of an end cap of the UA2 experiment. The photosensitive chamber was a Time Projection Chamber (TPC see Subsection 2.1.4) with a surface area of $20 \times 20 \, cm^2$. Charged tracks were reconstructed independently, hits in an area around the track direct impact point in the chamber, typically lying outside the ring, were discarded. All hits in a ring around the predicted focussing centre were subjected to a circle fit. The position of the circle centre and the radius corresponded to the expected performance of the device, and showed the existence of a clean signal from electrons from γ conversions (from π^0) inside the set-up. No more than about 5 photons per relativistic electron were found on average, but the signal was clean, and allowed the prediction of a 10^{-4} e/π rejection capability for this type of counter (Botner, *et al.* 1987).

2.6.2.2 Particle identification in OMEGA and in E605

Qualitatively similar RICH counters have been operated successfully in fixed-target experiments at CERN and FNAL, and used for general particle identification. The OMEGA device used for experiment WA69 (Fig. 2.29) had a 6 m long gas radiator, an array of hexagonal spherical mirrors, and 16 TPCs (see Subsection 2.1.4) with a total surface of $4.2 \times 2.6 \, m^2$, using TMAE as the photosensitive agent. The aim of the device was the identification of hadrons in photoproduction experiments at high beam momenta (up to $200 \, GeV/c^{-1}$) at the CERN SPS.

The RICH counter in E605 was a 15 m long container, filled with pure helium gas; 7.5 photons were produced on average for a relativistic particle. An array of 4×4 mirror segments each $63.5 \times 66.0 \, cm^2$ in size projected the light onto multistep proportional chambers, whose readout consisted of 2 mm spaced anode wires, and two cathode wire planes at $\pm 45°$, also with

2 mm spacing. Single hits were read out in space by requiring a triple coincidence, remaining ambiguities were resolved by amplitude matching and pulse shape analysis (McCarthy, *et al.* 1986). A radius reconstruction error of $\sigma_R = 0.71$ mm was achieved. Circle fits were used with as few as two photon hits.

In both experiments, the RICH information is used in the late analysis stage only. The position and momentum of the tracks are known from measurements in the spectrometer geometrically preceding the RICH. This allows the prediction, for each hypothesis of particle mass, of a circular search band around a calculated ring centre, and the track's direct impact point. Background in the bands is estimated, event by event, from measuring the hit density outside the search bands. If the population inside the bands corresponds to the predictions, a radius fit is performed, and the resulting radius and hit density are used to obtain a probability for the hypothesis under consideration. Results obtained correspond to design criteria, and demonstrate the ultimate power of the RICH device (Apsimon, *et al.* 1986).

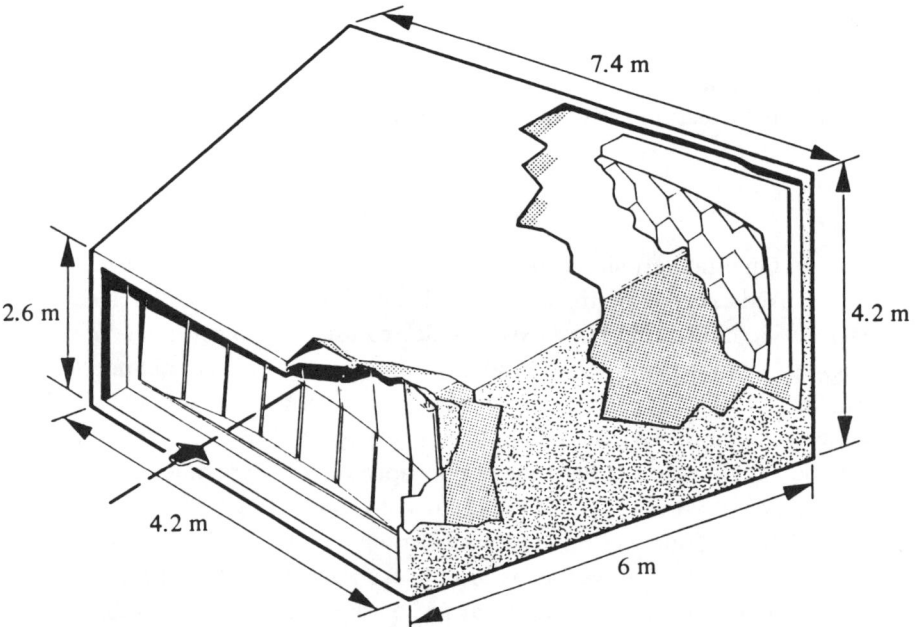

Fig. 2.29 The RICH detector in experiment WA69, at the OMEGA facility at CERN. TPCs are seen in the foreground, the array of hexagonal spherical mirrors is seen through the cutout.

2.6.2.3 Projected RICH is SLD, with two radiators

The SLD detector now under construction for running in the SLAC linear e^+e^- collider contains a Cherenkov Ring Imaging Detector (CRID) with two radiators, and is representative of other projects of the same type. The detector is described in the SLD Design Report (SLD 1984), and an overall picture of the CRID is shown in Fig. 2.30. The CRID uses a 1 cm layer of FC-72 (C_6F_{14}), contained by quartz windows, as a liquid radiator. FC-72 has (at 6.5 eV photon energy) a refractive index of 1.277, and hence a theoretical Cherenkov threshold of $\beta = 0.783$, or $p = 175$ MeV c^{-1} for pions. At $\beta = 1.0$, the Cherenkov ring in the photosensitive chamber will have a 17 cm radius (unfocussed image). The gaseous radiator is a 45 cm deep vessel of isobutane, with a refractive index of 1.0017, hence a pion momentum threshold of 2.4 GeV (the practically useful thresholds are somewhat higher). The focussed image from the gas radiator has a radius of 2.8 cm. The photosensitive chamber is of the TPC type (see Subsection 2.1.4), with drift times up to 20 µs (over 80 cm). The chamber gas

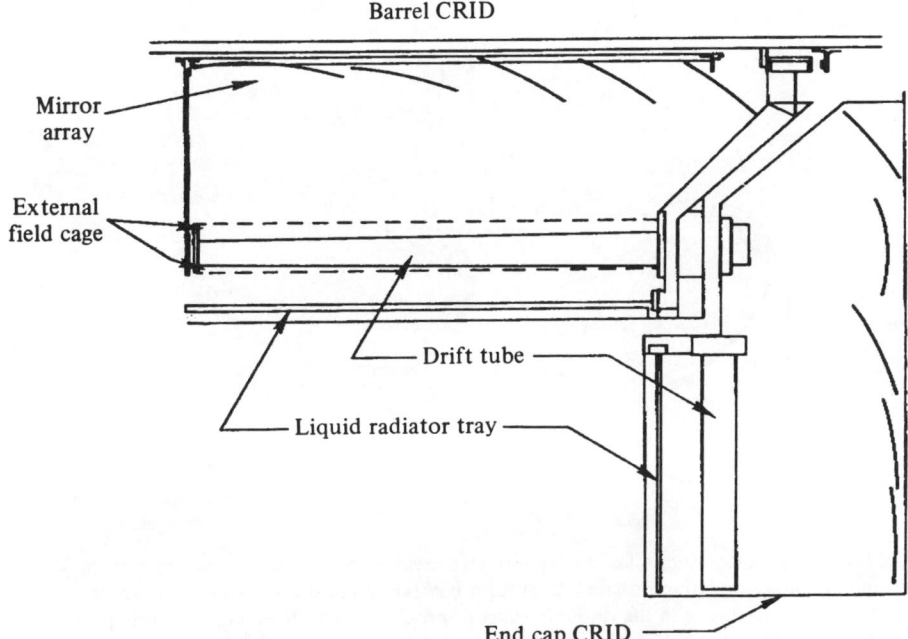

Fig. 2.30 The CRID detector in the SLD experiment at SLAC. Both barrel (top) and end cap part (right) operate with two radiators and a single chamber of the TPC type.

238 *Pattern recognition*

is a mixture of methane–isobutane, with TMAE as the photosensitive agent.

The following list gives the expected separation performances combining both radiators and assuming normal incidence and a 3σ level of separation.

e/π	0.2–7.0 GeV c^{-1}
μ/π	0.2–1.11 and 2.1–4.0 GeV c^{-1}
π/K	0.23–32 GeV c^{-1}
K/p	0.80–55 GeV c^{-1}

2.6.2.4 Projects for fast RICH devices

All analysed and nearly all planned experiments with RICH counters use these for a late and fine analysis in the final discrimination

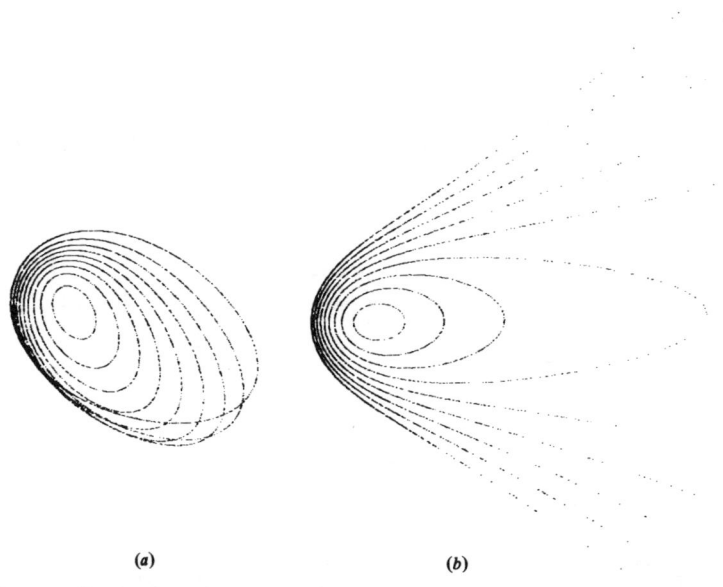

(a)   (b)

Fig. 2.31 Patterns from a fast RICH without focussing, generated at some angle to the normal in a simulation program. Curves for constant β are shown, with many tracks superimposed and with the causes for lateral spread removed, to enhance picture quality. (a) Pion–generated patterns in a magnetic field, at 15° incidence with respect to the normal. (b) Constant β without magnetic field, at 30°.

2.6 Identifying particles in RICH counters

stage, somewhere in the data reduction chain. In real time, full information is recorded on RICH-connected chambers, but no active use is made of this information. However this is not so in some experiments which are, at the time of writing, in an early preparatory phase. These plan to use RICH counters as active elements, ultimately in the trigger chain. One of these projects was studied for use at the LEAR antiproton storage ring at CERN (Bassompierre, *et al.* 1986).

Some simulated patterns arising from a RICH without focussing in the photosensitive detector are shown in Fig. 2.31. Other applications of this 'fast RICH' technique are in an advanced state of testing (Ypsilantis 1987).

3

Track and vertex fitting

3.1 The task of track fitting

Track fitting and the treatment of multiple scattering have had a long tradition in high-resolution *cosmic ray* and *bubble chamber experiments* (e.g. Rossi and Greisen 1941; Eyges 1948; Moorhead 1960; Gluckstern 1963; Laurikainen 1971a, b; Laurikainen, Moorhead, and Matt 1972). The importance and the feasibility of track and vertex fitting in the more complex environment of *experiments with electronics detectors* have been recognized over less than the last two decades. This was when detectors with good and well-defined resolution came into operation (Bouclier, et al. 1974; Charpak 1978; Charpak and Sauli 1987), and at the CERN ISR track and vertex fitting with rigorous attention to multiple scattering were successfully applied in the early seventies (e.g. Metcalf, Regler, and Broll 1973; Nagy, et al. 1978). Cross sections could be measured over more than seven decades with precise knowledge of the measurement errors on momentum transfer, knowledge of the losses due to χ^2 cuts, and a good estimation of the background. Important applications in set-ups with fixed targets were experiments at the CERN SPS, e.g. those in the OMEGA spectrometer (Lassalle, et al. 1980), or one where protons had been scattered on polarized protons (Fidecaro, et al. 1980). Since then trackfitting has been applied to several experiments at e^+e^- and $p\bar{p}$ storage rings where the selection of rare reactions in the presence of high multiplicities requires ultimate track separation and kinematic judgement, which can only be achieved by exact track and vertex fitting. On the other hand, the increased computing time required for complex events has sometimes discouraged the application of such rigorous reconstruction methods.

In the introduction to this book the experimental scenario of today's front-line high-energy physics experiments was described:

High multiplicities (20–50) due to the high collision energy obtained in

3.1 The task of track fitting

storage rings, and even much higher multiplicities in heavy ion collisions.

Momenta of particles in the final state ranging from a few hundred MeV up to several hundred GeV.

Very long spectrometers (up to 100 m) in fixed-target experiments, but with small spatial separation between tracks in the vertex region.

Complex modular track detectors combining different techniques and with different resolutions ranging from a few micrometres in microstrip detectors, allowing experiments to achieve the necessary precision for high momenta, to a few millimetres or even centimetres in calorimeters.

Resolutions which vary as a function of the impact point of the traversing particle.

Multiple scattering in the detector frames, supports, and cables, which competes with the detector resolution, at least, for lower energies and for secondary vertices, which lie inside the beam tube.

Large background from secondary activities of the particles to be measured, imposing optimal use of the information available for the final rejection of wrong associations of coordinates to tracks as well as of tracks to vertices.

And last, but not least, high event rates leading to a large amount of data, with a steadily increasing demand in triggering and mass storage devices.

From the physics point of view a few aspects should be mentioned again which are closely connected with the necessity for optimal track fitting:

Invariant masses are required to be determined with optimal precision and well-estimated errors, e.g. to determine the number of types of neutrinos from the width of the Z^0 mass ($p\bar{p}$ experiments), or to resolve the combinatorial problem when identifying short-lived particles from their decay products.

Secondary vertices must be fully reconstructed which requires optimal track bundling capability and optimal geometrical resolution in order to evaluate lifetimes of the order of 10^{-13} s (e.g. D mesons, B mesons, τ leptons).

Kinks must be located when a charged particle decays into a final state which again has only one charged particle, as in the case of escaping neutrinos (e.g. from the decay of charged pions or kaons).

Muon identification requires adequate treatment of the enormous amount of multiple scattering inside the muon filter, which is often made of ferromagnetic material (magnet yokes) from one to several metres thick.

The association of coordinates and sometimes directions into tracks and the bundling of tracks into vertices, with complete use of the information available from the track detectors to obtain the ultimate track resolution, also require the availability of a precise track model for the path of a charged particle in a magnetic field, and knowledge of the detector resolution of all the modules involved. Furthermore, the amount of material in the beam tube and the detector traversed by the particle must be known to a good approximation not only for correct and efficient treatment of multiple scattering (Metcalf, *et al.* 1973; Regler 1977) but also to account for the energy loss. Furthermore, appropriate matrix algorithms must be chosen, which perform the tasks mentioned above within a reasonable computing time.

Although a detailed track fitting is not be performed on all channels, even the selection of rare final states embedded in millions of similar topologies may require the detailed inspection of a high number of events if no clear trigger signal is available.

In this chapter first the use of the least squares method is discussed and justified (Section 3.2). Section 3.3 discusses the basic ingredients of trackfitting by the least squares method (track model, weight matrix, and minimization). Section 3.4 describes the bundling of individual tracks into vertices, and Section 3.5 gives some strategies for testing the track fitting module of the reconstruction program with Monte Carlo data and real data.

For the purpose of this chapter it is assumed that the problem of associating the often many hundreds of coordinates – as measured by different kinds of detectors – to track candidates (pattern recognition) has already been solved, with the exception of a few remaining ambiguities, although in special cases the tasks trackfitting and pattern recognition are interwoven. The same statement holds for vertex bundling. The task of reconstructing an interesting event out of a mass of measurements is, in general, split into several steps, depending on the complexity and modularity of the set-up (see also the previous chapter).

The final goal of this chapter is to help to obtain the ultimate geometrical resolution using the fastest and smallest program possible, and to confirm the hypothesis that the measurements which have been grouped together represent a particle's track up to the final confidence tests; the same also holds for track bundling into a common vertex. It should act as a guide for the physicist when he wants to adopt one or other method, with regard to speed, flexibility and trustworthiness (Eichiner and Regler 1981), and it should also help in the choice of an optimal detector arrangement during the experimental design. (Regler 1981), (Subsection 3.3.2.3).

Some symbols used in this chapter

(a, b), $[a, b)$ Interval
    $\{c\}$ Region
    $\langle c \rangle$ Expectation value (random vectors in italic)
    $\equiv$ Identical by definition
    $\cong$ Approximately equal
    $\sim$ Proportional
    $\leftarrow$ Is replaced by
    AB Matrix multiplication (but $\mathbf{A} \cdot (\mathbf{B}-\mathbf{C})$)
$(\mathbf{cc}^T)_{ij} = c_i c_j$

3.2 Estimation of track parameters

Before discussing track fitting in practice some theoretical estimation theory background will be reviewed.

3.2.1 Basic concepts

The task of track fitting, as described in Section 3.1, requires consideration of:

The behaviour of the detector, i.e. essentially its geometrical layout and accuracy ('resolution');

a mathematical 'model' which approximates the particle trajectories sufficiently well.

Knowledge of the detector behaviour allows an estimate of the track parameters with *minimum variance*. Furthermore, the resolution defines the scale in which decisions with respect to the track search must be taken, either to remove wrongly associated coordinates, or to reject a track string hypothesis as a whole.

The detector resolution can be estimated from theoretical considerations, from measurements in a **calibration experiment**, or from the tracks actually recorded when these have a sufficient number of constraints. In any case, the behaviour of the detector must be checked both during data acquisition and during off-line analysis by appropriate tests (Section 3.5).

In order to obtain the parameters (Equation (2.56)) defining a particle trajectory in a magnetic field (Subsection 2.2.3, Section 3.3), the path of the track must be known as a function of these parameters. In general it is appropriate to represent this 'track model' in *measurement space*: This is an n-dimensional space in the case of n measured coordinates along the track (e.g. if m planar detectors at fixed positions measure two coordinates each, then $n = 2m$). Of course, n is a widely varying function. A track is then given by a point on a *five-dimensional hyperplane* (a 'constraint surface'), according to the constraints given by the *equation of motion* of a charged particle in a magnetic field.

This 'measurement vector' deviates from the constraint surface due to randomly distributed *experimental errors* (Fig. 3.1):

$$\mathbf{f}: \mathbf{p} \to f_i(\mathbf{p}), \, i = 1, \ldots, n, \quad \text{or} \quad \mathbf{f}(\mathbf{p}) \tag{3.1}$$

i.e. $\mathbf{f}$ is a deterministic function of $\mathbf{p}$, and

$$c = \mathbf{f}(\mathbf{p}) + \varepsilon \tag{3.2}$$

where $\mathbf{p}$ is the 5-vector of track parameters (e.g. x, y at a reference surface $z_r = \text{const}$, and the momentum 3-vector), f_i is the ith coordinate of $\mathbf{f}$ corresponding to a track defined by $\mathbf{p}$, c is the random measurement vector (note that random quantities will only be denoted as such if necessary, e.g. for expectations, variances) and ε is the vector of the measurement errors. The task of track fitting is now to find a meaningful mapping $\mathbf{F}$ of the set of coordinate vectors, $\{c\}$, onto the set $\{p\}$ without bias and with minimum variance for the fitted parameters ($\overset{\shortmid}{\mathbf{p}}$ is the true value of $\mathbf{p}$, and $\tilde{p}$ the fitted value):

$$\tilde{p} = \mathbf{F}(c) \tag{3.3a}$$

$$\langle \tilde{p} \rangle = \overset{\shortmid}{\mathbf{p}} \tag{3.3b}$$

$$\sigma^2(p_i) \equiv \langle (\tilde{p}_i - \overset{\shortmid}{p}_i) \rangle \to \text{Minimum} \tag{3.3c}$$

where $\langle \ \rangle$ denotes an expectation value, or the average over many experiments.

For an individual track measured, $\overset{\shortmid}{\mathbf{p}}$ is a fixed but unknown quantity, while $\tilde{p}$ is a function of the random quantity c and is therefore also a random quantity. The variance has to be considered as describing the experimental errors for repeated measurement of the same track, and it may well differ for different $\overset{\shortmid}{\mathbf{p}}$. The fact that $\overset{\shortmid}{\mathbf{p}}$ itself usually has a random distribution from one track to another does not affect what follows. A simple example will be given at the end of subsection 3.2.2.

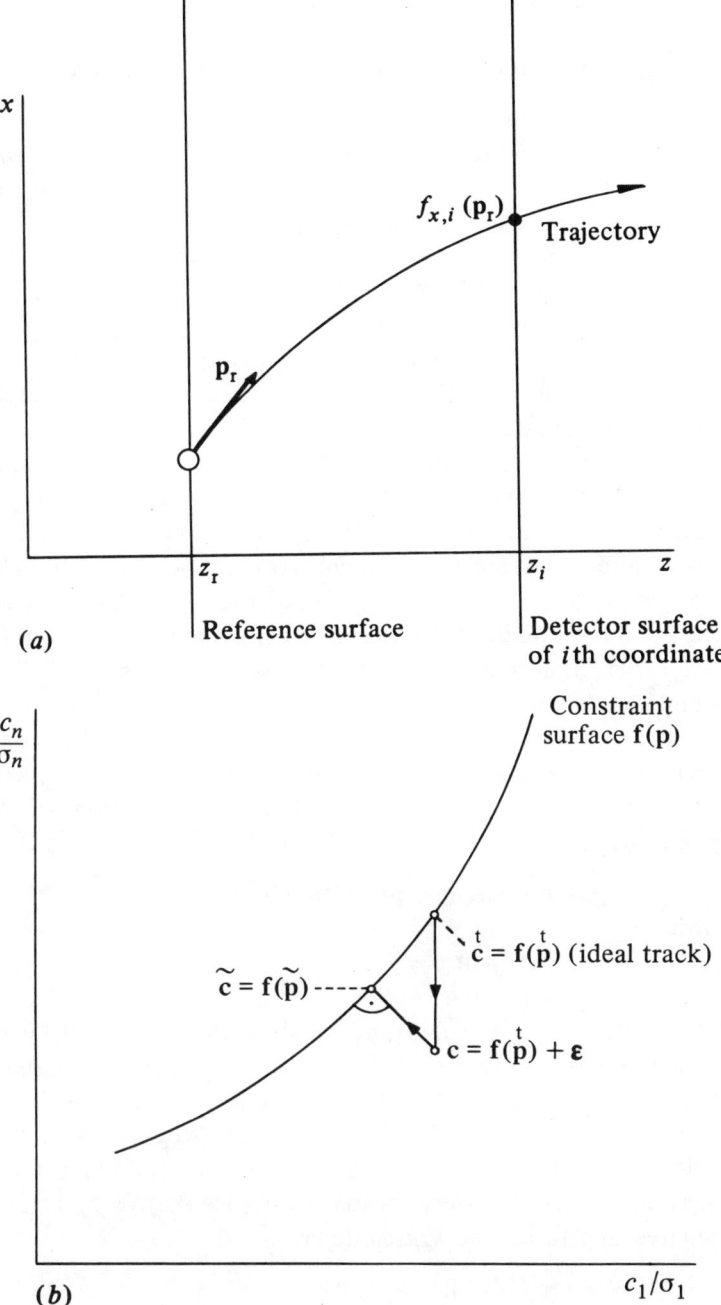

Fig. 3.1(a) Reference plane, the five-dimensional track parameter $\mathbf{p}_r$ and the 'impact functions' $f_i(\mathbf{p}_r)$, $i = 1, \ldots, n$. (b) The five-dimensional constraint surface; the fit minimizes the weighted Euclidian distance between a real measurement and an ideal track point lying on the constraint surface, $\mathbf{f}(\mathbf{p})$. The scale is the measurement error σ.

3.2.2 Global track fitting by the least squares method (LSM)

It turns out that the LSM best meets the requirements of track fitting, being simple, rather fast, and familiar to experimentalists. Its important statistical properties, together with its numerical simplicity, form the basis of the wide range of its application.

If the track model can be sufficiently well approximated by a *linear model* in the neighbourhood of the measurements, and if the errors vary sufficiently little with the track parameters that they can be considered as being constant in the neighbourhood of an individual track's path, then the LSM *estimation* has *minimum variance* among the class of linear and unbiassed estimates (for more details see Subsection 3.2.3).

In the linear approximation, the LSM is suitable for simple error propagation. Individual track fitting and track element merging can be done before vertex fitting, and in a third step the kinematical constraints can be imposed. If the measurement errors are not too far from 'Gaussian', or asymptotically in the limiting case of a large number of measurements, the square root of the variance can be considered as an interval estimation (see Table 3.1 (Subsection 3.2.3.1) and Table 3.2 (Subsection 3.2.3.2)), and a simple test quantity is obtained in parallel (the χ^2).

The track model is, in general, the set of solutions of the equations of motion, while the track model in the LSM is the *linear expansion* of the functions $\mathbf{f}(\mathbf{p})$ (Equation (3.1)) at a first approximation (the 'expansion point' $\overset{0}{\mathbf{p}}$):

$$\mathbf{f}(\mathbf{p}) = \mathbf{f}(\overset{0}{\mathbf{p}}) + \mathbf{A} \cdot (\mathbf{p} - \overset{0}{\mathbf{p}}) + O((\mathbf{p} - \overset{0}{\mathbf{p}})^2) \tag{3.4}$$

with

$$\mathbf{A} = \partial \mathbf{f}(\mathbf{p})/\partial \mathbf{p} \text{ at } \mathbf{p} = \overset{0}{\mathbf{p}}$$

The range of the interval $(\mathbf{p}_1, \mathbf{p}_2)$ in which the linear model is sufficiently close to the real track model *depends strongly on the appropriate choice of track parameters* (Subsection 3.3.1.2). This choice should be facilitated by a clever detector layout ('experimental design' (Regler 1981)).

In addition to the track model, the 'weight matrix' $\mathbf{W}$ must be evaluated (it is defined as the inverse of the covariance matrix $\mathbf{V}$), before the least squares 'ansatz' can be written down:

$$\mathbf{V} = \langle (\mathbf{c} - \langle \mathbf{c} \rangle)(\mathbf{c} - \langle \mathbf{c} \rangle)^\mathrm{T} \rangle \tag{3.5}$$

with

$$\mathbf{W} = \mathbf{V}^{-1}$$

3.2 Estimation of track parameters

where it is assumed that c has been corrected beforehand for a possible *bias*, (i.e. $\langle \boldsymbol{\varepsilon} \rangle = \mathbf{0}$ Equation (3.2)), or $\langle \boldsymbol{c} \rangle = \mathbf{f}(\overset{1}{\mathbf{p}})$.

In simple cases the measurement errors are *uncorrelated*, i.e. $\mathbf{W}$ is of the form

$$(\mathbf{W})_{ij} = \delta_{ij}/\sigma_j^2 \tag{3.6}$$

where σ_j is the standard deviation of the jth measurement ε_j, i.e. $\sigma_j = (\langle \varepsilon_j^2 \rangle)^{\frac{1}{2}}$, and δ_{ij} is the Kronecker delta. In general (mainly in the cases where *multiple scattering* must be considered) $\mathbf{W}$ will also have *off-diagonal* terms (see Subsection 3.3.2), and matrix inversion is necessary to obtain it numerically.

The LSM tries to minimize the function (see Subsection 3.2.3)

$$M = [\mathbf{f}(\overset{0}{\mathbf{p}}) + \mathbf{A} \cdot (\mathbf{p} - \overset{0}{\mathbf{p}}) - \mathbf{m}]^T \mathbf{W} \cdot [\mathbf{f}(\overset{0}{\mathbf{p}}) + \mathbf{A} \cdot (\mathbf{p} - \overset{0}{\mathbf{p}}) - \mathbf{m}] \tag{3.7}$$

where $\mathbf{m}$ is a '*realization*' of the random quantity c, i.e. a specific measurement. Differentiating M with respect to $\mathbf{p}$ and putting $\partial M/\partial \mathbf{p} = \mathbf{0}$ yields (with rank of $\mathbf{A} \geqslant$ dimension of $\mathbf{p}$, i.e. at least as many independent measurements as independent track parameters):

$$\tilde{\mathbf{p}} = \overset{0}{\mathbf{p}} + (\mathbf{A}^T \mathbf{W} \mathbf{A})^{-1} \mathbf{A}^T \mathbf{W} \cdot [\mathbf{m} - \mathbf{f}(\overset{0}{\mathbf{p}})] \tag{3.8}$$

As an example, the two track parameters will be evaluated in a Cartesian projection (y, z) for the case in which there is no magnetic field. Detectors are placed along the z axis, measuring coordinate y_i. Then the trajectory (Fig. 3.1a, with y instead of x) is a straight line:

$$y = a + zb \tag{3.9}$$

If the reference plane is chosen to be at $z = z_r = 0$, the linear track model (Equation (3.4)) is

$$\mathbf{f}(\mathbf{p}) = \mathbf{A}\mathbf{p} \tag{3.10a}$$

with

$$\mathbf{p} = \begin{pmatrix} a \\ b \end{pmatrix}, \quad A = \begin{pmatrix} 1 & z_1 \\ \vdots & \vdots \\ 1 & z_n \end{pmatrix}$$

or

$$f_i(p_1, p_2) = p_1 + z_i p_2 \tag{3.10b}$$

If multiple scattering is negligible, $(\mathbf{W})_{ij}$ is equal to δ_{ij}/σ_j^2 (Equation 3.6)) and Equation (3.8) has the form

$$\begin{pmatrix} \tilde{a} \\ \tilde{b} \end{pmatrix} = \begin{pmatrix} \sum 1/\sigma_i^2, & \sum z_i/\sigma_i^2 \\ \sum z_i/\sigma_i^2, & \sum z_i^2/\sigma_i^2 \end{pmatrix}^{-1} \begin{pmatrix} \sum m_i/\sigma_i^2 \\ \sum (m_i z_i)/\sigma_i^2 \end{pmatrix}$$

$$= \frac{1}{\det(\mathbf{A}^T \mathbf{W} \mathbf{A})} \begin{pmatrix} \sum z_i^2/\sigma_i^2, & -\sum z_i/\sigma_i^2 \\ -\sum z_i/\sigma_i^2, & \sum 1/\sigma_i^2 \end{pmatrix} \begin{pmatrix} \sum m_i/\sigma_i^2 \\ \sum (m_i z_i)/\sigma_i^2 \end{pmatrix}$$

(3.11)

where

$$\det(\mathbf{A}^T \mathbf{W} \mathbf{A}) = (\sum 1/\sigma_i^2)(\sum z_i^2/\sigma_i^2) - (\sum z_i/\sigma_i^2)^2.$$

If the σs do not depend on the impact coordinates, the matrices are constants of the set-up and, for efficient algorithms, must be evaluated only once, which can be an important feature for *fast real-time processing* (Chapter 1).

Note that if **W** is diagonal most of the matrix operations can be simplified by the following substitution ($a_{ij} \equiv (\mathbf{A})_{ij}$):

$$a_{ij} \leftarrow a_{ij}/\sigma_i$$
$$m_i \leftarrow m_i/\sigma_i \qquad (3.12)$$
$$f_i \leftarrow f_i/\sigma_i$$

However, some care has to be taken when calculating the 'pull quantities' (Equation (3.24)).

3.2.3 A few remarks on estimation theory

3.2.3.1 Generalities

In order to discuss the properties of the *estimators* appropriate for track fitting, it is assumed that the 'hypothesis' of the track model is correct (Subsection 3.3.1). This implies that:

the equation of motion can be solved with sufficient precision;
the magnetic field is measured sufficiently accurately and the information is easily accessible;
the material traversed is known, thus allowing the evaluation of energy loss and multiple scattering, which also require in some cases (low-energy particles, electrons) the identification of the particle;
no wrong measurements have been associated during the process of pattern recognition, a necessary condition for the consistency of the track model with the vector of measurements.

3.2 Estimation of track parameters

It is further assumed that the detector resolution is understood, in particular that the covariance matrix **V** of the measurement vector **c** is known (Subsection 3.3.2).

With the definition of the track model in coordinate space (Equation (3.1)) and the randomly displaced measurements (Equation (3.2)), one can obtain a *conditional probability density function* d describing the detector resolution, where the dominant variable is the difference between the measurement vector and the corresponding quantities of the 'undisturbed track' **f(p)**, namely $\varepsilon = \mathbf{m} - \mathbf{f(p)}$. However, this function quite often also depends explicitly on the measurement vector itself:

$$d(\mathbf{m}; \mathbf{\dot{p}}) = d'(\varepsilon, \mathbf{m}) = d'(\mathbf{m} - \mathbf{f(\dot{p})}; \mathbf{m}) \tag{3.13}$$

Before discussing the properties of the estimation of track parameters from a measured coordinate vector **m** we shall review a few of the requirements which one may ask of an estimator. The task is to estimate the track parameters themselves ('*point estimation*'), as well as an interval indicating how much the true parameters may possibly deviate from the estimated ones ('*interval estimation*'). Additional tasks will be to check whether the pattern recognition hypothesis associating this set of coordinates into a track segment was correct, and whether the detector behaviour (measurement error) has been understood properly.

In general, one distinguishes between *explicit* and *implicit* estimators. In the first case the estimator is defined by a function which has as the argument the measurement vector, while in the latter case the quantities to be determined are only implicitly defined.

The most common implicit estimator is given by the '*Maximum Likelihood Method*' (MLM). Its aim is to inspect the joint probability density function of the measurement vector (Equation (3.13)) and to compare the relative chance of obtaining a similar experimental result (measurement vector) for different assumptions for the track parameters **p** versus a 'zero hypothesis' $\mathbf{\overset{0}{p}}$. To establish this 'likelihood ratio' R, the quantity $\mathbf{\dot{p}}$ in Equation (3.13) is replaced by a running parameter **p**:

$$R_m(\mathbf{p}) = d(\mathbf{m}; \mathbf{p})/d(\mathbf{m}; \mathbf{\overset{0}{p}}) \tag{3.14a}$$

or, since the denominator is constant, just

$$L_m(\mathbf{p}) = d(\mathbf{m}; \mathbf{p}) \tag{3.14b}$$

The estimated parameters $\tilde{\mathbf{p}}$ will be obtained by the requirement that $L_m(\mathbf{p})$

(or, equivalently, $\ln(L_m)$) is a maximum or

$$\partial \ln L_m(\mathbf{p})/\partial \mathbf{p} = \mathbf{0} \tag{3.15}$$

which fulfills the requirements (Equation (3.3)).

If d' depends, at least *asymptotically*, only on the difference $\varepsilon = \mathbf{m} - \mathbf{f}(\mathbf{p})$, and if its distribution function can be approximated by a *Gaussian distribution*, then Equation (3.14) is fully determined by the corresponding covariance matrix $\mathbf{V}$ or by the *weight matrix* $\mathbf{W} = \mathbf{V}^{-1}$ (Equation (3.5)):

$$L_m(\mathbf{p}) = d_m(\varepsilon)$$
$$= (2\pi)^{-n/2}[\det(\mathbf{V})]^{-1/2} \prod_{i,j=1}^{n} \exp[\varepsilon_i(\mathbf{W})_{ij}\varepsilon_j/2] \tag{3.16}$$

where $\varepsilon = \mathbf{m} - \mathbf{f}(\overset{1}{\mathbf{p}})$ is a 'realization' of ε from Equation (3.2). In this case the MLM *coincides* with the LSM. Other, more general assumptions about how to perform track fitting by the MLM become impossible, in practice due to the computer time required.

If $\mathbf{W}$ is not constant but varies only slightly with $\mathbf{p}$, it may be sufficient to evaluate this explicit dependence on $\mathbf{p}$ once for a track at an approximate point $\overset{0}{\mathbf{m}}$, and to keep it constant for the remainder of the track fitting procedure.

There are cases such as the Multi-Wire Proportional Chambers (MWPC) where the main variable is the distance from the physical impact to the nearest wire, which is a deterministic step function rather than a probability density (see Subsection 3.3.2.1).

The linearized LSM yields an explicit estimator (Equation (3.8)). For most applications by an appropriate choice of $\mathbf{p}$, $\mathbf{f}(\mathbf{p})$ can be approximated by a linear expansion in the neighbourhood of a 'starting vector' $\overset{0}{\mathbf{p}}$ (Equation (3.4)). (For the nonlinear case an iteration procedure can be adopted, see Subsection 3.3.4.)

Recalling Equation (3.8)

$$\tilde{p} = \overset{0}{\mathbf{p}} + (\mathbf{A}^T\mathbf{W}\mathbf{A})^{-1}\mathbf{A}^T\mathbf{W} \cdot [c - \mathbf{f}(\overset{0}{\mathbf{p}})]$$

we will now discuss some properties of the LSM. (Those proofs which are not given here can be found in the literature (e.g. Kendall and Stuart 1967)):

(a) Bias

If the measurement vector $c = \mathbf{f}(\mathbf{p}) + \varepsilon$ (Equation (3.2)) is *unbiassed*, i.e. $\langle \varepsilon \rangle = \mathbf{0}$, then $\tilde{p}$ is also unbiassed (3.3a).

3.2 Estimation of track parameters

Proof: Since $\overset{0}{p}$ can be chosen arbitrarily, it can be set equal to $\overset{\downarrow}{p}$ without loss of generality

$$\langle \tilde{p} - \overset{\downarrow}{p} \rangle = \langle X\varepsilon \rangle = X\langle \varepsilon \rangle = 0 \qquad (3.17)$$

with

$$X = (A^T W A)^{-1} A^T W \quad \square$$

(b) Variance

The covariance matrix of $\tilde{p}$, i.e. the *error matrix* of the *fitted parameters*, is given by

$$C(\tilde{p}) = (A^T W A)^{-1} \qquad (3.18)$$

Proof:

$$\langle (\tilde{p} - \overset{\downarrow}{p})(\tilde{p} - \overset{\downarrow}{p})^T \rangle = \langle ((A^T W A)^{-1} A^T W \varepsilon)(\varepsilon^T W A (A^T W A)^{-1}) \rangle$$
$$= (A^T W A)^{-1} A^T W V W A (A^T W A)^{-1}$$
$$= (A^T W A)^{-1} A^T W A (A^T W A)^{-1} = (A^T W A)^{-1}$$

with

$$V = \langle (c - \langle c \rangle)(c - \langle c \rangle)^T \rangle$$
$$W = V^{-1}$$
$$\langle (X\varepsilon)(\varepsilon^T X^T) \rangle = X \langle \varepsilon \varepsilon^T \rangle X^T = X V X^T,$$
$$\langle \varepsilon \varepsilon^T \rangle = V = W^{-1} \qquad (3.5) \quad \square$$

(c) The 'Gauss–Markov theorem'

For a linear model, e.g. the track model of Equation (3.4), the LSM is the linear *unbiassed estimator* with *least variance* ('optimal') within this class if:

$\langle \varepsilon \rangle = 0$ (see above);
ε has a nonsingular covariance matrix;
the weight matrix W used to compute the minimum of M (Equation (3.7)) is the inverse of the covariance matrix, $V(\varepsilon)^{-1}$.

(d) Consistency

An estimator is called *consistent* if the following relation holds: $\lim_{n \to \infty} \tilde{p}_n = \overset{\downarrow}{p}$, where 'lim' means convergence in probability, i.e. for any $\eta > 0$ and any $\varepsilon > 0$ (not to be confused with the measurement error) and $N(\varepsilon, \eta)$ can be found such that prob $(|\tilde{p}_n - p| > \varepsilon) < \eta$.

Table 3.1. *Properties of the LSM*

Model	Errors	
	Non-Gaussian	Gaussian
Linear	Optimal among unbiassed linear estimators	Equivalent to the MLM with a sufficient statistic and therefore a minimum variance bound estimator even for a finite sample of measurements
Nonlinear	No general properties for a finite sample of measurements	Equivalent to the MLM and therefore asymptotically a minimum variance bound estimator

(e) The 'Cramér–Rao inequality'

For the variance of an unbiassed estimator $\tilde{p}$, the following inequality holds as a general lower bound for the precision which can be achieved in an experiment:

$$\sigma^2(\tilde{p}_i) \geq (\mathbf{I}_c^{-1})_{ii} \tag{3.19}$$

$$(\mathbf{I}_c(\mathbf{\acute{p}}))_{ij} = \left\langle \left. \frac{\partial \ln L_c(\mathbf{p})}{\partial p_i} \right|_{\mathbf{\acute{p}}} \times \left. \frac{\partial \ln L_c(\mathbf{p})}{\partial p_j} \right|_{\mathbf{\acute{p}}} \right\rangle_c$$

$$= \int \left\{ \left[\left. \frac{\partial L_m(\mathbf{p})}{\partial p_i} \right|_{\mathbf{\acute{p}}} \bigg/ L_m(\mathbf{\acute{p}}) \right] \times \left[\left. \frac{\partial L_m(\mathbf{p})}{\partial p_j} \right|_{\mathbf{\acute{p}}} \bigg/ L_m(\mathbf{\acute{p}}) \right] \right\} d(\mathbf{m}; \mathbf{\acute{p}}) \, d\mathbf{m} \tag{3.20}$$

where $\sigma^2(\tilde{p}_i) \equiv \text{var}(\tilde{p}_i) \equiv \langle (\tilde{p}_i - \acute{p}_i)^2 \rangle$ for an unbiassed estimator and $\langle \ \rangle_c$ denotes the expectation with respect to the random variable c. L comes from Equation (3.14b) and d from Equation (3.13). $I_c(\mathbf{p})$ is called the '*information*'. If the measurements are uncorrelated, d factorizes and so does L, and Equation (3.20) is just the sum over the information given by the individual measurements (thus justifying its name) together with Expression (3.19) for the variance. Equations (3.19) and (3.20) show that the information is just the square of the relative change (sensibility) of L due to a variation of $\mathbf{p}$, weighted with the probability density of the measurement vector, which accounts for the frequency of occurrence of a measurement. If in Expression (3.19) the equality is true, then $\tilde{p}$ is called a *minimum variance*

3.2 Estimation of track parameters

bound estimator (or the 'efficient estimator') (see Table 3.1). This is true only for a special class of probability density functions d, e.g. for Equation (3.16). Otherwise, one defines the relative efficiency of an unbiased estimator var $(\tilde{p})$/(minimum variance bound). If an estimator has minimum variance but a larger than the minimum variance bound, it is called *optimal*; this also applies when one is restricted to a special class of estimators (see the Gauss–Markov theorem above).

If the measurement errors are Gaussian (e.g. Equation (3.16)), the LSM is equivalent to the MLM. It follows that for the linear model the LSM is then a minimum variance bound (efficient) estimator even for a finite sample of measurements.

Table 3.1 summarizes the properties of the LSM. Asymptotically, in all four cases, the LSM is *normal, unbiassed* and *consistent* (e.g. Eadie, et al. 1971).

(f) 'Robustness'

An important property of an estimator is its *robustness*. An estimator is called robust if it is insensitive to measurements which deviate from the expected behaviour. There are two ways to treat such deviating measurements: one may either try to recognize them and then remove them from the data sample; or one may leave them in the sample, taking care that they do not influence the estimate unduly. In both cases robust estimators are needed.

In practice two types of deviating measurements have to be distinguished namely:

The measurement is generated by the particle under investigation, but with an error much larger than expected. Such a measurement, for which the probability of occurrence is low, is called an '*outlier*'. It can be caused by several processes, but the most common one is the creation of energetic electrons in gasous detectors (δ-rays). Sometimes it can be detected by a larger 'cluster size' (Subsection 3.3.2.1).

The measurement is a noise signal which was associated to the track by the pattern recognition; it can either be genuine detector noise, or picked up from another particle's set of measurements ('ghosts').

In the framework of the LSM the outlier problem could be handled globally by modifying the error matrix, thus ensuring the proper propagation of errors. However, in most cases this would spoil the overall resolution, and the χ^2 would be distorted, so that a χ^2-cut would be nontransparent and could spoil the overall normalization and introduce a bias (Subsection 3.2.4).

An algorithm should be made *robust* by taking some reasonable default action when 'normal proceeding' does not seem appropriate. Robust procedures. compensate for systematic errors as much as possible, and indicate any situation in which a danger of not being able to operate reliably is detected. As robust estimators are not normally fully efficient, a compromise must be found between robustness and the final resolution of the physical quantities to be estimated.

In particle physics many pragmatic approaches have been invented to solve the outlier problem such as removal of the largest absolute reduced residual (Equation (3.24)) if it exceeds a few standard deviations for position-sensitive devices ('truncation') and removal of a fixed percentage from the signal sample in the case of a pulse height analysis (e.g. for a dE/dx measurement, see Subsection 3.3.1.5). This latter method is called the 'censored mean'.

For a more systematic study on the detection of outliers in track fitting see (Frühwirth 1988).

3.2.3.2 The LSM in practice

Many modern particle track detectors collect very large samples of data, and the asymptotic properties of the LSM are therefore of real interest in practice. The convergence properties depend mainly on the approximate linearity of the track model and on the deviations of the measurement error density function from a Gaussian function (Equation (3.16)).

An important advantage of the LSM is that the covariance matrix of the fitted quantities (Equation (3.18)) is easily obtained for each individual track, thus allowing error propagation and subsequent fits with additional constraints (e.g. common vertex, relativistic energy–momentum conservation; see also Section 3.4).

Another interesting property in the linear model is the *equivalence* of the LSM to the class of Gaussian errors with respect to the properties of the forms of *quadratic averages* (i.e. of variances, $\langle \chi^2 \rangle$, etc.), which thus allows the definition of reduced test quantities with a variance of 1 or a 'pseudo χ^2 distribution' with an average equal to the corresponding number of degrees of freedom. Using the correct variances, and with the asymptotic property of being normal, in many cases reduced quantities are well approximated by normal distributions, and the effective pseudo χ^2

3.2 Estimation of track parameters

Table 3.2. *Confidence interval for Gaussian-distributed quantities*

n	prob($|\overset{t}{p}_i - \tilde{p}_i| < n\sigma_i$)(%)	Confidence level (%)
1	68.3	31.7
2	95.4	4.6
3	99.7	0.3

distribution is well approximated by a real χ^2 distribution. These quantities and their use for test purposes will be discussed shortly.

Use of this equivalence has already been made when calculating the error matrix of the fitted parameters by error propagation (Equation (3.18)). For a linear model and Gaussian errors, error propagation again gives Gaussian-distributed quantities (Fig. 3.2), and therefore this matrix is a direct measure of the interval limiting the deviation from the true value with a given probability (*'interval estimation'*), see Table 3.2 with $\sigma_i = (\mathbf{c}(\tilde{p}))_{ii}$ from Equation (3.18). The estimated interval $(\tilde{p}_i - \sigma_i, \tilde{p}_i + \sigma_i)$ is also called the *'confidence interval'*.

If no assumption is made about the distribution of the measurement

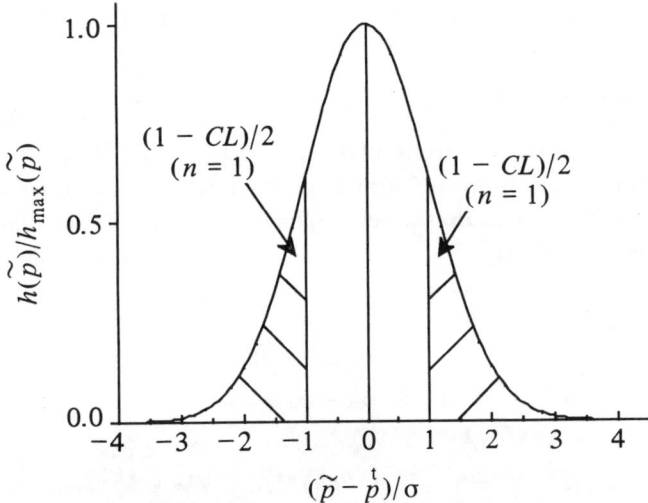

Fig. 3.2 The probability density function for $\tilde{p}$ (i.e. for repeated experiments) in the linear Gaussian case (or in the asymptotic limit). The shaded region is the complement to the confidence level for $\pm 1\sigma$. In practice σ is taken from the likelihood function, or, in the nonlinear finite case, from Monte Carlo simulation.

errors, only an upper limit can be given for the probability that $\overset{\iota}{p}_i$ is not within $\tilde{p}_i \pm \sigma_i$:

$$\text{prob}(|\overset{\iota}{p}_i - \tilde{p}_i| > n\sigma_i) \leqslant 1/n^2 \tag{3.21}$$

This is called the 'Chebycheff inequality'. In practice this limit is quite unsatisfactory compared with Table 3.2, and more restrictive limits exist; often the final meaning of σ must be gained from Monte Carlo or from the empirical 'χ^2' distribution of the experimental data. For a large number of measurements advantage can be taken of the asymptotic normality.

A classical check for the proper use of the LSM is the inspection of the *reduced residuals* ('stretch functions' or 'pull quantities'). Again it is assumed that the parameter dependence is not too far from being linear: During the fit, the parameters undergo an adjustment, and so does the track model in coordinate space. As the coordinates are the directly measurable quantities, it is advisable to inspect the differences between these measurements **m** and the readjusted or improved ones, i.e. $\mathbf{f}(\tilde{\mathbf{p}}) = :\tilde{\mathbf{c}}$. The difference $m_i - \tilde{c}_i$ is called the ith residual. The variance of $\tilde{c}$ is given by simple error propagation:

$$\mathbf{C}(\tilde{c}) = (\mathbf{A}(\mathbf{A}^\mathrm{T}\mathbf{W}\mathbf{A})^{-1}\mathbf{A}^\mathrm{T} \tag{3.22}$$

where **A** is given by Equation (3.4) and **W** by Equation (3.5).

The covariance of the residual vector $r = c - \tilde{c}$, $\mathbf{C}(r)$, is smaller than the measurement error $\mathbf{V}(c)$

$$\mathbf{C}(r) = \mathbf{V}(c) - \mathbf{C}(\tilde{c}) \tag{3.23}$$

where the minus sign in front of $\mathbf{C}(\tilde{c})$ reflects the positive correlation between c and $\tilde{c}$ (Equation (3.22)), and approaches the fitted values towards the measurements. (Note that the covariance matrix is always a symmetric, positive definite matrix.)

Proof

$$\begin{aligned}
r = c - \tilde{c} &= \mathbf{f}(\overset{\iota}{\mathbf{p}}) + \varepsilon - \mathbf{A} \cdot (\tilde{p} - \overset{\iota}{\mathbf{p}}) - \mathbf{f}(\overset{\iota}{\mathbf{p}}) = \varepsilon - \mathbf{A}\tilde{p} + \mathbf{A}\overset{\iota}{\mathbf{p}} \\
&= \varepsilon - \mathbf{A} \cdot [(\mathbf{A}^\mathrm{T}\mathbf{W}\mathbf{A})^{-1}\mathbf{A}^\mathrm{T}\mathbf{W}\varepsilon + \overset{\iota}{\mathbf{p}}] + \mathbf{A}\overset{\iota}{\mathbf{p}} \\
&= [1 - \mathbf{A})\mathbf{A}^\mathrm{T}\mathbf{W}\mathbf{A})^{-1}\mathbf{A}^\mathrm{T}\mathbf{W}]\varepsilon = \mathbf{Y}\varepsilon \\
\mathbf{C}(r) = \mathbf{Y}\langle \varepsilon\varepsilon^\mathrm{T}\rangle \mathbf{Y}^\mathrm{T} &= \mathbf{V} + \mathbf{A}(\mathbf{A}^\mathrm{T}\mathbf{W}\mathbf{A})^{-1}\mathbf{A}^\mathrm{T}\mathbf{W}\mathbf{V}\mathbf{W}\mathbf{A}(\mathbf{A}^\mathrm{T}\mathbf{W}\mathbf{A})^{-1}\mathbf{A}^\mathrm{T} \\
&\quad - \mathbf{A}(\mathbf{A}^\mathrm{T}\mathbf{W}\mathbf{A})^{-1}\mathbf{A}^\mathrm{T}\mathbf{W}\mathbf{V} - \mathbf{V}\mathbf{W}\mathbf{A}(\mathbf{A}^\mathrm{T}\mathbf{W}\mathbf{A})^{-1}\mathbf{a}^\mathrm{t} \\
&= \mathbf{V} + \mathbf{A}(\mathbf{A}^\mathrm{T}\mathbf{W}\mathbf{A})^{-1}\mathbf{A}^\mathrm{T} - 2\mathbf{A}(\mathbf{A}^\mathrm{T}\mathbf{W}\mathbf{A})^{-1}\mathbf{A}^\mathrm{T} \\
&= \mathbf{V} - \mathbf{A}(\mathbf{A}^\mathrm{T}\mathbf{W}\mathbf{A})^{-1}\mathbf{A}^\mathrm{T} = \mathbf{V} - \mathbf{C}(\tilde{c}) \quad \square
\end{aligned}$$

3.2 Estimation of track parameters

The reduced residuals can now be calculated (Equation (3.23)). As mentioned above they should be distributed with mean 0 and variance 1:

$$p_i = \frac{m_i - \tilde{c}_i}{[\sigma^2(c_i) - \sigma^2(\tilde{c}_i)]^{\frac{1}{2}}} \tag{3.24}$$

3.2.3.3 The χ^2 distribution

Finally the χ^2 *distribution* will now be discussed briefly. For measurement errors with a given n-dimensional Gaussian distribution (Equation (3.16)), the χ^2 distribution is the probability density function for the weighted sum of the squares of the measurement error. It is obtained by transformation into spherical coordinates $((\chi^2)^{\frac{1}{2}}, \omega)$ and by integrating over the $(n-1)$-dimensional solid angle:

$$g(\chi^2) = (2\pi)^{-n/2} [\det(\mathbf{V})]^{-\frac{1}{2}} \int e^{-\chi^2/2} \frac{\partial \varepsilon_1, \ldots, \varepsilon_n}{\partial \chi^2 \partial^{n-1} \omega} d^{n-1} \tag{3.25a}$$

where χ^2 is the square of the absolute value of the reduced radius vector (the weighted sum of squared measurement errors $= \sum_{i,j=1}^{n} \varepsilon_i(W)_{ij}\varepsilon_j$) and ω is the angular contribution. After a unitary transformation $\mathbf{T}$ diagonalizing $\mathbf{W}$ $((\mathbf{TWT}^{-1})_{ij} = \delta_{ij}/\sigma_j'^2, \sigma_j' = \sigma(\varepsilon_j'), \varepsilon_j' = (\mathbf{T}\varepsilon)_j)$ and with the reduced quantities $\varepsilon_j'' = \varepsilon_j'/\sigma_j'$ one obtains: $\chi^2 = \sum_{j=1}^{n} \varepsilon_j''$.

It follows from *spherical symmetry* and from *normalization* (with $\chi^{n-1} d\chi = (\chi^2)^{(n-2)/2} d\chi^2/2$ and $|\mathbf{T}| = 1$), or from the evaluation of the Jacobian matrix that

$$g_n(\chi^2) = e^{-\chi^2/2} (\chi^2)^{(n-2)/2} / 2^{n/2} / 2\Gamma(n/2) \tag{3.25b}$$

and

$$\langle \chi^2 \rangle = n \tag{3.25c}$$

For non-Gaussian errors, the χ^2 distribution is distorted with respect to Equation (3.25b), but in the linear model the expectation value is conserved. The following relation holds for the probability density function:

$$\chi^2 = \chi_{n_1} + \chi_{n_2} \text{ implies } g_{n_1+n_2}(\chi^2) \tag{3.26a}$$

Furthermore, in the LSM, the χ^2 distribution is defined by the number of degree of freedom (e.g. Blobel 1984b):

number of degrees of freedom = number of (independent) measurements (plus other constraints) − number of (independent) adjustable parameters   (3.26b)

The notion of 'number of degrees of freedom' can be misleading; here it means the number of degrees of freedom for adjusting $\tilde{c}$ other than just choosing **m** which would be the natural choice if no additional constraints were given by the track model ('zero-constraint fit').

3.2.4 Test for goodness of fit

The test for goodness of fit consists of two separate parts, but with strong interdependence in practice.

(a) Pull quantities In order to obtain a correct estimation of the track parameters, three conditions must be fulfilled (see above):

The track model must be correct (see also Subsection 3.3.1);
The covariance matrix of the measurement errors must be correct (Subsection 3.3.2);
The reconstruction program must work properly (Section 3.5).

A common way to test these requirements is to check the variances of the *pull quantities* (Equation (3.24)). A regular check of these quantitites also shows whether the detector behaviour is stable with time or not. Furthermore, the pull quantities also give a good indication of any individual track detector which is beginning to deteriorate, while if only the global χ^2 is known, this recognition may be difficult. A more global but less sensitive check is to observe the mean of χ^2 i.e. $\langle\chi^2\rangle$ and, in the case of Gaussian errors, even the whole χ^2 distribution itself. In this latter case it is convenient to plot the corresponding value of the cumulative distribution function $G(\chi^2)$ which is uniformly distributed (Fig. 3.2) between 0 and 1:

$$G_k(\chi^2) = \int_p^{\chi^2} g_k(x)\,dx \qquad (3.27)$$

When fitting the five parameters of a particle trajectory in a magnetic field, k is usually the number of measured coordinates minus five, because each independent coordinate contributes one constraint. $G_k(\chi^2)$ allows one to make a single plot independent of the number of degrees of freedom, i.e. independent of the number of measured coordinates of an individual track.

So far it has been implicitly assumed that no wrong coordinates were associated by the pattern recognition, which is, in fact, an additional condition for the correctness of the track model. In order to ensure this in practice, only selected event topologies with little background and low

3.2 Estimation of track parameters

track multiplicities (e.g. elastic scattering) should be chosen to tune the fitting program and to test and calibrate the detector. However, it must be guaranteed that such a sample is still representative: the detector behaviour usually deteriorates when there are high track multiplicities, and so does the behaviour of the pattern recognition program; or the low-energy behaviour (i.e. multiple scattering and energy loss) may not be tested well enough by such a sample; also the event rate might influence the resolution. It is only after all these conditions are fulfilled that one may have confidence in the covariance matrix of the fitted quantities, $C(\tilde{p}) = (A^T W A)^{-1}$.

(b) χ^2 *test* The second check concerns a proper test of the correctness of the association of the measured coordinates into a track. The 'Null hypothesis', H_0, assumes that all coordinates belong to the track. In track fitting the 'alternative hypothesis' H_1 is usually just the logical complement of H_0. However, in practice one distinguishes two classes of alternative hypotheses: (a) that several coordinates have been associated although there was no track (this is called background), and (b) that only one coordinate (or a small fraction of coordinates) has been associated wrongly to a real track.

A suitable quantity to decide if H_0 is tenable (if no alternative hypothesis has been estabilshed) such a test is called a 'significance test' is obtained automatically by the LSM i.e. the χ^2. When the null hypothesis is true, the expectation value of the χ^2 distribution is just the number of degrees of freedom (Equations (3.25b), (3.26b)); in the linear model and with Gaussian errors the minimum of M (Equation (3.7)) obeys a real χ^2 distribution (Equation (3.25a)) according to Cockran's theorem (Kendall and Stuart 1967, p. 84), and the χ^2 is *uncorrelated to the fitted track parameter $\tilde{p}$*.

As almost any kind of background has a tendency to have a larger χ^2 (this requires of course an overdetermined track, i.e. more independent measurements than independent parameters (Equation (3.26b))), a common way to test H_0 is to define a decision criterion making use of the χ^2 value of an individual track, e.g. to give up a certain percentage of good tracks with large χ^2, in order to eliminate the background (see also Section 4.2).

A function of random observables is called a 'statistic'. When using a test variable, e.g. χ^2, for the goodness of fit it is called a 'test statistic'. In practice one preassigns a *critical* χ^2 value, dividing the interval $(0, \infty)$ into two regions: and acceptance region $[0, \chi_c^2)$ and a critical (or rejection) region $[\chi_c^2, \infty)$ which implies the selection in the coordinate space of a certain region of acceptance, the set $\{m\}_{Acc}$ out of the set of all $\{m\}$ around the constraint surface (Fig. 3.1(b)), and its complement, the rejection region

$\{\mathbf{m}\}_{\text{Rej}}$. The loss α caused by this 'χ^2 cut' (i.e. the rejection of tracks with $\chi^2 \geq \chi_c^2$) is given by

$$\alpha = \int_{\chi_c^2}^{\infty} g_k(x)\,dx = 1 - G_k(\chi^2) = \text{prob}(\{\mathbf{m}\}_{\text{Rej}}) \quad (3.28)$$

with $G_k(\chi^2)$ as defined in Equation (3.27).

The probability α of rejecting real tracks, i.e. with H_0 true $(\text{prob}(\{\mathbf{m}\}_{\text{Rej}}|H_0 = \text{true}))$ – often quoted as percentage 100α – is called the '*level of significance*' or the '*size*' of the test. In practice α should be kept as small as possible; according to the level of background, it should lie between 10% and 0.1%. If, due to high background contamination, α must be chosen to be larger than this, the danger of a bias must be considered with great care. Tables for $\alpha(\chi_c^2, k)$ are given in most text books on statistics, and practical implementation of Formula (3.28) may be found in scientific computer program libraries (e.g. NAGLIB and CERNLIB). The quantity $1 - \alpha$ is called the '*confidence level*' or '*confidence coefficient*'. Rejecting the track hypothesis H_0 when it is true is called a 'type I error' or an error of *the first kind* (see also Section 4.2).

The experimental χ^2 distribution g_{exp} is the weighted sum of two probability densities: A real χ^2 distribution and a 'χ^2 distribution as obtained from the background', i.e. of wrongly associated coordinates (g_{bg}) where the χ^2 was evaluated using the LSM under the false assumption $H_0 = \text{true}$ (Fig. 3.3):

$$g_{\text{exp}}(\chi^2) = g(\chi^2|H_0 = \text{true})\,\text{prob}(H_0 = \text{true})$$
$$+ g_{\text{bg}}(\chi^2|H_0 = \text{false})\,\text{prob}(H_0 = \text{false}) \quad (3.29)$$

where bg denotes background.

Accepting H_0 when it is false is called a 'type II error' or an error of *the second kind*, and the corresponding probability is

$$\beta = \int_0^{\chi_c^2} g_{\text{bg}}(x|H_0 = \text{false})\,dx \quad (3.30)$$

The quantity $1 - \beta$ is called the '*power*' of the test. The total contamination is given by the ratio of background to accepted events:

$$\text{Contamination} = \frac{\beta\,\text{prob}(H_0 = \text{false})}{(1-\alpha)\,\text{prob}(H_0 = \text{true}) + \beta\,\text{prob}(H_0 = \text{false})} \times 100\% \quad (3.31)$$

H_1 can be more elaborate than just the complement of H_0, e.g. when a

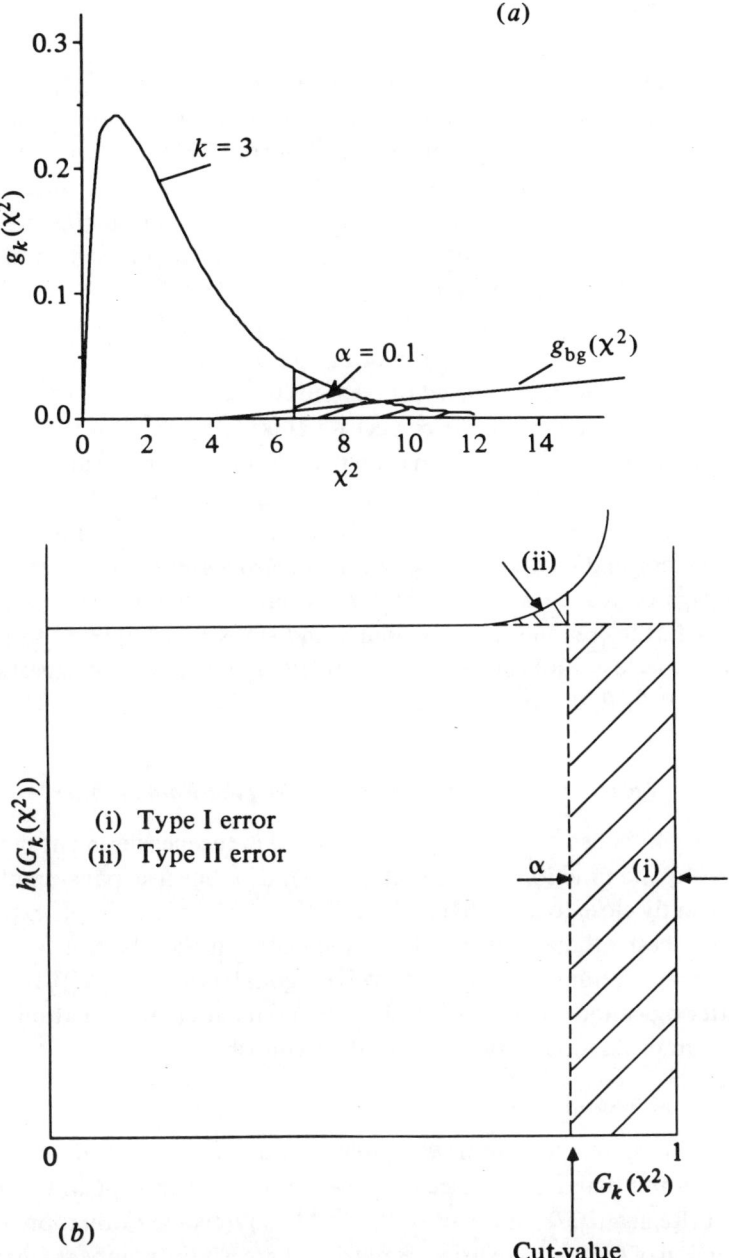

Fig. 3.3(a) χ^2 distribution for three degrees of freedom. The level of significance α of 10% corresponds to a confidence level of 90%. The experimental χ^2 distribution is the weighted sum of g_k and g_{bg}. (b) The density function as obtained by the inverse of the distribution function $G_k(\chi^2)$, with $G'_k(\chi^2) = g_k(\chi^2)$. The enhancement on the right side is due to wrongly associated coordinates, giving an excess of large χ^2s.

coordinate could be assigned to two different tracks or when looking for a primary and a secondary vertex, a second test quantity can be chosen for the LSM fit under the assumption that H_1 is true. This will reduce the risk of a wrong decision, but can easily cause a bias: namely, that the loss is no longer a *representative sample* for the all track topologies of good events, which means that differential cross sections can be biassed by different losses for different momenta.

If the background is such that its suppression would cause too important a loss of tracks, and if no other information is obtainable, the only solution is to improve the overall detector resolution, keeping the χ^2 distribution of good events stable when the correct weights are applied for the LSM, but shifting the pseudo χ^2 obtained from background to larger values. Another way to improve the background separation would be to add additional detectors, but there are fundamental limits to this due to multiple scattering energy loss, and secondary processes ('*overinstrumentation*'), and also more practical ones such as the space, money, and the manpower available.

For further reading on probability and statistics as applied to particle physics, see e.g. Blobel (1984b), Brandt (1984), Frodesen, Skjeggestad, and Tøfte (1979) Eadie, et al. (1971).

3.2.5 Recursive track fitting by the LSM (the Kalman filter)

In Subsection 3.2.1 the track model in the measurement space was given by the function $\mathbf{f}(\mathbf{p})$ (Equation (3.1)). If a linear expansion of $\mathbf{f}(\mathbf{p})$ is sufficiently close to an individual track, the LSM gives a simple expression for the best estimate of $\mathbf{p}, \tilde{\mathbf{p}}$ (Equation (3.8)). In the absence of multiple scattering, $\mathbf{V}(\varepsilon)$ and also therefore $\mathbf{W}$ is diagonal (Equation (3.6)). If multiple scattering cannot be neglected, the measurement error (Equation (3.2)) is the sum of essentially two independent contributions:

$$\varepsilon = \varepsilon_{\text{detector}} + \varepsilon_{\text{ms}} \qquad (3.32)$$

where ms stands for multiple scattering (Subsections 3.3.1.4 and 3.3.2.3). $\mathbf{V}(\varepsilon_{\text{ms}})$ is not diagonal because multiple scattering is an angular cumulative effect (Regler 1977). In this case the LSM requires the evaluation and the inversion of the $n \times n$ covariance matrix (where n is the number of measured coordinates). While the evaluation of this matrix is already a lengthy procedure, the computing time necessary for the inversion grows as n^3. In complex storage ring detectors, n can be as large as 100. The situation is even worse if ambiguities are still left or if outliers have to be removed.

3.2 Estimation of track parameters

There is also the disadvantage that the evaluated track corresponds to a best estimate of the track parameters $\mathbf{p}_r$ at the reference plane, while the calculated crossing points with the detectors, $\mathbf{f}(\tilde{\mathbf{p}}_r)$, correspond to an unscattered prolongation of a track given by $\tilde{\mathbf{p}}_r$, and not to an optimal description of the real scattered track. This deteriorates predictions, interpolations and error tuning, and the detection of outliers, which are hidden in multiple scattering.

If scattering only occurs on a few thin obstacles, track fitting with the inclusion of a few 'breakpoints' as additional pairs of parameters (changes in the direction) into the track model is a solution (Subsection 3.3.2.2).

If multiple scattering occurs at several places, or in continuous media, and if the measurements are sufficiently numerous and close to each other, it is more convenient to follow the track from one detector to the next and to update the estimated track parameters when a suitable additional measurement is available (Billoir 1984; Billoir, Frühwirth, and Regler 1985).

It is assumed that at a given initial surface the track parameter $\tilde{\mathbf{p}}_1$ is known with a given covariance matrix $\mathbf{C}(\tilde{\mathbf{p}}_1) \equiv \mathbf{C}_1$. The first step is now to propagate the parameter vector $\tilde{\mathbf{p}}_i$ and its covariance matrix to the position $i+1$ (starting with $i=1$):

$$\tilde{\mathbf{p}}_{i+1}^{(i)} = \mathbf{f}_{i+1}(\tilde{\mathbf{p}}_i) \tag{3.33a}$$

$$\mathbf{C}_{i+1}^{(i)} = \mathbf{D}_{i+1}^{(i)} \mathbf{C}_i (\mathbf{D}_{i+1}^{(i)})^T \tag{3.33b}$$

$$\mathbf{D}_{i+1}^{(i)} = \partial \mathbf{f}_{i+1} / \partial \mathbf{p}_i \tag{3.33c}$$

where $\mathbf{f}_{i+1}$ is the precise track model between i and $i+1$. The contributions of multiple scattering (the matrix **MS**) between i and $i+1$ are added to $\mathbf{C}_{i+1}^{(i)}$ to obtain the correct errors for the prediction:

$$\mathbf{C}_{i+1}^{(i)'} = \mathbf{C}_{i+1}^{(i)} + \mathbf{MS}_{i+1}^{(i)} \tag{3.33d}$$

Assuming a set of measurements at the position $i+1$, $c_k^{(i+1)}$, which corresponds to $\mathbf{p}_{i+1}$:

$$\mathbf{c}_{i+1} = \mathbf{h}_{i+1}(\mathbf{p}_{i+1}) + \boldsymbol{\varepsilon}_{i+1} \tag{3.34}$$

The measurements $\mathbf{c}_{i+1}$ are, in fact, now the subset of measurements at $i+1$. This subset can consist of an arbitrary number of measurements (also of a full 5-vector, or even of a vector with a dimension larger than 5).

Usually the function $\mathbf{h}_{i+1}(\mathbf{p}_{i+1})$ is either linear or can be well approximated by a linear expression (a possible constant is omitted):

$$\mathbf{h}_{i+1}(\mathbf{p}_{i+1}) = \mathbf{H}_{i+1} \mathbf{p}_{i+1} \tag{3.35}$$

p̃ᵢ
From module 1
mᵢ₊₁
mᵢ₊₂
mᵢ₊₃
mᵢ₊₄
p̃ᵢ₊₄
To module 3
Module 2
(a)
Parameter vector (i − 1) with errors
Detector surface (i − 1)
Scattering matter
Parameter vector (i − 1) propagated to surface (i)
Measurement (i) with errors
Detector surface (i)
Weighted mean of parameter vector (i − 1) and measurement (i) at surface (i) = parameter vector (i) with errors to be propagated to surface (i + 1)
(b)

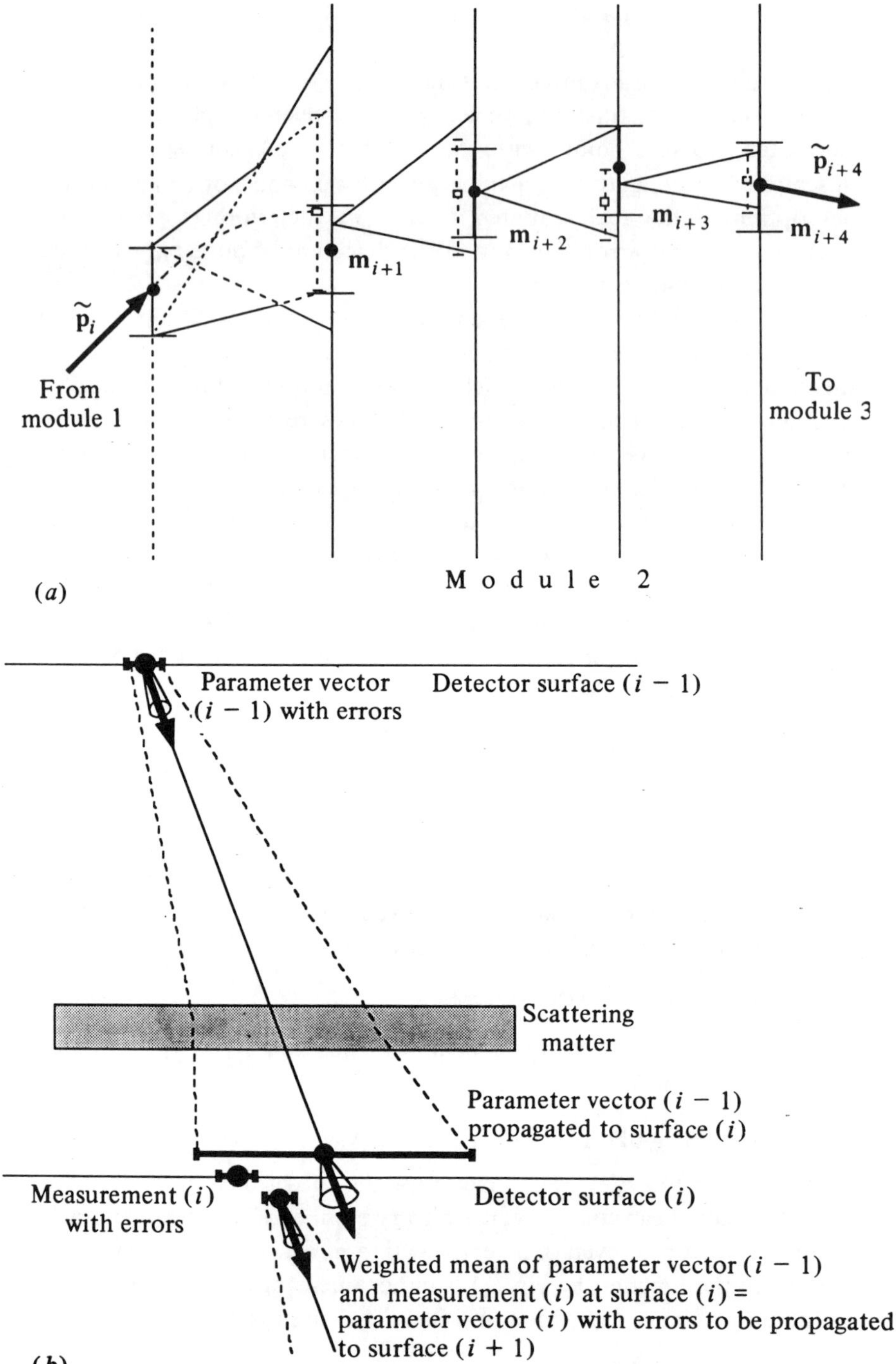

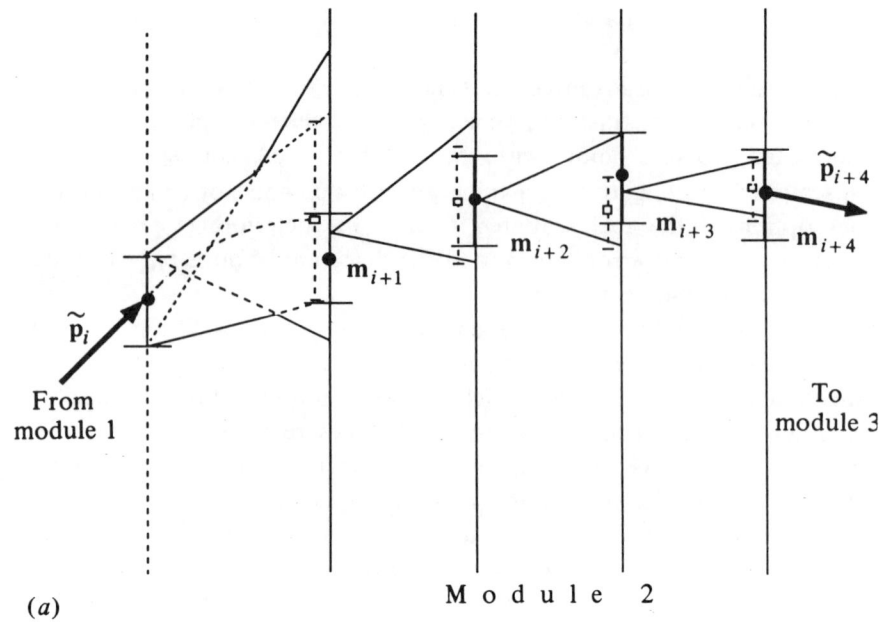

3.2 Estimation of track parameters

where some columns of **H** can be **0**.

If one now considers $\tilde{\mathbf{p}}_{i+1}^{(i)}$ (Equation (3.33a)) as 'direct measurements' of $\mathbf{p}_{i+1}$ (Fig. 3.4), one can make a joint *least squares ansatz* in order to combine $\tilde{\mathbf{p}}_{i+1}^{(i)}$ with the information from the measurement $\mathbf{c}_{i+1}$, with a realization $\mathbf{m}_{i+1}, \mathbf{V}_{i+1} = (\langle \boldsymbol{\varepsilon}_{i+1} \boldsymbol{\varepsilon}_{i+1}^T \rangle)$:

$$M(\mathbf{p}_{i+1}) = (\mathbf{p}_{i+1} - \tilde{\mathbf{p}}_{i+1}^{(i)})^T \mathbf{C}'^{-1} \cdot (\mathbf{p}_{i+1} - \tilde{\mathbf{p}}_{i+1}^{(i)})$$
$$+ (\mathbf{H}\mathbf{p}_{i+1} - \mathbf{m}_{i+1})^T \mathbf{V}^{-1} \cdot (\mathbf{H}\mathbf{p}_{i+1} - \mathbf{m}_{i+1}) \quad (3.36a)$$

(replace $\mathbf{C}' \leftarrow \mathbf{C}_{i+1}^{(i)'}, \mathbf{H} \leftarrow \mathbf{H}_{i+1}, \mathbf{V} \leftarrow \mathbf{V}_{i+1}$), with the solution for the minimum:

$$\tilde{\mathbf{p}}_{i+1} = (\mathbf{C}'^{-1} + \mathbf{H}^T \mathbf{V}^{-1} \mathbf{H})^{-1} \cdot (\mathbf{H}^T \mathbf{V}^{-1} \mathbf{m}_{i+1} + \mathbf{C}'^{-1} \tilde{\mathbf{p}}_{i+1}^{(i)}) \quad (3.36b)$$

The final estimate at the position $i+1$ is nothing but the properly weighted mean of the *information of the actual measurement* $\mathbf{m}_{i+1}$ and the *prediction based on the information of all preceding detectors*, with the covariance matrix

$$\mathbf{C}(\tilde{\mathbf{p}}_{i+1}) = [(\mathbf{C}_{i+1}^{(i)'})^{-1} + \mathbf{H}_{i+1}^T \mathbf{V}_{i+1}^{-1} \mathbf{H}_{i+1}]^{-1} \quad (3.37a)$$

and the residual vector

$$\mathbf{r}_{i+1} = \mathbf{m}_{i+1} - \mathbf{H}_{i+1} \tilde{\mathbf{p}}_{i+1} \quad (3.37b)$$

This allows the evaluation of the 'χ^2 increment'

$$\delta\chi_{i+1}^2 = \mathbf{r}_{i+1}^T \mathbf{V}_{i+1}^{-1} \mathbf{r}_{i+1} + (\tilde{\mathbf{p}}_{i+1} - \tilde{\mathbf{p}}_{i+1}^{(i)})^T (\mathbf{C}_{i+1}^{(i)'})^{-1} (\tilde{\mathbf{p}}_{i+1} - \tilde{\mathbf{p}}_{i+1}^{(i)})$$
$$(3.38a)$$

and of the *total* χ^2:

$$\chi_{i+1}^2 = \chi_i^2 + \delta\chi_{i+1}^2 \quad (3.38b)$$

$\delta\chi_{i+1}^2$ is a measure of the distance of the measurement $\mathbf{m}_{i+1}$ from the estimate $\tilde{\mathbf{p}}_{i+1}$.

This method, called the 'progressive fit', has several advantages:

It is suitable for combined trackfinding and track fitting.
No large matrices have to be inverted, and the number of computations increases linearly with the number of detectors crossed by the track.

Fig. 3.4(a) Progressive adjustment of the fitted parameters by propagation ('predictions') and the stepwise inclusion of further measurements. *(b)* A detailed outline of one step of the filtering procedure.

Therefore, it is relatively fast even when there is frequent multiple scattering and a large number of measurements.

The estimated track parameters closely follow the *real path* of the particle. The linear approximation of the track model does not need to be valid over the whole track length, but only from one detector to the next.

The method has, however, a fundamental drawback: the track parameters at any point include only the information from the preceding detectors, and are therefore known with optimal precision only after the last step of the fit. Consequently the ability to discriminate outliers or ambiguities in the early steps is low, and because of the track density the method must normally be applied from the outer detectors, where the track parameters are usually poorly known, towards the vertex region, in order to end up with the optimal estimate there.

This problem has recently been studied and successfully overcome (Frühwirth 1987; Regler and Frühwirth (1989): track fitting, which deals with the estimation of track parameters, is a special case of the 'Kalman filter', which deals with the analysis of (linear) dynamic systems (Gelb 1975; Brammer and Siffling 1975). The Kalman filter is the optimum solution of the problem mentioned above: a simple algorithm called '*smoothing*', using only weight matrices and track propagation matrices already evaluated during the progressive track fit, allows the recursive estimation of the track parameter vector **p** at all intermediate points *based on all measurements incorporated so far*:

$$\tilde{\mathbf{p}}_i^{(s)} = \tilde{\mathbf{p}}_i + \mathbf{A}_i \cdot (\tilde{\mathbf{p}}_{i+1}^{(s)} - \tilde{\mathbf{p}}_{i+1}^{(i)}) \tag{3.39a}$$

$$\mathbf{A}_i = \mathbf{C}(\tilde{p}_i)\mathbf{D}_{i+1}^{(i)}(\mathbf{C}_{i+1}^{(i)})^{-1}$$

$$\mathbf{C}(\tilde{p}_i^{(s)}) = \mathbf{C}(\tilde{p}_i) + \mathbf{A}_i \cdot (\mathbf{C}(\tilde{\mathbf{p}}_{i+1}^{(s)}) - \mathbf{C}_{i+1}^{(i)'})\mathbf{A}_i^T \tag{3.39b}$$

Smoothing is an optimal estimation of the track parameters anywhere along the track, and constitutes a substantial improvement of the simple progressive fit (Fig. 3.5):

It allows computation of optimal predictions in other detectors from both ends of a track segment, as well as optimal interpolation.

It is easy to remove the information contained in measurements $\mathbf{m}_i$ from the smoothed estimated $\tilde{p}_i^{(s)}$. This yields an optimal estimate of the track parameters based on all measurements with the exception of $\mathbf{m}_i$. Since this estimate normally gives a better definition of the track than the filtered estimate, it should be used for the final decision on outliers and

3.2 Estimation of track parameters

ambiguities, after a *less restrictive*, but similar procedure during the filter (progressive fit). This new estimate can also be used for the *checking* and *tuning* of the *detector alignment* and *resolution*.

The 'smoother' can also be used for efficient track segment merging.

To summarize: Smoothing is a useful and necessary complement of the Kalman filter, making the progressive method a powerful, flexible, and efficient tool not only for track fitting, but also for the computation of optimal predictions and interpolations, for outlier detection and rejection, and for the merging of track segments (see also Subsection 3.3.2.2).

Sometimes it is difficult to find a suitable starting element for the filter. If

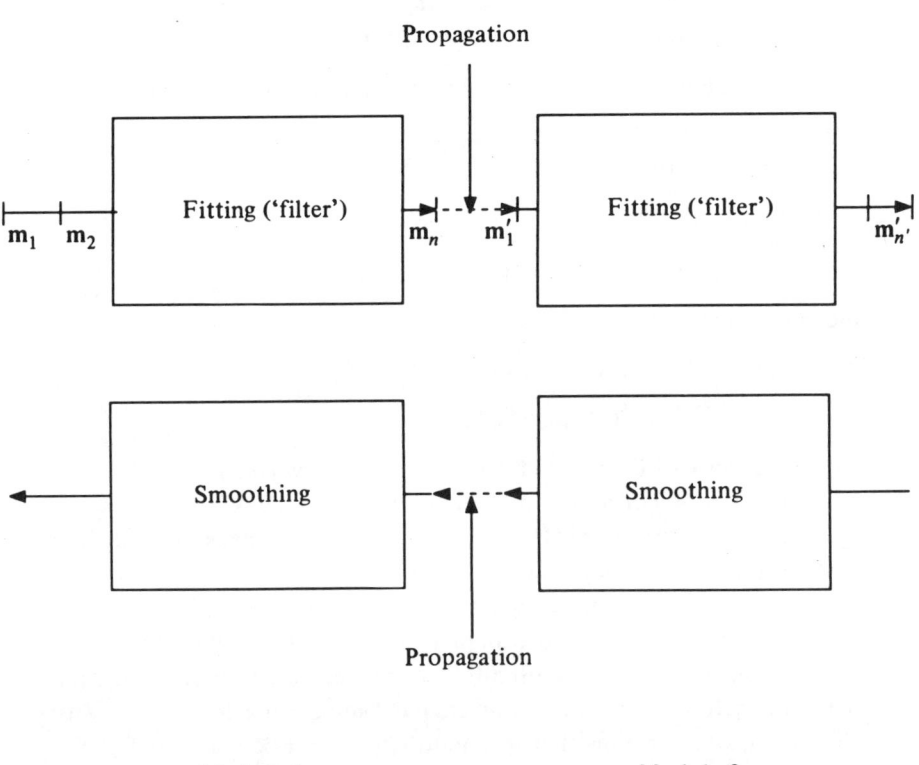

Fig. 3.5 After the complete filtering procedure (track fit) the *full information* is transmitted back to all intermediate points by the smoothing algorithm.

the angular and/or momentum resolution of the starting module is poor, and the next detector module is far away, it might be dangerous to make a prediction with a quite large error:

It is hard to compute properly the amount of material along the real trajectory, as the predicted path might not correspond to reality at all.
The linearity requirement for the track model might be overstressed; when predictions have large errors, this can also introduce a bias.

A solution is to take a 'reference track' as obtained from *pattern recognition* (*the matrix of derivatives is calculated along the reference track*) which defines the expansion point (P. Billoir, private communication). By replacing:

the state vector $\mathbf{p}_i$ by $\Delta\mathbf{p}_i = \mathbf{p}_i - \mathbf{p}_i^r$ (r denotes the reference track);
the prediction $\mathbf{p}_{i+1}^{(i)}$ by $\Delta\mathbf{p}_{i+1}^{(i)} = \mathbf{D}_{i+1}^{(i)} \Delta\mathbf{p}_i$;
the measurement vector $\mathbf{m}_{i+1}$ by $\Delta\mathbf{m}_{i+1} = \mathbf{m}_{i+1} - \mathbf{H}_{i+1}\mathbf{p}_{i+1}^r$.

Equation (3.36b) can be used without any change. The predicted residuals are:

$$\mathbf{r}_{i+1}^{(i)} = \Delta\mathbf{m}_{i+1} - \mathbf{H}_{i+1}\Delta\mathbf{p}_{i+1}^{(i)}$$

the filtered residuals:

$$\mathbf{r}_{i+1} = \Delta\mathbf{m}_{i+1} - \mathbf{H}_{i+1}\Delta\mathbf{p}_{i+1}$$

and the χ^2 increment:

$$\chi_+^2 = \mathbf{r}_{i+1}^T \mathbf{V}_{i+1}^{-1} \mathbf{r}_{i+1} + (\Delta\mathbf{p}_{i+1} - \Delta\mathbf{p}_{i+1}^{(i)})^T (\mathbf{C}_{i+1}^{(i)})^{-1} (\Delta\mathbf{p}_{i+1} - \Delta_{i+1}^{(i)})$$

The smoother can be adapted accordingly (Frühwirth, 1988).

A second problem arises, if the covariance matrix $\mathbf{C}_1$ of the starting vector $\mathbf{p}_1$ is singular, simply because there are too few measurements in $\mathbf{m}_1$ to allow a unique estimate of $\mathbf{p}_1$. In this case the inversion of the predicted covariance matrix in the second step of the filter will fail. One way of avoiding this is to propagate a singular weight matrix, which need not be inverted, but which will eventually become regular, if enough measurements are included in the fit. However, in the smoother a covariance matrix is required, which means that the smoother cannot be computed back to step 1, which is a serious drawback.

Another solution consists in making $\mathbf{C}_1$ artificially regular, by giving those elements, which are undefined, large diagonal errors. Note that these large quantities have to cancel again in the smoother. This might lead to numerical problems on a computer with a short word length.

3.3 Fitting the tracks of charged particles

In high-energy physics it is important to determine the momentum of a charged particle. The most common way to achieve this is to deflect it with a magnetic field and to measure the deflection by position-sensitive detectors. Such a device is called a *'magnetic particle spectrometer'*. In some special cases, other methods such as *'calorimetry'* are used (e.g. for very-high-momentum electrons from a Z^0 decay, Section 2.3), complementing the information given by the tracking detectors. However, the sign of the curvature at least must still be measured by the tracking devices.

3.3.1 The track model

3.3.1.1 The equations of motion

The trajectory of a particle in a *static* magnetic field $\mathbf{B}(\mathbf{x})$ must satisfy the equations of motion given by the *Lorentz force*. It is assumed that there is no electric field. Neglecting bremsstrahlung and material effects, this force $\mathbf{f}$ is derived from Maxwell's equations to be

$$\mathbf{f} \sim q\mathbf{v} \times \mathbf{B} \tag{3.40a}$$

where $\mathbf{v}$ is the velocity of the particle, $v = |\mathbf{v}|$ ($\mathbf{v} \equiv d\mathbf{x}/dt$) and q is the (signed) charge. This gives the equation of motion in a vacuum:

$$m\gamma \, d^2\mathbf{x}/dt^2 = c^2 \kappa q \, \mathbf{v}(t) \times \mathbf{B}(\mathbf{x}(t)) \tag{3.40b}$$

where κ is a proportionality factor, dependent on the choice of units (see below), $\mathbf{B}(\mathbf{x})$ is the static magnetic field (defined by its flux density), m is the rest mass and $\mathbf{x}$ is the position (a space point) of the particle, c is the velocity of light, and t is the time in the laboratory frame. The relativistic Lorentz factor γ is given by

$$\gamma = (1 - \beta^2)^{-\frac{1}{2}}$$

where

$$\beta = |\boldsymbol{\beta}| = v/c, \quad \boldsymbol{\beta} = \mathbf{v}/c = d\mathbf{x}/d(ct)$$

This equation can be rewritten in the form of geometrical quantities only:

$$d^2\mathbf{x}/ds^2 = (\kappa q/P)(d\mathbf{x}/ds) \times \mathbf{B}(\mathbf{x}(s)) \tag{3.41a}$$

where $s(t)$ is the distance along trajectory (path length), with $ds/dt = v$, and $\mathbf{p} = m\gamma\boldsymbol{\beta}c$ is the momentum of the particle, $P = |\mathbf{p}| = m\gamma\beta c$ (laboratory frame).

Proof:
$$(d^2\mathbf{x}/dt^2) \times (d\mathbf{x}/dt) \equiv 0 \quad \text{(see Equation (3.40b))}$$

and therefore:
$$|d\mathbf{x}/dt| = \text{constant} = v = \beta c$$
$$d\mathbf{x}/dt = (d\mathbf{x}/ds)(ds/dt) = (d\mathbf{x}/ds)\beta c$$
$$d^2\mathbf{x}/dt^2 = (d^2\mathbf{x}/ds^2)\beta^2 c^2 \quad \square$$

In particle physics, the following standard units are used (for more details on units see e.g. Jackson (1962), Bock, et al. (1984a)):

q in multiples of the positive elementary charge (dimensionless),
$\mathbf{x}$ and s in metres
P in GeV c^{-1}
$\mathbf{B}$ in tesla
κ is proportional to the velocity of light and is therefore defined as 0.299792458 (GeV c^{-1}) T^{-1} m^{-1}.

Sometimes, mainly in experiments with fixed targets and hence comparatively small P_T/P_L at least in the forward region (P_T and P_L being respectively the transverse and longitudinal components of $\mathbf{p}$ with respect to the beam), it might be advantageous to rewrite Equation (3.41a) choosing the coordinate along the beam (say z) as a variable (the primes denote derivatives with respect to z), giving

$$\left.\begin{array}{l} x'' = (\kappa q/P)(ds/dz)[x'y'B_x - (1 + x'^2)B_y + y'B_z] \\ y'' = (\kappa q/P)(ds/dz)[(1 + y'^2)B_x - x'y'B_y - x'B_z] \end{array}\right\} \quad (3.41b)$$

with
$$ds^2 = dx^2 + dy^2 + dz^2$$
$$ds/dz = (1 + x'^2 + y'^2)^{\frac{1}{2}}$$

(The third equation becomes a trivial identity.)

When integrating Equation (3.41a) there are six integration constants plus the unknown momentum P, but with the identity

$$(dx/ds)^2 + (dy/ds)^2 + (dz/ds)^2 \equiv 1$$

3.3 Fitting the tracks of charged particles

and an arbitrary choice of one coordinate (the 'reference surface'), there are, in fact, (for known mass) only *five free parameters* (see also Section 3.2) defining the track (e.g. two for the impact with a reference surface at $z_r = $ const., two for the direction at that point, and one for the momentum). Note that s can be given an arbitrary value on the reference surface.

In order to become familiar with these equations, the special case of a *homogeneous magnetic field* will first be discussed. Without loss of generality, **B** is chosen to be parallel to the z axis, $\mathbf{B} = B\mathbf{e}_z$, with $\mathbf{e}_z^T = (0,0,1)$;

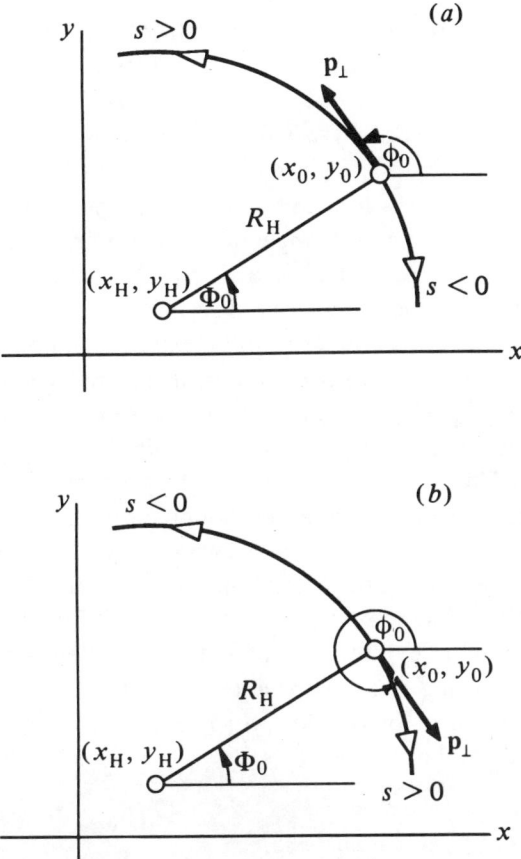

Fig. 3.6(a) For a constant magnetic field parallel to the z axis, the solution of the equation of motion is a helix. The projection on the xy plane is a circle, which is chosen here with a positive sense of rotation ($h = +1$). (b) Projected helix with negative sense of rotation ($h = -1$).

Equation (3.41a) then takes the form

$$\left.\begin{array}{l} d^2x/ds^2 = (\kappa q/P)(dy/ds)B \\ d^2y/ds^2 = -(\kappa q/P)(dx/ds)B \\ d^2z/ds^2 = 0 \end{array}\right\} \quad (3.41c)$$

and the solution is a *helix* with an axis parallel to z (Fig. 3.6):

$$\left.\begin{array}{l} x(s) = \overset{o}{x} + R_H[\cos(\Phi_0 + hs\cos\lambda/R_H) - \cos\Phi_0] \\ y(s) = \overset{o}{y} + R_H[\sin(\Phi_0 + hs\cos\lambda/R_H) - \sin\Phi_0] \\ z(s) = \overset{o}{z} + s\sin\lambda \end{array}\right\} \quad (3.42)$$

where s is the path length along the helix, which increases when moving in the particle's direction, $\overset{o}{x}$ is the starting point at $s = \overset{o}{s} = 0$, λ is the slope ('dip') angle ($= \arcsin(dz/ds)$, $-\pi/2 < \lambda \leqslant \pi/2$), and R_H is the radius of the helix ($= (P\cos\lambda)/(|\kappa qB|)$). The sense of rotation of the *projected helix* in the xy plane, h, is given by $h = -\text{sign}(qB) = \pm 1$ ($= \text{sign}(d\phi/ds)$, where ϕ is the track direction, and z is the polar axis parallel to the helix axis); Φ_0 is the azimuth angle of starting point (in cylindrical coordinates) with respect to the helix axis $= \phi_0 - h\pi/2$, ϕ_0 being arctan $(dy/dx)_{s=0}$ (the azimuth angle of the track direction at the starting point. For many applications, the assumptions for an explicit solution, namely a homogeneous **B** field and the absence of matter, are approximately fulfilled so that Equation (3.42) is precise enough to serve as a model for the track fit (Mitaroff 1987). This is the case in traditional bubble chambers and in the central region of track detectors at many storage ring experiments. In other cases, the approximation may still be good enough for pattern recognition.

In storage ring experiments, the form of the magnetic field can be a dipole field (as in the UA1 detector at the CERN p$\bar{\text{p}}$ collider or the Double Arm Spectrometer (DASP) detector at DESY) or a solenoidal field (like most of the existing or planned e^+e^- storage ring detectors, where synchrotron radiation makes a strong field perpendicular to the beam direction prohibitive). The treatment of **B** fields which occur in practice will be discussed in more detail later.

3.3.1.2 *The choice of track parameters*

Using the helix solution (Equation (3.42)), the choice of parameters to be fitted for a track model will now be discussed. Considering the usual set-up of a central detector consisting of coaxial cylinder surfaces with fixed

3.3 Fitting the tracks of charged particles

radii R_i, one measures for each track impact two coordinates, $(R\Phi)_i$ (Fig. 3.6) and z_i. The main magnetic field component is assumed to be parallel to the axis of the cylinder. The reference surface for defining the parameters may be chosen to be the cylindrical beam tube itself. Then the following choice for the parameters to be fitted will give a track model which is locally not too far from linear (Equations (3.1), (3.4)):

$$\left. \begin{array}{ll} p_1 = (R\Phi)_r & p_2 = z_r \\ p_3 = \phi_r & p_4 = \tan \lambda_r \\ p_5 = (1/R_H)_r \times \text{sense of rotation} \end{array} \right\} \qquad (3.43)$$

where the subscript 'r' denotes the actual value of the track parameters at the intersection with the reference cylinder $R = R_r$ (Fig. 3.7(a)). Note that p_4 and p_5 are constant for the helix track model (Equation (3.42)).

For the *downstream spectrometer arm* of a *fixed-target detector* (Fig. 3.7(b)), the choice of parameters will be inspired by the equations (Equation (3.41b)). Such a spectrometer often consists of a first 'lever arm' with position-sensitive detectors in a field-free region (where the solution of the equation of motion simply gives a straight line), followed by a bending magnet with a strong magnetic field perpendicular to the main direction (z was chosen in Equation (3.41b)), covering only a small region as compared to the total length of the spectrometer, finally followed by a second lever arm of position-sensitive detectors. (For different kinds of detector arrangements see also Chapter 1.)

If the magnetic field is quite homogeneous, $\mathbf{B} = B\mathbf{e}_y$ (the beam direction is again $\mathbf{e}_z$), and the length of the magnet is L, the following approximate solution of Equation (3.41b) can be given:

$$\left. \begin{array}{l} \Delta x' = x'_2 - x'_1 \cong -\kappa qBL/P \\ \Delta y' = y'_2 - y'_1 \cong 0 \end{array} \right\} \qquad (3.44a)$$

with

$$B_x, B_z \ll B_y$$
$$x', y' \ll 1, |ds/dz| \cong 1$$

From $p_x = P\,dx/ds \cong P\,dx/dz$ it follows that the change $|\Delta P_T|$ of the transverse momentum $P_T (= (p_x^2 + p_y^2)^{1/2})$ is approximately

$$|\Delta P_T| \cong \kappa |q|BL \qquad (3.44b)$$

and the deflection $\Delta\alpha$

$$|\Delta\alpha| \cong \kappa |q|BL/P \qquad (3.44c)$$

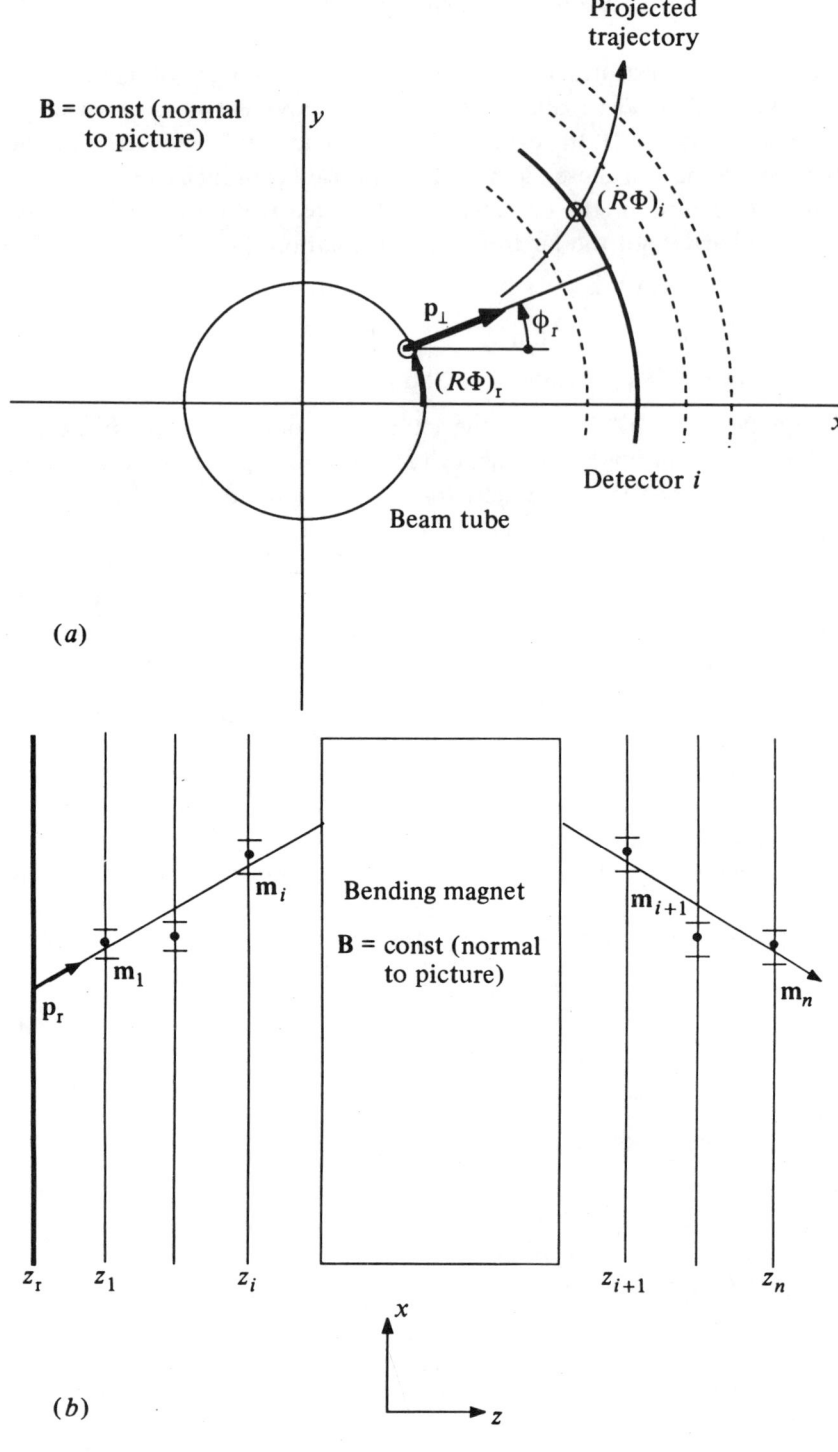

3.3 Fitting the tracks of charged particles

In practice, this formula gives an error of 10–20%.
The natural choice for the track parameters is now

$$\left.\begin{array}{ll} p_1 = x_r & p_2 = y_r \\ p_3 = (dx/dz)_r & p_4 = (dy/dz)_r \\ p_5 = (1/P) \times \text{sense of rotation} \end{array}\right\} \quad (3.45)$$

at a reference plane $z = z_r$.

3.3.1.3 Several types of track models

In the presence of an *inhomogeneous* magnetic field, appropriate algorithms are needed to allow a particle to be followed efficiently through a given detector set-up. The algorithms discussed here deal only with the *deterministic* part of the *track model*, i.e. with the solution of the equation of motion (Equations (3.41)).

Material effects (Subsection 3.3.1.5) are nondeterministic, and will be treated in two different ways: energy loss will be added in a *deterministic approximation* as an average, while multiple scattering will be treated either as a *stochastic perturbation* of the trajectory (noise contribution) or added to the measurement errors. The track model can be formulated either as the full trajectory from an analytical or numerical solution of the equation of motion (e.g. Equation(3.42)), or as the functional relation between the impact points on specific detectors and some initial parameters **p** (integration parameters), thus defining one single point in the measurement space for one set of initial parameters (Equation (3.1), Fig. 3.1). The set of all possible *undisturbed tracks* defines a *five-dimensional hyper-surface* in the measurement space.

To account for multiple scattering in obstacles between the detectors, one may also need the derivatives $\partial \mathbf{p}(s_j)/\partial \mathbf{p}(s_i)$. If there is no multiple scattering, it is sufficient to compute the functional dependence $\mathbf{f}(\mathbf{p})$ of the measurement vector on the initial parameters. Usually $\mathbf{f}(\mathbf{p})$ is approximated by a linear expansion which also requires the evaluation of $\partial \mathbf{f}/\partial \mathbf{p}$ in Equation(3.4).

Track models which do not exactly fulfill the equations of motion can still

Fig. 3.7(a) In many storage ring experiments, the magnetic field is rotationally invariant about the beam axis. In this case, the reference surface (at least for the barrel region) is chosen to be a cylinder. (*b*) In the forward arm of a fixed-target spectrometer, the natural choice of parameters is: two Cartesian coordinates at the reference plane, two directions (tangents or direction cosines) and $1/P$ (or the deflection angle).

give quite valuable parameter estimations. However, in such cases the statistical properties of the LSM are no longer the ones described above, and the loss of information depends on whether the deviation of the approximate track model from the ideal one is significant compared to the measurement errors. Special care is then advised to avoid a systematic bias.

A few of the common methods used in today's experiments are discussed below.

Helix tracking

In a *homogeneous* magnetic field, integration of the equation of motion gives a helix (Equation (3.42)). In a *strongly inhomogeneous field*, however, step-by-step tracking along small helix segments is very inefficient and *should be avoided*, as the derivatives of the field are totally neglected within one step.

Numerical integration

Two different cases must be distinguished: numerical integration of a bundle of trajectories, one 'zero trajectory' and five accompanying trajectories each corresponding to a slight variation of an initial parameter, in order to get the derivatives by numerical differentiation (e.g. Metcalf, et al. 1973; Eichinger and Regler 1981), and parallel integration of the derivatives.

(a) *The fourth-order Runge–Kutta method of Nyström* The equations of motion can be rewritten more generally for a given momentum:

$$\mathbf{u}'' = \mathbf{g}(\mathbf{u}', \mathbf{B}(\mathbf{u})) = \mathbf{f}(\mathbf{u}', \mathbf{u}) \tag{3.46}$$

where the prime denotes either the derivative with respect to the path length (Equation (3.40a)) and $\mathbf{u} = \mathbf{x}$, or the derivative with respect to z (Equation (3.14b)) and $\mathbf{u} = (x, y)^T$.

For tracking, the following recursive formula is used (e.g. Abramowitz and Stegun 1970 (formula 25.5.20))

$$\left.\begin{array}{l}\mathbf{u}_{n+1} = \mathbf{u}_n + h\mathbf{u}'_n + (h^2/6)(\mathbf{k}_1 + \mathbf{k}_2 + \mathbf{k}_3) + O(h^5) \\ \mathbf{u}'_{n+1} = \mathbf{u}'_n + (h/6)(\mathbf{k}_1 + 2\mathbf{k}_2 + 2\mathbf{k}_3 + \mathbf{k}_4)\end{array}\right\} \tag{3.47}$$

where

$$\mathbf{k}_1 = \mathbf{f}(\mathbf{u}'_n, \mathbf{u}_n) = \mathbf{g}(\mathbf{u}', \mathbf{B}(\mathbf{u}))$$
$$\mathbf{k}_2 = \mathbf{f}(\mathbf{u}'_n + (h/2)\mathbf{k}_1, \mathbf{u}_n + (h/2)\mathbf{u}'_n + (h^2/8)\mathbf{k}_1)$$
$$\mathbf{k}_3 = \mathbf{f}(\mathbf{u}'_n + (h/2)\mathbf{k}_2, \mathbf{u}_n + (h/2)\mathbf{u}'_n + (h^2/8)\mathbf{k}_2)$$
$$\mathbf{k}_4 = \mathbf{f}(\mathbf{u}_n + h\mathbf{k}_3, \mathbf{u}_n + h\mathbf{u}'_n + (h^2/2)\mathbf{k}_3)$$

3.3 Fitting the tracks of charged particles

In practice, it is quite often sufficient to take the same field values when evaluating $\mathbf{k}_2$ and $\mathbf{k}_3$ (this is Nyström's advantage), which corresponds to replacing $\mathbf{k}_2$ by $\mathbf{k}_1$ in the second argument of $\mathbf{k}_3$. Another frequent approximation is to use the field value in $\mathbf{k}_4$ of one step to evaluate the $\mathbf{k}_1$ of the subsequent one (see below).

(b) *The Runge–Kutta method of Simpson* A slightly different approach is to transform the equation of motion into a system of two first order, simultaneous differential equations:

$$\left.\begin{array}{l} \mathbf{u}' = \mathbf{v} \\ \mathbf{v}' = \mathbf{g}(\mathbf{v}, \mathbf{B}) \end{array}\right\} \tag{3.48}$$

Results of this approach are very similar – quite independent of the field inhomogeneities – to those obtained with the Nyström algorithm, depending slightly on the different algorithms applied to solve Equation (3.48), (e.g. Abramowitz and Stegun (1970)).

(c) *Parallel integration of the derivatives* In order to minimize the 'χ^2 ansatz' (Equation (3.7)), the functional dependence of the track interceptions with the detectors from the track parameters (Equation (3.1)) is needed. In practice, this is achieved either by numerical differentiation (see below) or by integrating the derivatives $\mathbf{A}(s)$ (Equation (3.4)) together with the 'zero trajectory'. However, it should be mentioned that the exact algorithm also requires knowledge of the field derivatives, except when the influence of the field gradient transverse to the trajectory integrated over one step length can be neglected for the derivatives. In practice this approach saves about a factor of 2 in computing time spent in the tracking module (Myrheim and Bugge 1979; Bugge and Myrheim 1981). It was successfully applied in several experiments, e.g. for the recoil particle in the WA6 experiment at CERN (pp↑ → pp) measuring elastic proton–proton scattering on a polarized target (Fidecaro, *et al.* 1980).

(d) *'Numerical differentiation'* If the gradient of the magnetic field transverse to the trajectory cannot be neglected and if it is not easy to obtain the field derivatives from the 'field model', numerical differentiation can be used:

$$(\mathbf{A})_{ik} = \frac{\partial f_i}{\partial p_k} = \frac{f_i(\overset{0}{\mathbf{p}} + \Delta_k \mathbf{p}) - f_i(\overset{0}{\mathbf{p}})}{\Delta p_k} \tag{3.49}$$

where $\Delta_k \mathbf{p}$ is a vector with all components being zero except the kth component which is Δp_k.

The variations should be small, but they must be large enough not to run into trouble with the machine precision of the computer. A lower limit is given by the word length of the computer used. As a rule of thumb typical variations are 0.1–1.0 mm for space coordinates and 10^{-4}–10^{-3} rad for angles. As, in practice, the fifth parameter is usually chosen to be 1/momentum or 1/transverse momentum ($1/P$ or $1/P_T$), the variation should be chosen accordingly. The lateral displacement being roughly (in metres) $0.3 \times Bs^2/P$ (where B is in teslas, s in metres, P in GeV c^{-1}), a displacement of 1 mm corresponds to $\Delta p_s = (10^{-4}$–$10^{-3})/(0.3 \times Bs^2)$, where some average detector dimension should be chosen for s (Equation (3.41)). A fast check of the proper choice of the variations is the stability of the derivative values when the variations are changed slightly.

To avoid discontinuities during numerical differentiation if the field is represented in separate volumes (see Subsection 3.3.1.4), the *same 'boxes' of the field model* (Subsection 3.3.1.4) should be used for the zero track and for the variations at the borders of a box.

Predictor–Corrector methods

Multistep integration has been applied for the AFS at the CERN-ISR. This method first makes a prediction for $\mathbf{u}'_{n+1}$, using a polynomial for $\mathbf{u}''$ through the previous points, and then a prediction for $\mathbf{u}_{n+1}$, before computing the final 'corrected' $\mathbf{u}_{n+1}$, again by integrating a polynomial. The method can be tuned by several parameters, but this makes the method less transparent. Considering the fact that at the beginning of the track no polynomials are yet available, the modest gain in computing time compared to the Runge–Kutta method does not really suggest this method should be used. However, if the method is used in connection with the analytical properties of the magnetic field, a slightly larger step length than with the Runge–Kutta method can be achieved for a required precision.

Taylor expansion

(a) *General case using derivatives of the field model* If not only the field values but also the derivatives can be evaluated by the field representation up to order $n-2$, an exact Taylor expansion can be evaluated up to order n. The Taylor expansion is obtained by comparing the coefficients of both

3.3 *Fitting the tracks of charged particles* 279

sides of the equations of motion at a given point:

$$\begin{aligned}
x_i''/\text{const} &= x_j' B_k - x_k' B_j \\
x_i'''/\text{const} &= x_j'' B_k - x_k'' B_j + \sum_\alpha \left[x_j' x_\alpha' \frac{\partial B_k}{\partial x_\alpha} + x_k' x_\alpha' \frac{\partial B_j}{\partial x_\alpha} \right] \\
x_i''''/\text{const} &= x_j''' B_k - x_k''' B_j \\
&\quad + \sum_\alpha \left[2\left(x_j'' x_\alpha' \frac{\partial B_k}{\partial x_\alpha} - x_k'' x_\alpha' \frac{\partial B_j}{\partial x_\alpha} \right) + x_j' x_\alpha'' \frac{\partial B_k}{\partial x_\alpha} - x_k' x_\alpha'' \frac{\partial B_j}{\partial x_\alpha} \right] \\
&\quad + \sum_{\alpha\beta} \left[x_j' x_\alpha' x_\beta' \frac{\partial^2 B_k}{\partial x_\alpha \partial x_\beta} - x_k' x_\alpha' x_\beta' \frac{\partial^2 B_j}{\partial x_\alpha \partial x_\beta} \right] \\
x_i'''''/\text{const} &= x_j'''' B_k - x_k'''' B_j + \cdots
\end{aligned} \quad (3.50\text{a})$$

where i, j, k denote the three coordinates cyclically, and α, β, γ are the summation indices running from 1 to 3 (x, y, z), from which, e.g.:

$$\begin{aligned}
\Delta x_i &= x_i' s + x_i'' \frac{s^2}{2!} + x_i''' \frac{s^3}{3!} + x_i'''' \frac{s^4}{4!} + x_i''''' \frac{s^5}{5!} \\
\Delta x_i' &= x_i'' s + x_i''' \frac{s^2}{2!} + x_i'''' \frac{s^3}{3!} + x_i''''' \frac{s^4}{4!}
\end{aligned} \quad (3.50\text{b})$$

The gain in precision when adding higher orders than those suggested by the field representation depends on the analytical properties of the field itself. For example in the case of a constant field all terms in Equation (3.50a) which contain derivatives of the field are equal to zero. The remaining terms with the field itself constitute the series expansion of a helix. This shows that in practice it might be meaningful to extend the series to orders higher than suggested by the field derivatives available (i.e. $> n$).

In general this method allows larger steps than the Runge–Kutta method (but needs more computer time per step for higher orders), and the step size is only limited by the volume of the box in which the local field model is valid. If the field computations are fast, and if the formulae for the derivatives are coded efficiently, the method is competitive with the other methods mentioned above, and has the advantage that the track's position and direction are known for any point along the path, *allowing the evaluation of material effects*, the *simulation of decays* etc.

A 'prediction–correction procedure', using the field values from the next step and correcting the derivatives of the present step in turn, starting with the highest derivative, would allow a jump with one single step from one field box to another, but this has never been field proven.

The method mentioned above has been successfully applied in the SFM at the CERN–ISR after careful tuning (Metcalf and Regler 1973; Metcalf, Regler, and Broll 1973). A general program performing the comparison of coefficients for a general field not fulfilling the 'Laplace equation' turned out to be of no practical use other than for a comparison of the precision which can be achieved by this method (Regler 1968).

If another method has been chosen, e.g. because of unknown field derivatives in a very inhomogeneous field, a second order or 'truncated' third order expansion (constant field approximation) may still be useful for several purposes: evaluation of multiple scattering and energy loss in inhomogeneous matter, simulation of electromagnetic processes such as ionization, or simulation of secondary reactions such as particle decays.

(b) Expansion in an axially symmetric magnetic field An interesting special case arises if the field expansion is still more restrictive, i.e. for a field of *axial symmetry and mirror symmetry about the median plane*:

$$\left.\begin{aligned} B_z &= B_z(R,z) = B_z(R, -z) \\ B_R &= B_R(R,z) = -B_R(R, -z) \\ B_\Phi &\equiv 0 \end{aligned}\right\} \quad (3.51a)$$

The vector potential **A** of the magnetic field in this special case has only one nonvanishing component – the azimuthal one $A_\Phi(R, z)$ – which can be obtained from

$$A_\Phi(R, z) = (1/R) \int_0^R B_z(r, 0)\,\mathrm{d}r - \int_0^z B_R(R, \zeta)\,\mathrm{d}\zeta \quad (3.51b)$$

The motion of a particle is then completely described by Equation (3.41)

$$\left.\begin{aligned} PR \sin\beta \cos\lambda + \kappa q R A_\Phi(R, z) &= \text{const} \\ P\,\mathrm{d}(\sin\lambda)/\mathrm{d}R - \kappa q \tan\beta\, \partial A_\Phi/\partial z &= 0 \end{aligned}\right\} \quad (3.51c)$$

where β is the angle between the radius vector and the direction of the particle projected onto the median plane. An efficient power series expansion can be obtained, providing the coordinates along the trajectory in terms of the inverse momentum of the particle, the distance of the particle from the symmetry axis, and the initial inclination with the median plane. For a small dip angle λ (Equation (3.42)), the series converges rapidly, and

3.3 Fitting the tracks of charged particles

the procedure of evaluating the intercepts with the detectors is very efficient when the detectors are cylindrical in shape and concentric to the symmetry axis of the field (Birsa, *et al.* 1977) (Fig. 3.8). By taking higher orders into account, any desired precision can be obtained.

For tracks going primarily along the direction of the symmetry axis, corrections to a helix are given in Billoir (1987a), which handles the inhomogeneities $\Delta \mathbf{B}$ of the field as a perturbation and computes the first order of this perturbation with integrals depending linearly on $\Delta \mathbf{B}$; the integration path is the helix approximation. (For the derivatives see Billoir (1986) and Billoir (1987b)).

(c) Magnetic quadrupole fields In some experiments, when measuring very forward particles i.e. those at a small angle with respect to the beam, it may be necessary to place tracking detectors behind the last 'magnetic quadrupole lens' of the beam line. This must be taken into account in the analysis program. The vector potential $\mathbf{A}$ of the magnetic field of a pure

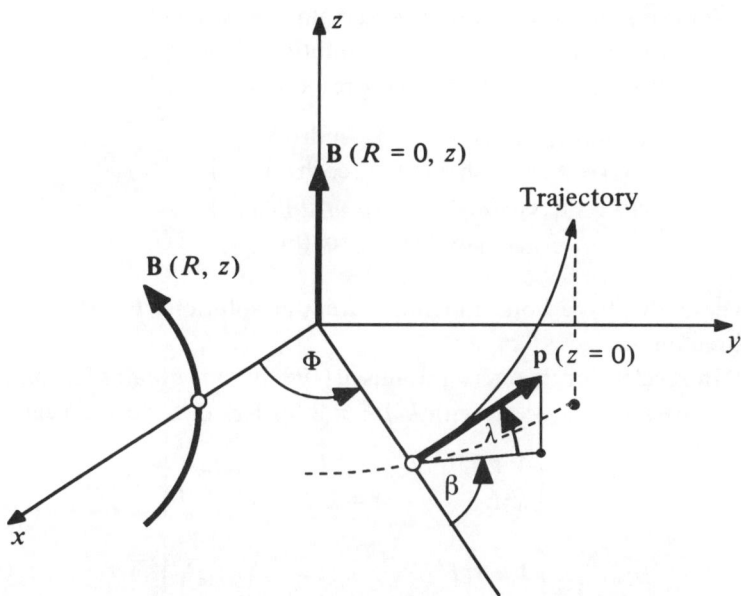

Fig. 3.8 A field of axial symmetry and mirror symmetry about the median plane. Φ is the azimuthal angle of the intersect of the trajectory with the median plane, β is the difference between the direction of the projected tangent and Φ, λ is the angle between the tangent and its projection χ, and p is the momentum.

quadrupole is particularly simple (G is the field gradient):

$$\mathbf{A}(\mathbf{x}) = \pm \frac{G}{2} \begin{pmatrix} 0 \\ 0 \\ x^2 - y^2 \end{pmatrix} \tag{3.52a}$$

and

$$\mathbf{B}(\mathbf{x}) = \mp G \begin{pmatrix} y \\ x \\ 0 \end{pmatrix} \tag{3.52b}$$

The equations of motion (Equation (3.41b)), with $\mathrm{d}x/\mathrm{d}z \ll 1$, $\mathrm{d}y/\mathrm{d}z \ll 1$, $\mathrm{d}s/\mathrm{d}z \cong 1$, give

$$\begin{aligned} \mathrm{d}s/\mathrm{d}z &= \pm (\kappa q/P) G x \\ \mathrm{d}y/\mathrm{d}z &= \mp (\kappa q/P) G y \end{aligned} \tag{3.52c}$$

where $\pm$ defines the polarity of the quadrupole and together with the sign of q, defines the focussing resp. defocussing direction: if the factor is positive, the lens is defocussing in this direction, and focussing in the other one.

If the quadrupole field is traversed along a path of length $z-z_{\mathrm{in}}$, and with $|\kappa q G/P| = k_0^2$ and $k = k_0 (z - z_{\mathrm{in}})$, the solution of the equations of motion is (a prime denotes the derivative with respect to z)

$$\left. \begin{aligned} x(z) &= x_{\mathrm{in}} \cosh(k) && + x'_{\mathrm{in}} \sinh(k)/k_0 \\ x'(z) &= x_{\mathrm{in}} k_0 \cosh(k) && + x'_{\mathrm{in}} \cosh(k) \\ y(z) &= y_{\mathrm{in}} \cos(k) && + y'_{\mathrm{in}} \sin(k)/k_0 \\ y'(z) &= -y_{\mathrm{in}} k_0 \cos(k) && + y'_{\mathrm{in}} \cos(k) \end{aligned} \right\} \tag{3.52d}$$

where the hyperbolic functions are the solutions for the defocussing direction.

In accelerator theory (e.g. Johnsen (1987b)) one obtains for *thin lenses* of thickness $L = z_{\mathrm{out}} - z_{\mathrm{in}}$, with $k_0 L \ll \pi/2$ (orders of $L \geqslant 3$ are neglected)

$$\left. \begin{aligned} \begin{pmatrix} x_{\mathrm{out}} \\ x'_{\mathrm{out}} \end{pmatrix} &= \begin{pmatrix} 1 + k_0^2 L^2/2 & L \\ k_0^2 L & 1 + k_0^2 L^2/2 \end{pmatrix} \begin{pmatrix} x_{\mathrm{in}} \\ x'_{\mathrm{in}} \end{pmatrix} \\ \begin{pmatrix} y_{\mathrm{out}} \\ y'_{\mathrm{out}} \end{pmatrix} &= \begin{pmatrix} 1 - k_0^2 L^2/2 & L \\ -k_0^2 L & 1 - k_0^2 L^2/2 \end{pmatrix} \begin{pmatrix} y_{\mathrm{in}} \\ y'_{\mathrm{in}} \end{pmatrix} \end{aligned} \right\} \tag{3.52e}$$

Note that without the magnetic field, the matrix has the form $\begin{pmatrix} 1 & L \\ 0 & 1 \end{pmatrix}$. If L

3.3 Fitting the tracks of charged particles

tends to 0 with $k_0^2 L = $ const., the remaining term is

$$\begin{pmatrix} 1 & 0 \\ k_0^2 L & 1 \end{pmatrix}_{x,x'}; \quad \begin{pmatrix} 1 & 0 \\ -k_0^2 L & 1 \end{pmatrix}_{y,y'} \tag{3.52f}$$

'Parametrization'

One way of obtaining a track model suitable for rigorous track fitting is the evaluation of a *local linear track model*. Many attempts have been made to find a more *'global* – in general nonlinear – *parametrization'*, and some of them have been successful, mainly if some of the five independent parameters vary over only a small range (e.g. limited interaction region in storage rings, small fixed target, limited momentum range in elastic scattering). Furthermore invariance of the magnetic field under rotation can reduce the dimension of the parameter vector (precisely speaking it leads to a trivial functional relation for one of the parameters), and symmetry planes of the magnetic field can reduce the number of possible coefficients for the track model. The technique consists of making a direct 'ansatz' for the coordinate functions of the undisturbed track as a linear combination of (if possible orthogonal) functions $\boldsymbol{\Phi}(\mathbf{p})$:

$$\mathbf{F}(\mathbf{p}) = \sum_{i_1,\ldots,i_5}^{m_1,\ldots,m_5} \mathbf{a}_{i_1,\ldots,i_5} \times \prod_{j=1}^{5} \Phi_{i_j}(p_j) \tag{3.53}$$

The functions Φ_i may be powers of p_j, trigonometric functions, Chebycheff polynomials, etc.

The choice of the functions Φ_i determines the properties of the approximation (Equation (3.53)). For instance, the choice of Chebycheff polynomials guarantees the minimization of the maximal linear (absolute) residual, if the appropriate 'training sample' of tracks is chosen $(\mathbf{p}^{(k)}, \mathbf{f}(\mathbf{p}^{(k)}); k = 1,\cdots,M)$. The coefficients $\mathbf{a}$ are then evaluated by requiring the quadratic form

$$Q^2 = \sum_{k=1}^{M} [\mathbf{f}(\mathbf{p}^{(k)}) - \mathbf{F}(\mathbf{p}^{(k)})]^2 \tag{3.54}$$

to be a minimum, leading to a set of $m_1 \times \cdots \times m_5$ linear equations.

If the functions Φ_i are *orthogonal* for the appropriate training sample, some advantage can be taken of an efficient choice of coefficients. First, the matrix to be inverted is *diagonal*, and second, each coefficient contributes *independently* to the total sum of Q^2, immediately allowing the selection of the most relevant coefficients. However, this requires a choice of the training sample according to the type of functions chosen, which is not always feasible in practice, due to limitations of the field map, of nonphysical regions etc. In the latter case, an iterative procedure has to be chosen, using a 'Gram–Schmidt transformation' for orthogonalization. An efficient program performing the choice of coefficients is available at CERN (Brun, et al. 1979). The algorithm is described in Wind (1972). Finally an *independent* uniform 'test sample' must be generated to test the quality of the approximation, and both the averaged squared and the maximal residual must be examined carefully, in comparison with the detector resolution, for a possible distortion of the χ^2 distribution, and a possible bias. However, it should be mentioned that the addition of higher order functions makes sense only if the averaged squared residual of the test sample *and* the maximal residual, together with the 'Q^2 ansatz', are still decreasing.

The method was successful in several experiments, a few of which should be mentioned. In the R401 experiment (using the SFM at CERN-ISR) the reaction $pp \to pp$ and $pp \to p(n\pi^+)$ have been measured. The result of parametrization was typically to keep 40 coefficients after having subdivided the phase space into several cells (Aubert and Broll 1974). For regions of a very inhomogenous field, the classical numerical interpretation was kept as back-up. The computer time used in the tracking routine was reduced by an order of magnitude for this large detector. A sophisticated correction procedure had to be applied when a detector was remounted in a slightly different position after having been removed for maintenance. In the $pp\uparrow \to pp$ elastic scattering experiment with a fixed target (CERN WA6, see above) the forward particle's trajectory could be parametrized by 30 coefficients for the deflection plane and by 15 for the perpendicular one, for a total spectrometer length of 50 m and a variation of deflection power '$\Delta BL/BL$' $\cong 20\%$ (Fidecaro, et al. 1980). The computer time for evaluating these functions in the track fitting program could almost be neglected.

A drawback of this method is the necessity to reevaluate the coefficients when the position of a detector inside the magnetic field is changed during the overall lifetime of the experiment (outside the field simple corrections can be applied).

Spline approximation

Another way to represent a trajectory is to connect the coordinates in the two projections by a 'spline curve'. A first approximation would be obtained by interpolating between two subsequent coordinates along the

3.3 Fitting the tracks of charged particles

track by quadratic parabolas, such that the following continuity constraints hold for the resulting curve of subsequent parabola segments:

The curve passes through the measured coordinates and is therefore continuous;

the slope of the curve is continuous at the measured coordinates (this assumes that multiple scattering is not concentrated in the dense material in the vicinity of these measurements (detector frames) or anywhere else (walls)).

Cubic splines are also often used (Subsection 2.2.1).

Such a model allows the evaluation of the *direction* at each measured point. After computation of the *magnetic field* at these points, the *second derivatives* ('curvature') can be calculated up to an overall proportionality factor $(1/P)$ from the equations of motion. If continuity is also requested for the second derivatives, a *cubic spline* to the measurements can be used.

However, a spline is an approximation to the scattered path and also includes the measurement errors; it does not correspond to an exact solution of the equations of motion and is therefore not suitable for a rigorous track fit in the sense of this chapter. The more appropriate way of using splines for the final track fit is to assume a *cubic spline* for *the track's second derivatives*, i.e., a quintic spline model for the track as given by the equations of motion (Wind 1974, 1978); the most elegant case is when the equations of motion are written in the form of Equation (3.41b) (see also Bugge and Myrheim 1981):

$$\left.\begin{array}{l} x'' = (\kappa q/P) X''(\mathbf{x}, \mathbf{x}') \\ y'' = (\kappa q/P) Y''(\mathbf{x}, \mathbf{x}') \end{array}\right\} \quad (3.55\text{a})$$

It is assumed that from pattern recognition (possibly combined with a cubic spline or a parabolic fit to the measured coordinates) the full space point for each measured coordinate is available as well as the direction vector. Knowing the field, the right hand sides of the Equations (3.55a) can now be calculated at each space point corresponding to a coordinate:

$$\left.\begin{array}{l} X''(z_i) = [x_i' y_i' B_x(x_i) - (1 + x_i'^2) B_y(x_i) \\ \qquad + y' B_z(x_i)](1 + x_i'^2 + y_i'^2)^{1/2} \\ Y''(z_i) = [(1 + y_i'^2) B_x(x_i) - x_i' y_i' B_y(x_i) \\ \qquad - x_i' B_z(x_i)](1 + x_i'^2 + y_i'^2)^{\frac{1}{2}} \end{array}\right\} \quad (3.55\text{b})$$

where i denotes the value at $z = z_i$. In order to get a continuous interpolation between these points of curvature but to keep the freedom to account for measurement errors, a cubic spline interpolation is applied to the second derivatives, and then twice integrated along the preliminary reconstructed path of the particle:

$$\left. \begin{array}{l} X''(z) = S_x^{(3)}[X''(z_1), \cdots, X''(z_n)] \\ Y''(z) = S_y^{(3)}[Y''(z_1), \cdots, Y''(z_n)] \end{array} \right\} \quad (3.55c)$$

where $S_x^{(3)}, S_y^{(3)}$ are cubic spines interpolating the second derivatives of x and y.

If energy loss has to be taken into account $X''(z_i)$, $Y''(z_i)$ have to be replaced by $X''(z_i)/(1 - \varepsilon(z_i))$, $Y''(z_i)/(1 - \varepsilon(z_i))$, with $P(z_i) = P_0(1 - \varepsilon(z_i))$, and with a first guess of P_0 from pattern recognition.

The ansatz for the track model is (Equation (3.1), Fig. 3.1a):

$$\left. \begin{array}{l} f_{x,i} = x(z_i) = a_1 + a_2 z + (\kappa q/P) \int_{z_1}^{z_i} \left[\int_{z_1}^{v} X''(u) \, du \right] dv \\ f_{y,i} = y(z_i) = b_1 + b_2 z + (\kappa q/P) \int_{z_1}^{z_i} \left[\int_{z_1}^{v} Y''(u) \, du \right] dv \end{array} \right\} \quad (3.56)$$

where the integration parameters a_1, b_1, a_2 and b_2, and the inverse of the momentum $1/P$ can be considered as the track parameter vector **p**, giving in many cases a reasonable five-dimensional track model to be used in the LSM (Corporaal 1979; Wind 1979; Zupančič 1986), and also allowing the evaluation of a χ^2, and pull quantities, and the application of error propagation.

In order to be able to construct these cubic splines passing through $X''(z_i), Y''(z_i)$, one needs continuity constraints and appropriate boundary conditions (Cox 1982; Wind 1984):

(a) Ansatz: A cubic spline is a polynomial expression of order 3 (at most) for each interval (z_i, z_{i+1}).
(b) Continuity condition: If the spline is of order k, $k-1$ derivatives must be continuous. This also basically holds at the points z_i, the 'knots'.
(c) Boundary conditions: From the continuity condition, $k-1$ free parameters are left open, i.e. two for a cubic spline, giving one degree of freedom for each boundary. The boundary conditions should be chosen according to the spectrometer set-up, assuming either a constant field or a field-free region at the end points. This can be taken into consideration by requiring no change in the curvature at the end points $(S'_x(z_1) = S'_y(z_1) = S'_x(z_n) = S'_y(z_n) = 0)$.

3.3 Fitting the tracks of charged particles

Having performed the integration interval-by-interval, a particle's trajectory is now represented by a *doubly integrated cubic spline* (with the continuity conditions of a quintic spline but without a strict interpolation constraint). The five parameters can then be determined from a linear least squares fit. The evaluation of the cubic spline expressions $S_x^{(3)}(z)$, $S_y^{(3)}(z)$ could also be based on more data points than have actually been measured; these 'additional knots' then, of course, have to be excluded from the least squares fit. This is of importance if the magnetic field varies strongly between the measured coordinates.

The method has been successfully applied in many experiments, giving good results, χ^2 distributions and pull quantities (SFM and AFS detectors at the CERN-ISR, OMEGA detector at the CERN-SPS and some others). However, some drawbacks should be mentioned:

Only the global track fitting method can be used.

The track model depends in a hidden way on the measurements. Therefore it is difficult to determine the rigorous weight matrix of the least squares fit.

If multiple scattering is significant compared to the detector resolution, the method follows the physical path of a particle rather than the ideal one too closely, giving in practice pull quantities and χ^2s which are too small (Subsection 3.2.3). This is because multiple scattering is a discontinuous term superimposed on the curvature as given by the equation of motion (Equation (3.41b)), and is therefore not well described by s-spline interpolation. Some preliminary smoothing is already done by the model itself before the proper fit procedure.

The trajectory of the particles should have a common 'drift direction' (z), and the deflection angle should not be large ($<\pi/2$). The detectors should all be arranged perpendicular to this drift direction.

The parameters chosen give a very simplified derivative matrix, corresponding to just a variation of a rigid curve as far as the first four parameters are concerned; the field gradient transverse to the trajectory is completely neglected.

Several attempts have been made to overcome these drawbacks (e.g. more additional points, piecewise track reconstruction, break points, iterative procedures, etc.), but – reliability and clarity being major requirements in large collaborations – it seems advisable to limit the application of this method to tracks with the following properties:

similar order of magnitdue for the resolution of the individual detectors;
detector errors which are not too large;
little multiple scattering;
small transverse field gradient;
general 'main direction' perpendicular to the detectors;
limited deflection.

3.3.1.4 The field representation

When integrating the equations of motion (Equations (3.40) and (3.41)), knowledge of the magnetic field (defined by its induction **B**) is needed at some known points (Equation (3.46)). Other tracking methods require the knowledge of higher derivatives, e.g. Equation (3.50).

Only the use of the field representation as needed for tracking will be discussed here. The problem of preevaluation of the field during the design of a spectrometer magnet will not be covered, nor will the algorithms for controlling or smoothing field measurements in practice be considered.

The most straightforward way to establish a field representation, which gives a fast response when looking for the field at a certain point **x**, is a dense grid of points at each of which the field vector is stored:

$$\mathbf{B}_{ijk} = \mathbf{B}(x_i, y_j, z_k) \tag{3.57}$$

where i, j, k are obtained by rounding down the values x, y, z according to the grid; for an equidistant grid: $i =$ integer $(x/\Delta x + 0.5 \,\text{sign}(x))$ etc. This method is frequently used with a possible improvement of the precision by simple interpolation formulae.

However this field representation is only efficient if the map is stored in a fast access memory, leading to some limitation of the number of values which can be stored. For 'virtual memory' computers, the size is not limiting, but some care must be taken to avoid excessive 'page faults', making it highly inefficient in time and limiting the portability of the map. Further optimization may be gained by using sophisticated grid search algorithms. In a very inhomogenous field as many as 100 steps in each direction can be needed, leading to some 3×10^6 field values.

If some symmetry planes exist, the mechanical precision and homogeneity of the magnet yoke and the coils should be good enough that the dimension of the field map could be reduced by a factor 2, 4 or even 8.

A measure of the required precison is again the detector resolution, with respect to which the momentum-dependent lateral displacement along the

3.3 Fitting the tracks of charged particles

full track due to the field approximation should be negligible. Whether this displacement, summed over all the boxes along a track, cancels on average and can be treated as a pseudostochastic process $\sigma(\Delta x) \sim$ (number of boxes)$^{-\frac{1}{2}}$ or whether it corresponds to a biassed summation, must be checked by a Monte Carlo calculation. Sometimes $\Delta P/P$ originating from the detector errors versus $\Delta|\mathbf{B}|/|\mathbf{B}|$ is examined instead of the error in the lateral displacement.

A way of avoiding numerical tables which describes magnetic field components in sufficient detail for simple interpolation formulae is to subdivide the volume into sufficiently small fractions (e.g. nonoverlapping 'boxes') and also to store the derivatives. Then the box size can be enlarged by an important factor. Although the higher order coefficients must be stored in addition (three for the field itself plus $3 + 2n$ for the n'th derivative), the number of values to be stored can be reduced considerably. Note that fulfilling the Laplace equation (magnetic fields in spectrometer magnets are constant in time) does not always give the minimum number of coefficients for a given precision, but it has the advantage of always selecting the same set for a given order of the polynomial, speeding up those tracking methods which use the field derivatives. However, an interpolation polynomial must be evaluated, limiting the usefulness of a higher order representation due to the time needed for interpolation. This limit depends on whether tracking makes direct use of the derivatives, and is of the order $n = 1$ to $n = 3$. (For more details of this and also symmetry operations on the derivatives, see Metcalf et al. (1973) and Metcalf (1974)).

Another approach is an expansion in terms of low-order orthogonal polynomials for each field component inside each fractional volume. An implementation in terms of Chebycheff polynomials has frequently been used for experiments at CERN (Louis and Verkerk 1975). The output generated by this program is a FORTRAN source code, containing a loop-free evaluation of the polynomial, with all insignificant terms eliminated. No precision is lost by the polynomials not satisfying the Laplace equation. For example, in the UA1 experiment at CERN, the entire magnetized zone was represented by these polynomials (including all three components, and both high-precision tracking zones and intrumented iron volumes). Again the highest order of polynomials was no more than 3.

If the field has rotational invariance within the precision limits discussed above, the field map grid is only *two-dimensional*, allowing more freedom for the choice of the field representation. If these symmetries are also observed when designing the detector, some additional computation can be

290   *Track and vertex fitting*

avoided, e.g. how to reach the detector surface exactly during numerical integration, or to ease parametrization (Equation (3.51) and (3.53)). The field shapes for several storage ring experiments are given in Fig. 3.9.

In practice it is often convenient to establish a *global field* model in parallel, allowing a check of the measurements on the boundary versus those inside the field volume, to smooth out the field measurements and to detect outliers. Such a model should be kept as a back-up, allowing a change in the field representation of the tracking program if needed. Such a model, although slow for the field evaluation, can also be a useful tool for evaluating the training sample for parametrization purposes, avoiding discontinuities which can occur between adjacent field map boxes. These discontinuities can cause difficulties during numerical differentiation (Equation (3.49)); a solution is to choose the boxes when tracking the zero track, and to keep them fixed as the track varies, even if a slight extrapolation is needed, although in general, extrapolation should be avoided.

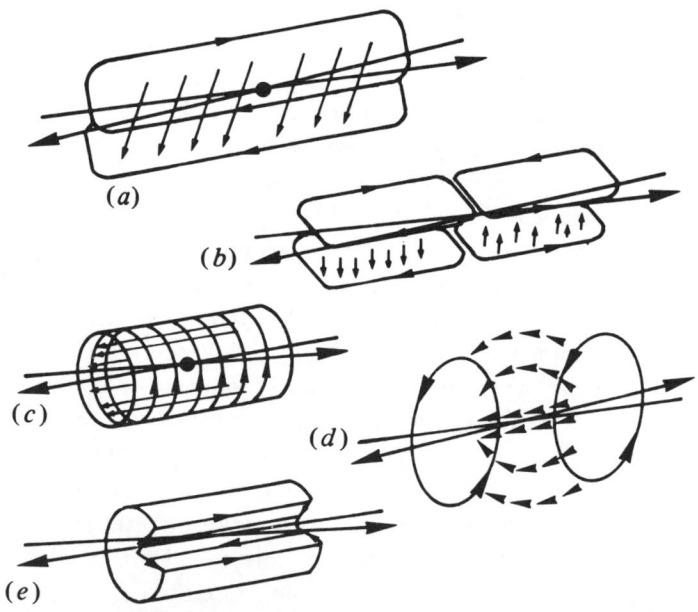

Fig. 3.9 Several field shapes from storage ring experiments (from C. Fabjan): (*a*) Dipole (UAI CERN, DASP DESY); Dipoles (*H*-magnets, *C*-magnets) are also used in the forward spectrometer of fixed-target experiments. (*b*) split field magnet (SFM CERN-ISR; (*c*) solenoid (DELPHI CERN-LEP); (*d*) axial field magnet (AFS CERN-ISR); (*e*) toroid (MARK II SLAC).

3.3 Fitting the tracks of charged particles

To summarize: when speaking about the field representation one must distinguish between the *general field map* which allows an evaluation of a field value at any point with low speed, but high precision, and the usual *boxwise field representation for the tracking program*, where the kind of field model must be chosen in conjunction with the tracking algorithm. A trade-off is necessary between memory utilization, speed, and precision.

3.3.1.5 The effects of matter on the trajectory

So far it has been assumed that the trajectory of the charged particle is not affected by any material. In reality, however, several types of secondary interactions between particle and material may occur: multiple scattering and energy loss due to electromagnetic interactions (also, the external 'bremsstrahlung' may be not negligible for high-energy electrons deflected by strong magnetic fields), elastic nuclear scattering, and processes of any kind which create additional new particles.

In this subsection, only *multiple scattering and energy loss due to electromagnetic interactions* are discussed. The theory of these processes is well understood, and good descriptions of these phenomena are given in text books, e.g. Allison and Wright (1987), Fernow (1986), and Kleinknecht (1986); for numerical values see also Particle Data Group (1986). A standard text book on this subject is Rossi (1965). For multiple scattering see also Gluckstern (1963).

Multiple scattering

When an electrically charged particle traverses a layer of matter, it is deflected from the path determined by the equations of motion in a vacuum by elastic scattering off the electrons or the nuclei of that matter.

The transverse momentum transfer of this process can be calculated by integrating the transverse components of the force acting between two charged particles, one moving (a projectile) and the other a *target* at rest (e.g. Jackson 1962) (the average of the longitudinal contribution is 0):

$$P_T = \int f_T \, dt = 2q_1 q_2 e^2 / vb \tag{3.58a}$$

or

$$\theta \cong \sin \theta \equiv P_T / P = 2q_1 q_2 e^2 / Pvb \tag{3.58b}$$

where t is the time (i.e. integration over the interval of interaction), f_T the transverse force component between the particles, e the positive elementary charge, q_1 and q_2 the charges of the projectile and target, respectively, in multiples of e, v the velocity along the path of the projectile, b the impact parameter of the undisturbed path of the projectile with respect to the target, P the momentum of the projectile, P_T the transverse momentum transfer caused by the interaction, and θ is the deflection angle.

The *differential cross-section* can be calculated from Equation (3.58b) for small deflection angles θ (note that $\cos\theta$ runs from $+1$ to -1):

$$\frac{d\sigma}{d\Omega} = \frac{1}{2\pi\theta}\frac{d\sigma}{d\theta} = \frac{1}{2\pi\theta}\frac{d\sigma}{db}\left|\frac{db}{d\theta}\right| \tag{3.59a}$$

and with

$$d\sigma/db = 2\pi b, \quad b = \text{const}/\theta, \quad |db/d\theta| = |\text{const}/\theta^2|$$

one obtains

$$\frac{d\sigma}{d\Omega} = \left(\frac{2q_1 q_2 e^2}{Pv}\right)^2 \frac{1}{16}\left(\frac{2}{\theta}\right)^4 \tag{3.59b}$$

This is an approximation of the more general *Rutherford formula* for the scattering of *point-like spinless particles*:

$$\frac{d\sigma_R}{d\Omega} = \frac{1}{4}q_1^2 q_2^2 \left(\frac{e^2}{Pv}\right)^2 \frac{1}{\sin^4(\theta/2)} \tag{3.59c}$$

If scattering occurs on an electron, $q_2 = -1$, and the cross section per atom increases with Z (the atomic number which is also the number of electrons in the atom). On the other hand, for scattering off a nucleus, $q_2 = Z$, and the cross section per atom increases with Z^2 (Fig. 3.10). Therefore setting the charge of the projectile $q_1 \equiv z$,

$$\frac{d\sigma_R}{d\theta}d\theta = \frac{d\sigma_R}{d\Omega}2\pi\sin\theta\, d\theta \cong 8\pi z^2\left(\frac{e^2}{Pv}\right)^2\frac{d\theta}{\theta^3}(Z + Z^2) \tag{3.60a}$$

The *cross-section per unit length* is given by

$$\frac{d^2\sigma}{d\theta dx}d\theta dx = n8\pi z^2 Z(Z+1)\left(\frac{e^2}{Pv}\right)^2\frac{d\theta}{\theta^3} \tag{3.60b}$$

where n is the density of atoms per unit volume ($= N\rho/A$, N is Avogadro's number ($\cong 6.022 \times 10^{23}$ atoms per mol), ρ is the density (grams per unit volume of matter) and A is the number of nucleons (protons and neutrons) in the nucleus).

The *variance of θ per scattering process* can then be evaluated by

3.3 Fitting the tracks of charged particles

choosing suitable $\theta_{\min}$ and $\theta_{\max}$ (see below):

$$\langle \theta^2 \rangle = \int_{\theta_{\min}}^{\theta_{\max}} \theta^2 (1/\theta^3) \, d\theta \Big/ \int_{\theta_{\min}}^{\theta_{\max}} (1/\theta^3) \, d\theta \tag{3.61a}$$

The *variance of θ per unit length* is therefore

$$\langle \theta^2 \rangle = (N\rho/A) 8\pi z^2 Z(Z+1)(e^2/Pv)^2 \ln(\theta_{\max}/\theta_{\min}) \tag{3.61b}$$

$\theta_{\min}$ can be derived from screening effects when b is large, and $\theta_{\max}$ from a more sophisticated model which takes into account the projectile's Compton wavelength:

$$\theta_{\max}/\theta_{\min} \approx 183 Z^{-\tfrac{1}{3}}$$

Considering the *variance per unit length of the projections of θ* in a local Cartesian coordinate system with one axis pointing in the projectile's direction ($\theta^2 = \theta_1^2 + \theta_2^2, \theta \, d\theta \, d\phi = d\theta_1 \, d\theta_2$) yields an extra factor of $\tfrac{1}{2}$:

$$\langle \theta^2 \rangle_{\text{proj}} = (N\rho/A) 4\pi z^2 Z(Z+1) \left(\frac{e^2}{Pv}\right)^2 \ln(183 Z^{-\tfrac{1}{3}})$$

or

$$\langle \theta^2 \rangle_{\text{proj}} = \tfrac{1}{2} z^2 \left(\frac{E_s}{\beta Pc}\right)^2 \frac{\rho}{X_0} \tag{3.61c}$$

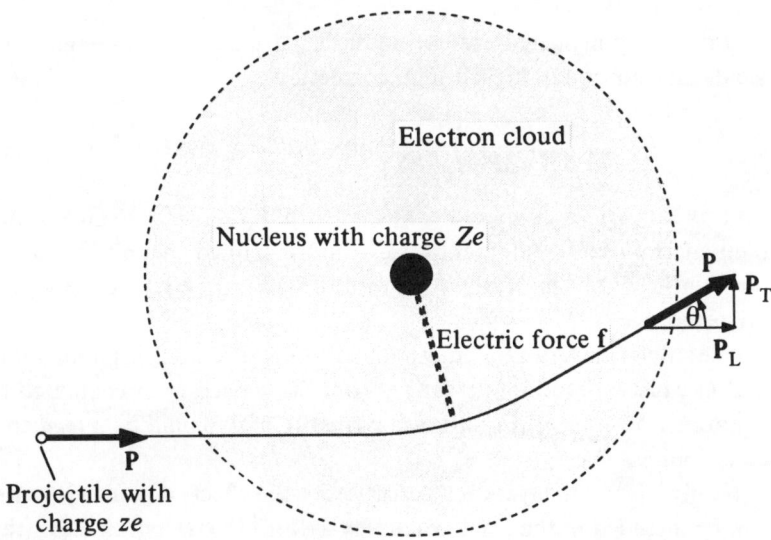

Fig. 3.10 Scattering of an incident particle with momentum **p** and charge ze on a nucleus of charge Ze. The electric force is transmitted by 'virtual photons'.

where

$$E_s \equiv m_e c^2 (4\pi/\alpha)^{\frac{1}{2}} \cong 0.0212 \text{ GeV}$$

$\beta = v/c$ (see Subsection 3.3.1.1)

m_e is the electron mass ($\cong 0.5 \times 10^{-3}$ GeV ($\cong 9.11 \times 10^{-28}$ g)), α is *Sommerfeld's fine structure constant* ($\cong 1/137$), and L_0 ($\equiv X_0/\rho$) is a geometrical (density-dependent) scaling variable of dimension unit length, called the *'radiation length'*. X_0 is a density-independent scaling variable (grams per unit area) (see also Equation (2.60)) and is given by

$$X_0 = [(4N\alpha/A)Z(Z+1)r_e^2 \ln(183 Z^{-\frac{1}{3}})]^{-1}$$

where $r_e (\equiv e^2/m_e c^2)$ is the classical (electromagnetic) electron radius $\cong 2.818 \times 10^{-13}$ cm. If scattering occurs sufficiently often in a layer of thickness x, the probability density function for the projected scattering angles (θ_1, θ_2) can be approximated by a Gaussian distribution:

$$f(\theta_1, \theta_2) = \frac{1}{2\pi x \langle \theta^2 \rangle_{\text{proj}}} = \exp[-\theta_1 \theta_2/(2x\langle \theta^2 \rangle_{\text{proj}})] \quad (3.61\text{d})$$

However, since the differential cross section for individual scattering has a large tail, convergence to a Gaussian distribution is poor, and for thin layers of matter these tails have to be taken into account (e.g. Fernow (1986); Particle Data Group (1986)).

A more sophisticated evaluation of the variance per unit length $\langle \theta^2 \rangle_{\text{proj}}$ yields an additional 'logarithmic correction':

$$\langle \theta^2 \rangle_{\text{proj}} = \tfrac{1}{2} \left(\frac{E_s}{\beta Pc}\right)^2 z^2 \frac{\rho}{X_0} [1 + 0.12 \log_{10}(x/L_0)]^2 \quad (3.61\text{e})$$

This is known as *Molière's formula* (with $E_s/\sqrt{2} = 0.015$ GeV). Since in high-energy physics momenta are usually defined in units of GeV, βPc (originating from the Rutherford formula) has to be replaced by βP, with $1/\beta = (m^2 + P^2)^{\frac{1}{2}}/P$.

The stochastic nature of multiple scattering of a particle passing through matter causes a random deviation from the trajectory as evaluated by the equations of motion (Equations (3.41a, b)), and *should be added* to these equations.

If only a few thin layers act as scatterers, the effects of multiple scattering can be included in the track model by a pair of extra parameters (the two scattering angles) for each layer ('break points'). These can be considered as

3.3 Fitting the tracks of charged particles

an unbiased measurement with mean $\langle \theta_{1,2} \rangle = 0$, and with variance $\langle \theta_{1,2}^2 \rangle$ given by Molière's formula (Equation (3.61e)).

The general treatment of multiple scattering in the context of the LSM will be discussed in Subsection 3.3.2.2. (For more detailed calculations see Scott (1963) and Bichsel (1970)).

Energy loss

In the equations of motion (Equations (3.41)), the absolute value of the momentum of the particle is assumed to be constant. In reality, when a charged particle passes through matter, some energy is transmitted to this medium, and the 'constant' must be readjusted as a function of the path length parameter.

The energy transmitted to the medium can be calculated from the transverse momentum transfer of a single scattering process; for a nonrelativistic recoil electron, it is

$$\Delta E = \frac{(\Delta P)^2}{2m_2} = \frac{P_T^2}{2m_2} \tag{3.62a}$$

where ΔE and ΔP are respectively energy and momentum transmitted to the recoil particle ($\Delta P = P_T$) and m_2 is the mass of the recoil particle. It follows that energy loss due to elastic scattering on electrons dominates, by orders of magnitude, the energy loss induced by the heavier nucleus, so the latter can be neclected (and $m_2 = m_e$).

For a given impact parameter b, the energy loss in a scattering process follows from Equation (3.58a) with $P_T = \Delta P$:

$$\Delta E(b) = P_T^2/2m_e = 2z^2 e^4/b^2 v^2 m_e \tag{3.62b}$$

The *energy loss per unit length* in the matter is given by:

$$\frac{dE}{dx} = Z(N\rho/A)\frac{2z^2 e^4}{v^2 m_e} 2\pi \int_{b_{min}}^{b_{max}} \frac{1}{b^2} b \, db$$
$$= \{[4\pi(N\rho/A)Zz^2 e^4/(\beta^2 m_e c^2)] \ln(b_{max}/b_{min})\} \tag{3.63a}$$

(Remember that $n = N\rho/A =$ density of atoms per unit volume (see Equation (3.60b)) and $d\sigma = 2\pi b \, db$ e.g. Fernow (1986)). In the literature, b_{max}/b_{min} is sometimes replaced by E_{max}/E_{min}, with the intuitive argument that E_{min} should be chosen as the average ionization potential and E_{max} from kinematics. However, this would create an additional factor $\frac{1}{2}$ in front of the logarithm (with $db/b = -dE/(2E)$) which is not found in a more rigorous calculation which includes proper treatment of bound states of electrons.

The semiclassical treatment which considers the electrons as nonrelativistic harmonic oscillators yields the *Bethe–Bloch formula* (Ahlen 1980; Jackson 1962; Rossi 1965):

$$\frac{dE}{dx} = 4\pi(N\rho/A)Ze^4 \frac{z^2}{m_e\beta^2c^2}\left[\ln\frac{2m_ec^2\beta^2\gamma^2}{I}\right] \qquad (3.64a)$$

where I is the *mean ionization*. For historical reasons, the 'mean ionization' is retained in the Bethe–Bloch formula, although it is, in fact, the *geometrical mean*. All oscillators with frequency ω contribute in an additive way *(arithmetic mean)* to the energy loss via the *logarithmic term* $\ln I(= -\ln 1/I)$, and therefore:

$$\langle \ln I_j \rangle = \sum_j (m_j/Z)\ln \hbar\omega_j = \ln \hbar\prod_j \omega_j^{m_j/Z} = \ln \hbar\langle\omega\rangle$$

$\hbar(\equiv h/2\pi)$ is 'Planck's constant' ($\cong 6.58 \times 10^{-25}$ GeV s), m_j is the number of bound electrons with oscillation frequency ω_j (with $\sum_j m_j = Z$), and $\gamma \equiv (1-\beta^2)^{-\frac{1}{2}} = E/m_1c^2$, where m_1 is the mass of projectile (see also Subsection 3.3.1.1). The energy loss depends on β (or $\beta\gamma$), but not explicitly on the mass m_1 of the incident particle. Therefore, the behaviour of the Bethe–Bloch formula is discussed in the variable β:

For small β (but still large compared to $v(\omega)/c$ of the orbital electrons), the energy loss decreases proportionally to $1/\beta^2$.

All incident particles have a region of minimum ionization with

$$dE/dx \cong 0.002 \text{ GeV (g cm}^{-2})^{-1} \text{ around } \beta\gamma \equiv Pc/(m_1c^2) \cong 3.$$

For the purpose of experimental data analysis, energy loss can often be approximated by a constant for $\beta\gamma > 1$. Even for a more precise treatment it is usually sufficient to calculate β and γ only once per track. However, the constant must be readjusted for each change of medium because of the different average ionization potential I (Fernow 1986; Particle Data Group 1986):

$$I/Z \cong \begin{cases} 21 \text{ eV for helium} \\ 13\text{–}15 \text{ eV for light atoms} \\ 10 \text{ eV for heavy atoms} \end{cases}$$

For very high incident particle energies ($\beta\gamma > 10$), energy loss increases

3.3 Fitting the tracks of charged particles

proportionally with $\ln(\beta\gamma) \simeq \ln(E/mc^2)$ (the region of 'relativistic rise') (Fig. 3.11). This effect plays an important role in particle identification (the 'dE/dx method') (Allison and Wright 1987; Kleinknecht 1986).

Equation (3.64a) describes the energy loss in matter composed of isolated atoms. For real media the interatomic space is small, and dielectric screening reduces energy loss for collisions with large impact parameters (Sternheimer 1952):

$$\frac{dE}{dx} = \text{const} \times \left[\ln\left(\frac{2m_e c^2 \beta^2 \gamma^2}{I}\right) - \beta^2 - \frac{\delta(\gamma)}{2} \right] \quad (3.64b)$$

The correction term $\delta(\gamma)$ (Fernow 1986) causes the energy loss to become constant again for $\beta\gamma > 100$ (the *Fermi plateau*). This limits the range of application of the 'dE/dx method' for particle identification.

The ratio between the energy loss at the Fermi plateau and at minimum ionization depends upon the medium. This ratio is only $\simeq 1.1$ for very dense media, thus allowing an approximation of energy loss by a constant value up to highest $\beta\gamma$. For argon at normal pressure, the ratio is $\simeq 1.6$; in this case, the energy loss is negligible and may be omitted for the reconstruction of particle tracks.

The energy loss of incident electrons in material see, for example Allison

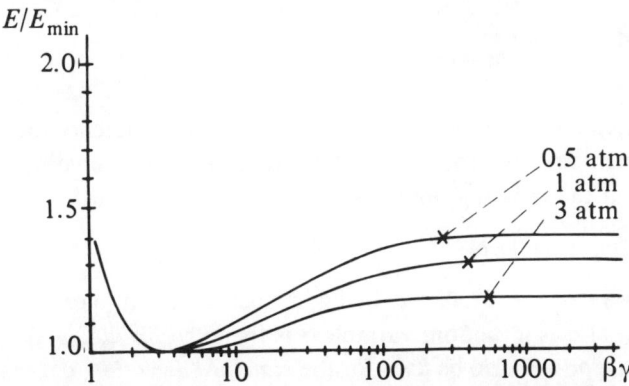

Fig. 3.11 Measured average mean energy loss in propane. If the momentum P is chosen as the abscissa, particles of different masses can be separated by the dE/dx method in the two 'slope regions'. However, statistical fluctuations are large ('Landau tail'), and the ionization path must be sufficiently long for the separation of individual particles (0.5–4.0 m).

and Wright (1987) and Fernow (1986) is given by:

$$\frac{dE}{dx} = 2\pi(N\rho/A)Ze^4 \frac{1}{m_e c^2}\left[2\ln\left(\frac{2m_e c^2}{I}\right) + 3\ln\gamma - 1.95\right] \quad (3.64c)$$

For electrons, the stochastic emission of synchrotron radiation due to the bending force of the magnetic field in a spectrometer must also be taken into account (Jackson 1962).

As mentioned above, energy loss must be included in the track model (Subsection 3.3.1.3) when solving the equations of motion (Equation (3.41)); in this context, however, it is treated as a deterministic process, the variance of energy loss being small. In this way energy loss is easily implemented into the Runge–Kutta and the spline methods, but it is less obvious for explicit track models like a helix. Some care is also necessary when propagating fitted track parameters and errors by numerical calculation of the derivatives.

3.3.2 The weight matrix

3.3.2.1 The measurement error of a detector

When a particle traverses a position-sensitive detector, the *measured position*, m, will, in practice, be different from the position of the real crossing point, $\overset{t}{c}$. This deviation can be described, in terms of statistics, by a conditional probability density function (Equation (3.13)), the *resolution function*:

$$\frac{d}{d\overset{t}{c}} P(c < \overset{t}{c}|m) = d(m; \overset{t}{c}) \quad (3.65a)$$

In practice, the dominant variable will be the difference between the measurement and the crossing point; however, this function quite often also depends on the crossing point itself:

$$d'(\varepsilon; \overset{t}{c}) \equiv d(m; \overset{t}{c}) \quad (3.65b)$$

where $\varepsilon = m - \overset{t}{c}$, with $m \equiv c$ being a measurement, i.e. an *individual realization* of c as a random variable; ε is the *experimental error*, although the word *error* will also be used for the *standard deviation* of ε, $\sigma(\varepsilon)$, namely the square root of the variance of ε, $\sigma^2(\varepsilon)$, or var (ε), which for an unbiassed measurement is the *expectation value* of ε^2: $\langle\varepsilon^2\rangle$.

Sometimes also the word 'resolution' (also experimental or detector resolution) is used for the standard deviation. This has historical reasons,

3.3 Fitting the tracks of charged particles

when in multichannel analysis the σ of a peak was also a measure of the two-peak separation. However, in some tracking detectors the two-particle separation is often determined by quantities other than σ (e.g. electronics dead time in drift chambers).

If the word resolution is used at all, it should not be used with confusing quantitative attributes (a tiny resolution would certainly not suggest a small σ; it is better to use the terms 'good resolution' or 'small error'). In the context of separation of the signals from two different particles, 'two-particle resolution' should be used.

An estimate of the error is the RMS:

$$\text{RMS} = \left(\frac{1}{N}\sum_{\alpha=1}^{N}\varepsilon_\alpha^2\right)^{\frac{1}{2}} \tag{3.66a}$$

If $\overset{t}{c}$ is constant but unknown, for instance in an ideal test beam, the variance should be estimated by the formula:

$$\text{RMS}_N^2 = \frac{1}{N-1}\sum_{\alpha=1}^{N}\left(m_\alpha - \frac{1}{N}\sum_{\beta=1}^{N}m_\beta\right)^2 \tag{3.66b}$$

which is an *unbiassed estimator* of the variance even for finite N. (The gain of information (Equation (3.19)) is proportional to the inverse of the variance, $1/\sigma^2$, when the model related to this measurement is linear.)

A tracking detector can measure a single coordinate (a space point), a space point and a direction, or the full track parameter vector including the momentum. Here, measurement of only a coordinate will be discussed, but most of what is said can be trivially extended to the more general cases.

As examples, three types of detectors will be discussed:

An idealized detector with a constant and genuine random error.
The drift chamber with a genuine but variable random error.
The MWPC with a quasi-random error.

An idealized detector

In an idealized detector the resolution function (Equation 3.65b) depends only on the difference between the measurement and the particle impact point:

$$d'(m - \overset{t}{c}; \overset{t}{c}) = d'(m - \overset{t}{c}) = d'(\varepsilon) \tag{3.67a}$$

In this case the error does not depend on the location of the impact point itself:

$$\sigma^2 = \int \varepsilon^2 d'(\varepsilon)\, d\varepsilon \qquad (3.67b)$$

Furthermore, it is assumed that the measurement is unbiassed. If it were not, the bias would be constant and one could easily correct for it by replacing ε by $\varepsilon - \langle \varepsilon \rangle$. If the errors are Gaussian (Equation (3.16)), the LSM applied to the track fit would coincide with the MLM (Subsection 3.2.3.1).

The drift chamber

In a drift chamber (Subsection 2.1.1), the coordinate is calculated from the drift time of the first electrons arriving at a sense wire after the particle has crossed the detector – trailing electron/ion pairs behind. The measurement error is generated by several effects:

The random behaviour and energy/space distribution of the primary electrons including outliers from energetic electrons (δ-rays). This contribution is constant for a fixed gas mixture, but can depend on the angle between the sense wire and the magnetic field.

The diffusion of the drifting electrons along the path. This contribution to the variance is proportional to the drift length. It is the dominant effect for large drift distances.

Different path lengths for different primary electrons due to inhomogeneities in the drift field and the 'avalanche field' near the sense wire. Special attention has to be paid to electrons coming from the region near the drift cell boundary where the field changes its sign at the adjacent cell.

Jitter effects in connection with the detector and readout electronics. The TDC clock is stopped when the signal is higher than a threshold given by a discriminator. Because of the variable shapes and variable integrated charges of the avalanches in a wire chamber, the threshold is passed at different relative times for different particles crossing the detector at the same position. This jitter depends on the rise time of the pulse, the relative variation of the pulse height, and the frequency of occurrence and shape of the noise signals, all putting constraints on the choice of the threshold. Pulse shaping at an early time may reduce jitter effects to their *intrinsic minimum*, while the integrating effect of long cables contributes to an increase of jitter. The jitter effects can be considerably reduced if the *pulse shape* is measured by a FADC.

3.3 Fitting the tracks of charged particles

The error due to the binning of the TDC. This contribution to the resolution function is of rectangular, triangular, or trapezoidal shape according to the bin sizes for the 'start' and 'stop signals', but is a real stochastic contribution due to the random relation between the arrival time and the clock status.

A typical time bin is in the order of a few nanoseconds; this is to be compared with a typical drift velocity of 50 μm ns^{-1}. The variance for a time bin of width Δt is:

$$\langle c^2 \rangle = \int_{-\Delta t/2}^{+\Delta t/2} (c - \bar{c})^2 f(c)(dc/dt) \, dt = (v\Delta t)^2/12 \tag{3.68}$$

or

$$\sigma(c) = (v\Delta t)/\sqrt{12}$$

where $f(c) = 1/\Delta t$ for t in the interval $(-\Delta t/2, +\Delta t/2)$, $f(c) = 0$ otherwise, and $dc/dt = v$ is the drift velocity.

Special behaviour must be expected for the pulse produced at the anode wire when the particle crosses in the avalanche region. In addition, the problem of how to decide on which side of the wire the particle has crossed the detector ('left–right ambiguity') is more difficult to resolve in this case.

For typical resolutions of a few tens to a few hundreds of micrometres, the mechanical tolerances from machining and mounting, as well as from gravitational sagging and electrostatic deflections must also be considered. Some of these effects can be corrected systematically (see below), but sometimes they must be considered 'on average' by a 'pseudostochastical treatment', smeared out for different trajectories of different particles due to the distances between detectors. However, as much as possible should be done at the systematic level, in order to avoid an ununnecessary loss of information.

Special problems arise inside a magnetic spectrometer, where the drift direction is determined both by the electric and the magnetic fields. If the magnetic field is constant, its effect can be compensated by inclining the direction of the electric field. Otherwise, quite complicated 'isochronous lines' (space curves of equal drift times) have to be expected. The magnetic field acts also as a focussing or defocussing force for the avalanche electrons, significantly influencing the measurement error.

Systematic effects can be due to mechanical tolerances, but also to changes in the drift velocity. Special care is needed when working in the nonsaturated region of drift velocity: the gas mixture, impurities, temperature, and pressure all influence the drift velocity and must be permanently monitored. This, of course, requires a reasonable number of calibration tracks for each monitoring period. For details on drift velocity see, e.g. Fernow (1986), Leo (1987), and Sauli (1978).

MWPCs

One of the milestones in the history of particle detectors and experimental particle physics was the invention of the MWPC (Bouclier et al. 1974; Charpak 1978; Sauli 1978). It features good spatial resolution, and therefore also provides good momentum resolution even for highly energetic particles. In addition, its good two-track separation allows the detection of 'jets' (bundles of particles close to each other). The MWPC also has a good efficiency even for large multiplicities, which is necessary for detecting complex event structures. It has only local dead time, thus allowing high event rates, and it can easily be incorporated into a fast decision trigger, which is its most outstanding advantage compared to the drift chamber (see above). Last, but not least, it is mechanically simple, not very sensitive to small changes of the gas mixture or the high voltage, and it is easy to calibrate.

A crude estimate of the measurement error perpendicular to the wires is straightforward: from Equation (3.68) it follows for a flat distribution that

$$\sigma = d/\sqrt{12} \tag{3.69}$$

where d is the distance between two adjacent wires (0.5–3.0 mm in practice). A *cluster* is defined as a set of adjacent fired wires belonging to one coordinate measurement. By considering odd and even wire cluster sizes for particles crossing in the neighbourhood of a wire or in a region near to the midpoint between two wires, respectively, this value could be reduced theoretically by a factor of 2, but in practice only a factor of 1.1 (i.e. about 10%) has been achieved for a particle path normal to the chamber surface.

For slightly inclined tracks the error goes through a minimum ($\cong 0.8d/\sqrt{12}$ in practice), becoming larger than the value in Equation (3.69) for angles in the region of 45°. This is due to effects such as 'cluster decays' (a missing wire in the set of wires covered by the projection of the particle's path in the chamber) (Fig. 3.12).

Using only information from single-wire clusters, the resolution function is an *ad hoc* deterministic step function when firing the wire nearest to the

3.3 Fitting the tracks of charged particles

impact point of the particle's trajectory. This function is randomized by the stochastic character of the position of the impact point, and correlations between different detectors are smeared out by the different directions and curvatures due to the phase space population, deflection in the magnetic field, and multiple scattering. Therefore, the standard LSM can be applied. In a dense stack of chambers, however, the correlations between measurement errors cannot be neglected, but so far no attempt to take them into account in a rigorous way has been successful. Replacing the LSM by a 'Chebycheff norm' did not result in a significant improvement (James 1983).

A parametrized formula for the resolution is $\sigma^{\frac{1}{2}} = C(d/\sqrt{12})f(\beta)$ (with $C = 0.8$–0.9), where β is the angle between the track projected onto a plane perpendicular to the wires and the normal to the wire plane; $f(\beta) = 1$ for $\beta = 0$, has a minimum around 15–20°, and rises sharply for larger angles. In practice, however, the dependence of the measurement errors on the track parameters (mainly position and angle) is sufficiently smooth that the errors used for the fit can be evaluated from the less precise parameters resulting from the previous pattern recognition (usually after a suitable parametrization). Only in exceptional cases must the error be reevaluated during the track fit. For the estimation of the variances see e.g. Frühwirth (1986).

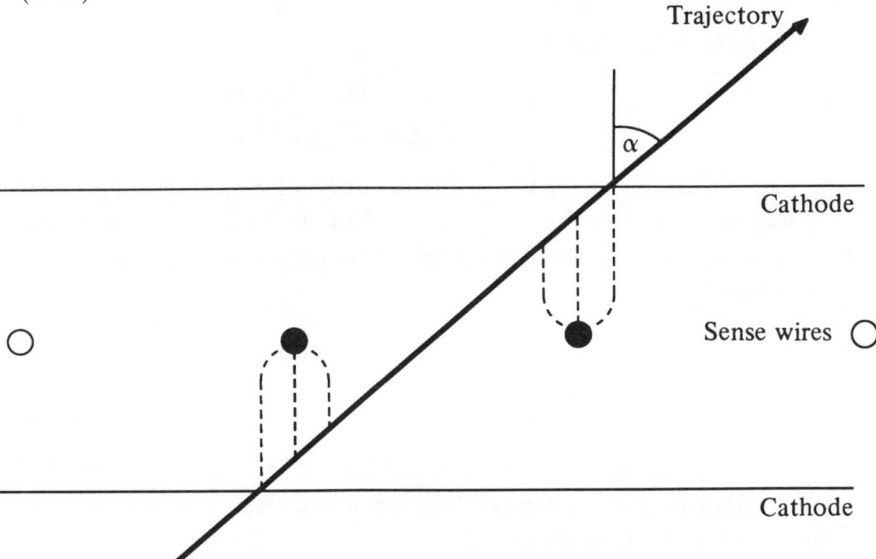

Fig. 3.12 Particles crossing the MWPC with a projected angle α can hit several wires, increasing the resolution for small α. The electrons and ions follow the field lines around the anode wires.

An interesting application of a standard MWPC for very high resolution was performed in the European Hybrid Spectrometer (EHS) at CERN. A standard MWPC was turned through 45° and the drift time correlations between the individual wires measured, giving a resolution of better than 50 μm and a 100% background rejection (Pernicka, Regler, and Sychkov 1978; Aquilar-Benitez *et al.* 1983).

3.3.2.2 Weight matrix formalism for multiple scattering

Multiple Coulomb scattering acting on a particle sums up to relatively small but random change of the direction of flight. The resulting effect is a *stochastic process*, in particular a *Markov process*, and one can only evaluate the 'probable' amount of influence of a scatterer on a particle trajectory (Equation (3.61e)).

Within the approximation that the multiple scattering can be described in two perpendicular planes by two random variables, the probability that a particle of momentum P travelling along the z axis leaves a scatterer of a length L in the interval of the projected angle $(\theta, \theta + d\theta)$ with a lateral displacement $(\epsilon, \epsilon + d\epsilon)$ can be approximately described by the distribution function, (Equation (3.61d))

$$dF(\epsilon, \theta; L) = 2\pi (L^4 \theta_s^2/24)^{-\frac{1}{2}}$$

$$\times \exp\left\{-\left[\frac{4}{\theta_s^2}\left(\frac{\theta^2}{L} - \frac{3\theta\epsilon}{L^2} + \frac{3\epsilon}{L^3}\right)\right]\right\} d\epsilon\, d\theta \quad (3.70a)$$

where $\theta_s^2/2$ is the projected mean squared angle of the scattering per unit length as calculated by Equation (3.61e). (Note that in Subsection 3.3.2.1, following the notation in the classical text books, x was the particle direction.)

Proof:

$$\langle \theta^2 \rangle = \left\langle \left(\int_{z_r}^{z} \frac{\partial \theta}{\partial z'} dz'\right)\left(\int_{z_r}^{z} \frac{\partial \theta}{\partial z''} dz''\right) \right\rangle \quad (3.70b)$$

θ is a random quantity, and so is $\partial \theta / \partial z$. If multiple scattering is regarded as 'white noise' one can write with $L = z - z_r$:

$$\left.\begin{array}{l} \langle (\partial \theta/\partial z')(\partial \theta/\partial z'') \rangle\, dz'\, dz'' = (\theta_s^2/2)\delta(z' - z'')\, dz'\, dz'' \\ \langle \theta^2 \rangle = (\theta_s^2/2)L \end{array}\right\} \quad (3.70c)$$

3.3 Fitting the tracks of charged particles

Furthermore

$$\langle \theta \varepsilon \rangle = \left\langle \left(\int_{z_r}^{z} \frac{\partial \theta}{\partial z'} dz' \right) \left(\int_{z_r}^{z} (z'' - z_r) \frac{\partial \theta}{\partial z''} dz'' \right) \right\rangle$$
$$= (\theta_s^2/2) L^2/2 \tag{3.70d}$$

$$\langle \varepsilon^2 \rangle = \left\langle \left(\int_{z_r}^{z} (z' - z_r) \frac{\partial \theta}{\partial z'} dz' \right) \left(\int_{z_r}^{z} (z'' - z_r) \frac{\partial \theta}{\partial z''} dz'' \right) \right\rangle$$
$$= (\theta_s^2/2) L^3/3 \tag{3.70e}$$

or in matrix notation

$$\left\langle \begin{pmatrix} \theta^2 & \theta\varepsilon \\ \varepsilon\theta & \varepsilon^2 \end{pmatrix} \right\rangle = \theta_s^2/2 \begin{pmatrix} L & L^2/2 \\ L^2/2 & L^3/3 \end{pmatrix} \tag{3.70f}$$

with the inverse

$$(8/\theta_s^2) \begin{pmatrix} 1/L & -(3/2)/L^2 \\ -(3/2)/L^2 & 3/L^3 \end{pmatrix} \quad \square \tag{3.70g}$$

For the LSM it is necessary to know the covariance matrix of the measurement error vector $\varepsilon^T = (\varepsilon_1, \ldots, \varepsilon_n)$ (Subsection 3.2.3) or Equation (3.32). It can easily be evaluated from Equation (3.70e) (with $k < l$):

$$(cov(\varepsilon))_{kl} = \langle \varepsilon_k \varepsilon_l \rangle = (\theta_s^2/2)(z_k - z_r)^2 (z_l - z_r)/2$$
$$- (z_k - z_r)^3/6 \tag{3.71}$$

For the general case of a coordinate vector c one can define a generalized geometry factor

$$\langle c_k c_l \rangle = (\theta_{s,r}^2/2) I_{kl}$$

$$I_{kl} = \int_{s_r}^{\min(s_k, s_l)} g(s) \left(\frac{\partial f_k}{\partial \theta_1(s)} \frac{\partial f_l}{\partial \theta_1(s)} + \frac{\partial f_k}{\partial \theta_2(s)} \frac{\partial f_l}{\partial \theta_2(s)} \right) ds \tag{3.72a}$$

where

$$\left\langle \frac{df_k}{d\theta(s')} \frac{d\theta(s')}{ds'} \frac{df_l}{d\theta(s'')} \frac{d\theta(s'')}{ds''} d\theta(s') d\theta(s'') \right\rangle$$
$$= \frac{\partial f_k}{\partial \theta(s')} \frac{\partial f_l}{\partial \theta(s')} \frac{\theta_s^2}{2} \delta(s' - s'') ds' ds''$$

f is the model corresponding to c (see equation (3.1)) and θ_1, θ_2 are the two uncorrelated and orthogonal scattering angles from Equation (3.61d), see Fig. 3.13 (note that the parameters θ_1, θ_2 have to be taken at s). The function $g(s)$ takes account of a possible change of medium and of energy loss. It is 0 in vacuum, usually 1 when entering the first medium (reference medium) and increases with energy loss (Equation (3.64b)) according to Molière's formula (Equation (3.61c)), where a change in radiation length must also be built in for the evaluation of $g(s)$. The particle's path is usually determined

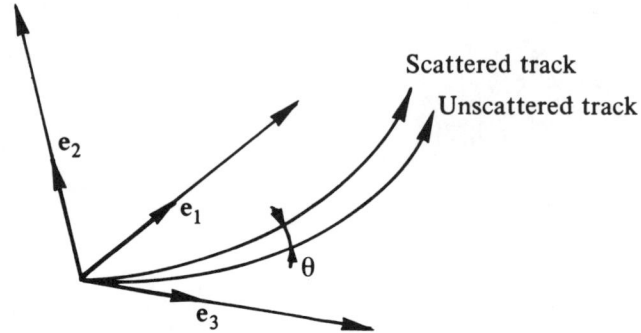

Fig. 3.13 The coordinate system with the axis e_3 parallel to the tangent of the trajectory. If multiple scattering is small it is sufficient to take into account only first order effects in the projected scattering angles θ_1 and θ_2 describing the change in the direction unit vector.

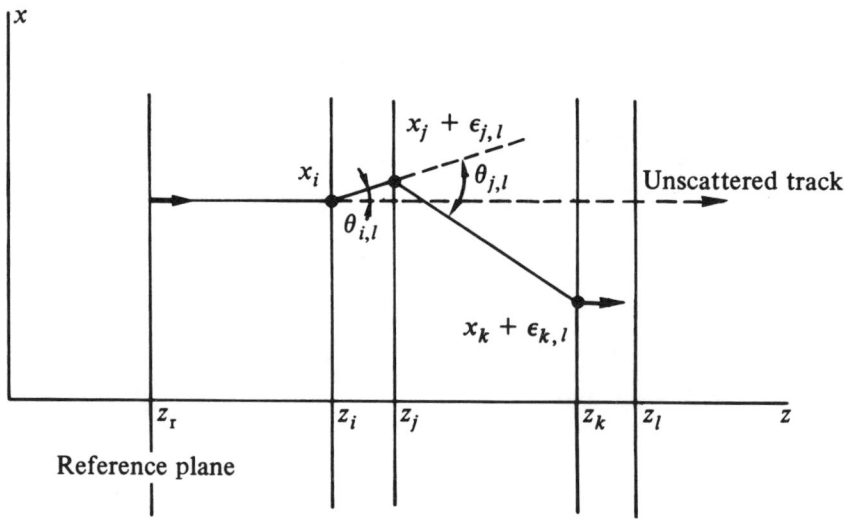

Fig. 3.14 The path of a track nearly parallel to the z-axis traversing discrete scatterers.

3.3 Fitting the tracks of charged particles

from pattern recognition information. If $g(s)$ only differs significantly from 0 when the particle passes through discrete detector layers, Equation (3.72a) can be approximated by a sum (Fig. 3.14):

$$\langle c_k c_l \rangle = \sum_i^{\min(k-1,l-1)} \langle \theta_i^2/2 \rangle \left(\frac{\partial f_k}{\partial \theta_{1,i}} \frac{\partial f_l}{\partial \theta_{1,i}} + \frac{\partial f_k}{\partial \theta_{2,i}} \frac{\partial f_l}{\partial \theta_{2,i}} \right) \quad (3.72b)$$

where $\langle \theta_i^2 \rangle = (\theta_s^2/2)[(\Delta L)_i/\cos \alpha]$ is the variance of the projected scattering angle at detector layer i. It is proportional to the length of material along the path and is therefore direction dependent. Each layer is counted once, regardless of the number of coordinates measured at this layer. Here α is the angle between the particle direction and the normal to the layer. For an efficient evaluation of the geometry factors using the standard track parameters see Regler (1977).

For the direction cosines the following relation holds (with $p_3 = dx/ds$, $p_4 = dy/ds$):

$$\left.\begin{aligned}\langle p_3 p_3 \rangle &= (1 - p_3^2)\langle \theta_{\text{proj}}^2 \rangle \\ \langle p_3 p_4 \rangle &= - p_3 p_4 \langle \theta_{\text{proj}}^2 \rangle \\ \langle p_4 p_4 \rangle &= (1 - p_4^2)\langle \theta_{\text{proj}}^2 \rangle\end{aligned}\right\} \quad (3.72c)$$

and for the direction tangents (with $p_3 = dx/dz, p_4 = dy/dz$) (Eichinger and Regler 1981):

$$\left.\begin{aligned}\langle p_3 p_3 \rangle &= (1 + p_3^2 + p_4^2)(1 + p_3^2)\langle \theta_{\text{proj}}^2 \rangle \\ \langle p_3 p_4 \rangle &= (1 + p_3^2 + p_4^2) p_3 p_4 \langle \theta_{\text{proj}}^2 \rangle \\ \langle p_4 p_4 \rangle &= (1 + p_3^2 + p_4^2)(1 + p_4^2)\langle \theta_{\text{proj}}^2 \rangle\end{aligned}\right\} \quad (3.72d)$$

Several approaches have been invented for efficient use of the LSM. They can roughly be summarized in three classes:

the global method;
the break point method;
the Kalman filter.

The '*global method*' has been discussed in this chapter to illustrate the weight matrix formalism. It is the appropriate method in the absence of multiple scattering. In the presence of multiple scattering, its contribution to the track uncertainty (Equations (3.72a, b)) is added to the error matrix (Equation (3.5)). The following ansatz has to be minimized (Equation (3.7)):

$$M(\mathbf{p}) = [\mathbf{f}(\mathbf{p}) - \mathbf{m}]^{\text{T}}(\text{cov}(c))^{-1}(\mathbf{f}(\mathbf{p}) - \mathbf{m}) \quad (3.73a)$$

The advantage of this method is the fact that all the information is used at once. Its disadvantages are:

It is not suitable for pattern recognition.

Inversion of a $n \times n$ matrix (number of operations $\sim n^3$).

The fitted track follows the ideal extrapolation of the starting vector and not the scattered path of the real track. The residuals are dominated by multiple scattering, and therefore the pull quantities show the real meaning of multiple scattering rather than of the measurement errors.

The '*break point*' method (e.g. Billoir (1984)) is adequate in the presence of a limited number of strong scatterers, e.g. plates and frames. An additional term is added to the least squares ansatz for each scatterer:

$$M(\mathbf{p}, \mathbf{\theta}_{bp}) = [\mathbf{f}(\mathbf{p}, \mathbf{\theta}_{bp}) - \mathbf{m}]^T [\text{cov}(c)]^{-1} [\mathbf{f}(\mathbf{p}, \mathbf{\theta}_{bp}) - \mathbf{m}] \\ + \mathbf{\theta}_{bp} [\text{cov}(\mathbf{\theta}_{bp})] q^{-1} \mathbf{\theta}_{bp} \qquad (3.73b)$$

The main advantage of this method is not only that all the information is used at once, but that the real path of the particle is closely followed. This is of great importance for error checking and tuning where optimal interpolation is needed, and for extrapolation.

For a large number of break points the method is computer-intensive because the derivatives $\partial \mathbf{f}/\partial \mathbf{\theta}$ at each break point have to be calculated. It can be shown that the global method and the break point method are equivalent as far as the estimate of the initial parameter $\tilde{\mathbf{p}}_r$ is concerned. All formulae are given in Billoir, *et al.* (1985).

'*Recursive track fitting*' by the LSM (the *Kalman filter*) is described in Subsection 3.2.5 (Frühwirth 1987). The contribution of multiple scattering between two subsequent steps can now be obtained from Equation (3.72a), and from discrete layers from Equation (3.72b), including inherently the break points at detector surfaces.

The advantage of this method is an efficient combination of pattern recognition and track fitting when a large number of measurements closely follow each other.

3.3.2.3 Resolution of magnet spectrometer

In most cases the errors in track reconstruction, as given by the LSM theory, are quite representative for the precision achievable. In order to arrive at a formula for a first guess of the resolution of a spectrometer, the two main types of spectrometers must be treated separately (Bock, *et al.* 1984b):

3.3 Fitting the tracks of charged particles

(a) A spectrometer consisting of a central 'bending magnet' and two position detectors 'lever arms'. This set-up is typical in *fixed-target experiments* (Fig. 3.7(b)).
(b) A compact spectrometer consisting of a set of equidistant detectors, all inside a magnetic field. This is typical for detectors in *colliding beam experiments* (Fig. 3.7(a)).

Case (a)

If one considers a charged particle moving parallel to the z axis, being subject to a small deflection by the spectrometer magnet's quasi-homogeneous field with only one significant component B_y, the deflection angle $\Delta\alpha$ can be obtained by integrating the equations of motion (Equation 3.41b)):

$$\Delta\alpha \cong \Delta x' \cong -(\kappa q/P)(1 + x'^2)\int_L B_y \, dz \qquad (3.74a)$$

where x' is the derivative of x with respect to z ($dx/dz \ll 1$), $\Delta\alpha$ is the deflection angle, and $\int_L B_y dz \equiv \bar{B}L$ where $\bar{B}$ is the average field along the z axis; $\bar{B}L$ is often called the deflection power or $B\,dl$(T m). Neglecting x'^2, Equation (3.74a) can be rewritten:

$$(P\Delta\alpha)/P \cong \Delta P_T/P = -(\kappa q/P)\bar{B}_L)$$
$$|\Delta P_T| \cong |\kappa q\bar{B}L| \qquad (3.74b)$$

with P, P_T, κ and q defined as in Subsection 3.3.1.1. Differentiation of the logarithm of Equation (3.74a) yields

$$\left|\frac{\delta\Delta\alpha}{\Delta\alpha}\right| = \left|\frac{\delta P}{P}\right| = \left|\delta\Delta\alpha \frac{P}{\kappa q\bar{B}L}\right| \qquad (3.74c)$$

With M position detectors available and a *symmetric spectrometer of length l*, the best theoretical angular resolution is obtained by placing $M/4$ detectors at the end of each arm and $M/2$ detectors at the centre (i.e. near the spectrometer magnet). The resolution obtainable is then:

$$\sigma(P)/P = P(|\kappa q\bar{B}L|)^{-1} 8\sigma/(lM^{\frac{1}{2}}) \qquad (3.74d)$$

where $\kappa = 0.299792458$ (GeV c^{-1})T^{-1}m^{-1} and σ is the error of an individual detector measuring the x coordinate (same units as l). This configuration of detectors, whilst optimizing geometrical precision, is a particularly unsuitable arrangement for correctly associating measured points into tracks and, therefore, can be used only in experiments with trivial track recognition problems (e.g. elastic scattering). At low energies, a limit on the precision is set by multiple scattering. Whereas the relative error in the momentum from position measurements is given by $\sigma(P)/P \sim P/l$, i.e. is proportional to P, another term arising from multiple scattering contributes $\sigma_{MS}(P)/P \sim m/(\beta \bar{B} L)$, which is *large for small β and constant for high momenta* ($\beta \cong 1$). These two contributions must be added quadratically in order to obtain the error on the momentum. With Equations (3.61e) and (3.74c) one obtains:

$$|\sigma_{MS}(\Delta\sigma)/\Delta\alpha| = |(M/2)^{\frac{1}{2}} q\, 0.015 (d/L_0)^{\frac{1}{2}} (1/\beta Pc)(P/\kappa q \bar{B} L)| \\ = (M/2)^{\frac{1}{2}} 0.015 (d/L_0)^{\frac{1}{2}} (1/\beta c \kappa \bar{B} L) \quad (3.75)$$

where P is the momentum (GeV c^{-1}), $\bar{B}L$ is the deflection power (T m), $M/2$ is the number of detectors placed around the magnet, and d/L_0 is the thickness of one individual detector layer in units of radiation length (note that one detector layer can measure either one or several coordinates; the latter case can be taken into account by reducing σ).

Case (b)

In central spectrometers, as used in most of the storage ring experiments, all tracking detectors are located inside a quasi-homogeneous magnetic field. This set-up was extensively studied for the first time for bubble chambers (Gluckstern 1963), both with and without multiple scattering.

The error in the momentum reconstructed in any projection is *inversely proportional to the field* normal to this projection, B_p, and to the square of the projected track length, l_p^2. In many storage ring experiments the field is of cylindrical symmetry, with its main component parallel to the beams (z axis). Therefore, the central detectors are designed such that they can measure the azimuthal coordinate $R\Phi$ precisely and therefore determine $P_T \equiv P_p\ (= P\cos\lambda = P\sin\theta)$ with ultimate precision (Fig. 3.7(a)): for the polar angle, θ, and the 'dip angle' $\lambda = \pi/2 - \theta$, see Fig. 3.8 and Equations (3.42).

Assuming a set-up consisting of M equidistant surfaces (concentric cylinders) for measurements of the coordinate $R\Phi$, and $M \gg 3$, one obtains asymptotically

$$\sigma(P_p)/P_p = P_p(|\kappa q B l_p|)^{-1}(\sigma/l_p)[720/(M+6)]^{\frac{1}{2}} \quad (3.76a)$$

Note that comparing Equation (3.76a) with Equation (3.74d), and with $L = l = l_p$, the relative error $\sigma(P_p)/P_p$ is proportional to $1/l_p^2$. This is an obvious disadvantage for the design of compact storage ring detectors (and the reason why highly energetic particles need an additional determination of energy by calorimeters surrounding the tracking detector Section 2.6). For a given relative error, the spectrometer must grow in size with $P^{\frac{1}{2}}$ if M is fixed. If M also grows linearly l_p, $\sigma(P)/P$ is asymptotically proportional to $P_p M^{-\frac{5}{2}} \sim P_p l_p^{-\frac{5}{2}}$.

If the set-up is not equidistant, but such that half of the measurements are at the centre of the track and one quarter at each of the ends, the momentum error is substantially improved to

$$\sigma(P_p)/P_p = P_p(|\kappa q B l_p|)^{-1}(\sigma/l_p)[256/(M+2)]^{\frac{1}{2}} \tag{3.76b}$$

This can easily be shown by evaluating the sagitta S:

$$S \cong l_p^2/(8R_p), R_p = P\cos\lambda/(|\kappa q B|),$$
$$\sigma(S) \cong \sigma/(M/4)^{\frac{1}{2}}, \sigma(S)/S = \sigma(P_p)/P_p = [\sigma/(M/4)]^{\frac{1}{2}}$$
$$\times [8P_p/(l_p^2 \kappa |qB|)],$$

with $\sqrt{256} = 16$, to be compared with $\sqrt{720} \cong 27$. This is, again, only a hypothetical arrangement of detectors, as it is unsuitable for recognizing tracks and difficult to install. Note that even this formula gives – for fixed $\overline{BL}$ and fixed M – a precision worse than that of the two lever arm spectrometer by a factor of 2.

For the precision of the initial direction of the track one obtains (again in the asymptotic limit) for equidistant detectors (Gluckstern 1963):

$$\sigma(\phi_0) = (\sigma/l_p)[192/(M+6)]^{\frac{1}{2}} \tag{3.76c}$$

and for the correlation between ϕ_0 and $1/P_p$ a constant i.e.

$$|\sigma(\phi_0, 1/P_p)| = 0.968 \tag{3.76d}$$

Note that ρ *vanishes in the centre* of the track segment. To evaluate the error in P, the error in θ must also be taken into account:

$$P = P_T/\sin\theta$$
$$\sigma(P) = [\sigma^2(P_T)/\sin^2\theta + \sigma^2(\theta)P_T^2 \cos^2\theta/\sin^4\theta]^{\frac{1}{2}} \tag{3.77a}$$

the covariance between P_T and θ being negligible in practice. For $\sigma(P)/P = P\sigma(1/P)$ one obtains

$$\sigma(P)/P = [\sigma^2(P_T)/P_T^2 + \sigma^2(\theta)\cot^2\theta]^{\frac{1}{2}} \tag{3.77b}$$

At *low energies*, a limit of the precision is again set by *multiple scattering* and the optimization becomes definitely more complicated, as the final resolution depends not only on the number of coordinates measured, but also on the amount of matter and on the momentum spectrum of the particles. This limitation is also important for high precision vertex detectors.

Modern storage ring spectrometers try to minimize the amount of matter inside the detectors themselves, and scattering occurs mainly in the supporting frames and in the beam tube. For such discrete layers the formula derived for homogeneously scattering matter can only be used as a guideline.

The contribution of *multiple scattering* to the *relative error* in the momentum measured by a set of equidistant detectors (Gluckstern 1963), after correction of the track length for the dip angle λ (subscript p' in contrast to p for the projection), is:

$$\sigma_{MS}(P_p)/P_p = 1.2 P_p/(\kappa|qB|)(\langle \theta_s^2/2 \rangle_{p'} l_p)^{\frac{1}{2}} \qquad (3.78a)$$

where $P_p = P \cos \lambda$ (GeV c^{-1}) is the momentum component in azimuthal plane perpendicular to **B**, $l_p = l \cos \lambda$ is the track length projected to azimuthal plane (l = track length in space in the detector module), $\lambda = \pi/2 - \theta$ is the dip angle (θ = polar angle), see Fig. 3.8, $\langle \theta_s^2/2 \rangle_{p'}$ is the variance per projected unit length by multiple scattering with $\langle \theta_s^2/2 \rangle$ as defined in Equation (3.70b), and

$$\langle \theta_s^2/2 \rangle_p = \langle \theta_s^2/2 \rangle / \cos^2 \lambda, \langle \theta_s^2/2 \rangle_{p'} = \langle \theta_s^2/2 \rangle_p / \cos \lambda$$

The term $1/\cos^2 \lambda$ accounts for the projection of the scattering angle, while the term $1/\cos \lambda$ accounts for the additional matter traversed in space.

Using the Molière formula (Equation (3.70b)) for $\langle \theta_s^2/2 \rangle$ yields:

$$\sigma_{MS}(P_p)/P_p = \frac{1.2}{\kappa|B|} \frac{0.015}{\beta c} \left(\frac{\rho}{x_0 \cos \lambda} \frac{1}{l_p} \right)^{\frac{1}{2}} \qquad (3.78b)$$

where $X_0/\rho = L_0$ is the radiation length (same unit as l_p). For a set-up which has coaxial cylindrical detectors parallel to **B**, any significant change of the angle β (Fig. 3.8) between the projected trajectory and the radius vector must be considered for a more accurate calculation.

It should be noted that $\sigma_{MS}(P_p)/P_p$ does not depend explicitly on the number of detectors. However, a set-up often consists of M detector layers, separated by gaps without multiple scattering. This can be accounted for by replacing ρ in Equation (3.78b) by $\rho_{eff} (= \rho_{mat} l_{mat}/l)$ where ρ_{mat} is the density of the detector material, and l_{mat}/l is the fraction of track length through the layers of matter.

If, in a cylindrical set-up, a total radial space $R_{max} - R_{min} = \Delta R$ is available for placing M detector layers, each with a radial thickness d, then

3.3 Fitting the tracks of charged particles

$l_{mat}/l = Md/\Delta R$ and (Fig. 3.7(a))

$$\sigma_{MS}(P_p)/P_p = \frac{1.2}{\kappa|B|}\frac{0.015}{\beta c}\left(\frac{l}{X_0}\frac{\rho_{mat}d}{\Delta R \cos\lambda}\right)^{\frac{1}{2}}\left(\frac{M}{\Delta R}\right)^{\frac{1}{2}} \quad (3.78c)$$

Considering $\sigma_{MS}(P_p)/P_p$ as a function of M and ΔR for fixed $\rho_{mat}\,d$, one finds it to be:

Decreasing $\sim 1/\Delta R$ for $M = $ const
Decreasing $\sim 1/(\Delta R)^{\frac{1}{2}}$ for $M/\Delta R$ (i.e. layers per unit length) = const
Increasing $\sim M^{\frac{1}{2}}$ for $\Delta R = $ const

This contribution, $\sigma_{MS}(P_p)/P_p$, has to be added quadratically to the errors in the relative momentum due to the measurements, $\sigma_{pos}(P_p)/P_p$ (see Subsection 3.3.2.3):

$$\sigma(P_p)/P_p = \{[\sigma_{pos}(P_p)/P_p]^2 + [\sigma_{MS}(P_p)/P_p]^2\}^{\frac{1}{2}} \quad (3.79a)$$

In the case of a cylindrical set-up, for a given ΔR, $\sigma_{pos}(P_p)/P_p \sim P_p/(M+6)^{\frac{1}{2}}$ (see Equation (3.76a)), and

$$\sigma(P_p)/P_p = \{[c_1 P_p/(M+6)^{\frac{1}{2}}]^2 + (c_2 M^{\frac{1}{2}})^2\}^{\frac{1}{2}} \quad (3.79b)$$

This shows clearly that an optimal M exists for an available space ΔR, for each value of P_p. Overinstrumentation will be counterproductive, unless additional measurements can be included *without further increasing the scattering matter* (e.g. by using a TPC).

These considerations also have an impact on the precision of vertex reconstruction (Section 3.4). Because of the strong correlation between the track errors in the direction and the momentum (Equation (3.67d)), the error in the vertex position is largely due to the lever arm between the innermost detector (usually just outside the beam tube) and the vertex region; therefore, precise spatial vertex evaluation also requires z and θ to be measured with high precision. However, the effects of multiple scattering within the beam tube and the innermost detector (a high-precision vertex detector) can be considerable, thus also limiting the precision which can be achieved in the (x, y) projection. The only solution is a beam pipe with a minimum of material (radiation lengths) at a small radius (Mitaroff 1986).

3.3.3 Track element merging

In a complex detector it is often necessary for track segments to be fitted separately. The problem of combining the information is discussed in this subsection. It will be assumed that there are two detector modules with two estimates ($\tilde{\mathbf{p}}_i$, $i = 1, 2$) of the track parameters at the respective reference planes $z_r = z_{r,1}$, $z_{r,2}$. The covariance matrices of the $\tilde{\mathbf{p}}_i$ are denoted by $\mathbf{C}_i$. The reference plane of the combined track information is assumed to be equal to $z_{r,2}$ (Fig. 3.15).

The principle is to propagate the estimate $\tilde{\mathbf{p}}_1$ and its covariance matrix $\mathbf{C}_1$ to $z_{r,2}$ (Equation (3.33)):

$$\tilde{\mathbf{p}}_2^{(1)} = \tilde{\mathbf{p}}_2(\tilde{\mathbf{p}}_1)$$
$$\mathbf{C}_1' = \mathbf{D}\mathbf{C}_1\mathbf{D}^T$$

with $\mathbf{D} = \partial \mathbf{p}_2 / \partial \mathbf{p}_1$.

Contributions to multiple scattering between $z_{r,1}$ and $z_{r,2}$ (including material at $z_{r,1}$ are added to $\mathbf{C}_1'$, to give the final covariance matrix $\mathbf{C}_2^{(1)}$ of $\tilde{\mathbf{p}}_2^{(1)}$

$$\mathbf{C}_2^{(1)} = \mathbf{C}_1' + \sum_i \frac{\partial \mathbf{p}_2}{\partial (p_{i,3}, p_{i,4})} \frac{\partial (p_{i,3}, p_{i,4})}{\partial (\theta_{i,1}, \theta_{i,2})} \mathbf{C}(\theta_i)$$
$$\times \frac{\partial (p_{i,3}, p_{i,4})^T}{\partial (\theta_{i,1}, \theta_{i,2})} \left(\frac{\partial \mathbf{p}_2}{\partial (p_{i,3}, p_{i,4})} \right)^T \quad (3.80)$$

+ contributions due to continuous scattering
(Equation (3.33d))

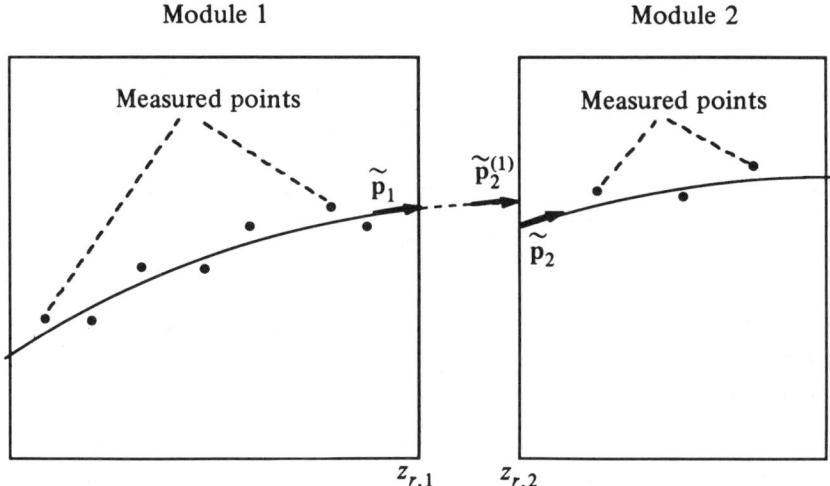

Fig. 3.15 In track element merging, the track parameters $\tilde{\mathbf{p}}_1$ are propagated to a new reference plane. $\tilde{\mathbf{p}}_2^{(1)}$ and $\tilde{\mathbf{p}}_2$ are considered as direct measurements of the true track parameters at the reference plante $z_{r,2}$.

3.3 Fitting the tracks of charged particles

where $\mathbf{C}(\theta_i)$ is the covariance matrix of the scattering angles $\theta_{i,1}, \theta_{i,2}$, and i is the index for all scatterers in the interval $[z_{r,1}, z_{r,2}]$. If one now considers $\tilde{\mathbf{p}}_2^{(1)}$ and $\tilde{\mathbf{p}}_2$ as direct measurements of $\mathbf{p}_2$, a joint least squares ansatz can be made

$$M(\mathbf{p}_2) = (\mathbf{p}_2 - \tilde{\mathbf{p}}_2)^T \mathbf{C}_2^{-1}(\mathbf{p}_2 - \tilde{\mathbf{p}}_2) + (\mathbf{p}_2 - \tilde{\mathbf{p}}_2^{(1)})^T (\mathbf{C}_2^{(1)})^{-1}(\mathbf{p}_2 - \tilde{\mathbf{p}}_2^{(1)}) \quad (3.81a)$$

The final estimate is a weighted mean:

$$\tilde{\tilde{\mathbf{p}}}_2 = [(\mathbf{C}_2^{(1)})^{-1} + \mathbf{C}_2^{-1}]^{-1} [(\mathbf{C}_2^{(1)})^{-1} \tilde{\mathbf{p}}_2^{(1)} + \mathbf{C}_2^{-1} \tilde{\mathbf{p}}_2] \quad (3.81b)$$

and $M(\tilde{\tilde{\mathbf{p}}}_2)$ is χ^2-distributed with five degrees of freedom.

Ansatz (3.80) contains an implicit assumption, namely that $\tilde{\mathbf{p}}_2^{(1)}$ and $\tilde{\mathbf{p}}_2$ are independent. This is, however, only true if the reference plane $z_{r,2}$ is at the near end of the module 2, as seen from module 1. In this case, the combined estimate is between the two modules and therefore not of great interest, except for a χ^2 test. In the more interesting case of $z_{r,2}$ being at the far end of module 2, one notes first that the difference

$$\mathbf{p}_2(\hat{\mathbf{p}}_1) - \hat{\mathbf{p}}_2$$

is a random quantity: it is *correlated to the contribution of multiple scattering* to the measurement errors in module 2 and hence also to $\tilde{\mathbf{p}}_2 - \hat{\mathbf{p}}_2$. Therefore, $\tilde{\mathbf{p}}_2^{(1)} - \hat{\mathbf{p}}_2$ is also correlated (Billoir, et al. 1985).

It is important to note that, in the presence of multiple scattering, the following rule has to be observed: if an optimal estimate should be achieved on one side of the set of detectors, all estimates for the individual sets should be made on the corresponding side of each detector module (e.g. for vertex fitting), and vice versa (e.g. for extrapolation to the muon chambers). Otherwise a complicated smoothing algorithm must be used (Subsection 3.2.5).

If the information from individual detector modules with poor resolution is going to be merged, a 'reference track' as obtained from the global track search (pattern recognition) can be used, as described in Subsection 3.2.5.

Adding nongeometrical information

In addition to the coordinates, some other information such as the 'TOF' or 'dE/dx' is sometimes available. Usually, however, this information, although very useful for particle identification, triggering, etc., must be treated with great care when used in a geometry program. We demonstrate this with TOF measurements.

If the velocity of a particle is significantly different from the velocity of light (e.g. for a recoil particle in diffractive processes), then the TOF yields valuable information about the momentum and the mass of this particle:

$$ct \cong cL/\beta = c(L/P)(P^2 + m^2)^{\frac{1}{2}} \quad (3.82a)$$

where m and P are in GeV, and L is the path length. Using $p_5 = 1/P$ one obtains:

$$ct \cong cLp_5[m^2 + (1/p_5)^2]^{\frac{1}{2}} \quad (3.82b)$$

It is usually possible to make a rough estimate of L and to consider it as a constant during the fit. Only for very slow particles and large curvature must the length L also be varied during the iteration process; but slow particles always cause several problems (important multiple scattering, energy loss, nuclear scattering).

There are now two possibilities:

either the TOF information is used during the geometrical fit;
or, preferably, this information is first used only for the mass assignment and added in an additional fit step, or even later in the kinematical fit if enough geometrical information is available for a first track fit without using additional information.

In either case, the additional term to be added to the least squares ansatz is:

$$M'(\mathbf{p}) = M(\mathbf{p}) + [t(p_5) - t_m]^2/\sigma^2(t) \quad (3.82c)$$

3.3.4 Numerical minimization technique

Data analysis in high-energy physics is confronted with a variety of problems which require the optimization of some function with respect to a set of parameters. The MLM (where usually the negative logarithm of the likelihood function is going to be minimized) and the LSM are examples. The methods used all have in common the fact that the *optimal estimate* of the parameters is defined by the *minimum of a function* which depends explicitly on these parameters. The measurements, together with their covariance matrix, are fixed parameters in these functions. The numerical technique will be outlined below only for the case of the LSM. The only assumption is an approximate quadratic behaviour of the least squares function in the immediate vicinity of its minimum. In the linear model, a sufficient condition for this assumption is that, within the measurement

3.3 Fitting the tracks of charged particles

space $\{c\}$ there exists a subspace with dimension $n \geq$ dimension of $\mathbf{p}$ (i.e. 5 in magnetic spectrometers) for which the covariance matrix is not singular, and the corresponding rows of the matrix of derivatives, $\mathbf{A}$, are at least of the rank of the dimension of $\mathbf{p}$. If this is not the case, part of the parameter vector may still be defined (e.g. the three-dimensional projection $R\Phi, \phi, 1/P_T$, without the knowledge of z and θ).

If the model for the LSM is linear, the minimization procedure is trivial: the least squares ansatz (Equation (3.7)) for the function $M(\mathbf{p})$ is of second order, therefore, the minimum condition $\partial M/\partial \mathbf{p} = \mathbf{0}$ yields a system of linear equations for the estimate $\tilde{\mathbf{p}}$ with the explicit solution (Equation (3.8)):

$$\tilde{\mathbf{p}} = \overset{0}{\mathbf{p}} + (\mathbf{A}^T \mathbf{W} \mathbf{A})^{-1} \mathbf{A}^T \mathbf{W} \cdot [\mathbf{m} - \mathbf{f}(\overset{0}{\mathbf{p}})]$$

where $\overset{0}{\mathbf{p}}$ is the expansion point (Equation (3.4)) of $\mathbf{f}(\mathbf{p})$. For the least squares ansatz, in addition to the track model, the covariance matrix $\mathrm{cov}(\varepsilon)$ must also be evaluated and inverted, which, in the presence of multiple scattering causing a nondiagonal covariance matrix (Subsection 3.3.2.2), might become a time consuming procedure if the number of measurements is large.

If, however, the track model is far from being linear, the linearization will give a result, $\tilde{\mathbf{p}}$, strongly dependent on the choice of $\overset{0}{\mathbf{p}}$, and an *iterative procedure* has to be applied, using the 'Newton method' or the quasi-Newton method, for iteratively approximating $\overset{0}{\mathbf{p}}$ towards $\tilde{\mathbf{p}}$. This is usually successful for the 'least squares function' $M(\mathbf{p})$. First and second derivatives of M with respect to $\mathbf{p}$ can be expected to exist everywhere in the region of interest of $\mathbf{p}$. One can then write the Taylor series expansion of M around some point $\mathbf{p}$, say $\overset{0}{\mathbf{p}}$

$$M(\mathbf{p}) = M(\overset{0}{\mathbf{p}}) + \mathbf{q} \cdot (\mathbf{p} - \overset{0}{\mathbf{p}}) + \tfrac{1}{2}(\mathbf{p} - \overset{0}{\mathbf{p}})^T \mathbf{G} \cdot (\mathbf{p} - \overset{0}{\mathbf{p}}) \tag{3.83a}$$

where $\overset{0}{\mathbf{p}}$ is an *ad hoc* estimate (e.g. from a previously performed pattern recognition). (As mentioned earlier (Subsection 3.3.2) the covariance matrix $\mathbf{V} = \mathrm{cov}(\varepsilon)$ varies only slowly with $\mathbf{p}$ in most cases and can therefore be kept constant; only when $\overset{0}{\mathbf{p}}$ must be corrected drastically, i.e. $|\overset{0}{\mathbf{p}} - \tilde{\mathbf{p}}_1|$ large, must $\mathbf{V}(=\mathbf{W}^{-1})$ be reevaluated for the second iteration, sometimes for both contributions V_{detector} and V_{MS} (Subsection 3.3.2).)

For the derivatives of the LSM ansatz, one gets for $\mathbf{q}$ and $\mathbf{G}$ from Equation (3.83a)

$$\mathbf{q} = -2\mathbf{A}^T \mathbf{W} \cdot [\mathbf{m} - \mathbf{f}(\overset{0}{\mathbf{p}})] \tag{3.83b}$$

and the 'Hessian matrix' **G**, i.e. the matrix of second derivatives, is

$$\mathbf{G} = 2\{\mathbf{A}^T\mathbf{W}\mathbf{A} - [\partial^2 \mathbf{f}/(\partial \mathbf{p})^2]\mathbf{W}\cdot[\mathbf{m}-\mathbf{f}(\overset{0}{\mathbf{p}})]\} \tag{3.83c}$$

where

$$[\partial^2 \mathbf{f}/(\partial \mathbf{p})^2]_{i;kl} \equiv [\partial^2 f_i(\mathbf{p})/(\partial p_k \partial p_l)]_{\overset{0}{\mathbf{p}}}$$

for summation over i.

The calculation of the second derivatives of **f** would cause prohibitive computing time if done numerically together with step-by-step tracking. Fortunately the second term of Equation (3.83c) can be neglected as long as *both* the second derivatives *and* all $m_i - f_i(\overset{0}{\mathbf{p}})$ are small (note that in the case of linear functions this term vanishes anyway). This leads to the same formula as for the linear case, see above and Equation (3.8). If the functions **f(p)** are explicitly known, there is no reason to suppress the second derivatives and thus reduce the convergence properties of the iterative procedure. The *convergence point* $\tilde{\mathbf{p}}$ *remains the same*, but is approximated faster with the second derivatives retained.

Neglecting the second term of Equation (3.83c) results in **G** being necessarily positive definite, but does not necessarily yield a meaningful curvature at $\overset{0}{\mathbf{p}}$ and may cause 'divergent oscillations' of the procedure, whereas with the second derivatives retained in the least squares ansatz, the Newton method, may diverge when the initial parameter values, $\overset{0}{\mathbf{p}}$, are outside the domain where M is parabolic, i.e. beyond an inflection point. This is not always a disadvantage, because it may be a useful test of the *ad hoc* method for initialization and, for instance, indicate an error in the previous pattern recognition.

In practice, two or three iterations are necessary in the nonlinear case until the minimum is reached, i.e. until the estimated parameters become stable; then the *change* in 'χ_i^2' $\equiv M(\tilde{\mathbf{p}}_i)$ for the last iteration becomes negligibly small.

$$0 \leqslant M(\tilde{\mathbf{p}}_{i-1}) - M(\tilde{\mathbf{p}}_i) \ll 1$$

usually less than 10^{-3}. Therefore, it is sufficient in most cases to reevaluate the derivatives only once. The decision criteria must be deduced from practical experience; they may depend on $\overset{0}{\mathbf{p}}$ itself. During the tuning period of the fit program, however, one extra iteration of the track fit at the expansion point $\overset{0}{\mathbf{p}}_{i+1} = \tilde{\mathbf{p}}_i$ yielding $\tilde{\mathbf{p}}_{i+1}$, may be needed. For more details see Blobel (1984a), Eichinger and Regler (1981), Frodesen, *et al.* (1979), James (1972) and James and Roos (1986).

3.4 Association of tracks to vertices

3.4.1 Basic concepts

The vertex fit serves two purposes. The first is to estimate the position of the point of interaction and the momentum vectors of the tracks emerging from this point (with improved precision due to the *vertex constraint*). The second is to check the association of tracks to a vertex, i.e. the decision of whether the track does indeed originate from this vertex. The following discussion is applicable both to the primary interaction vertex and to an eventual secondary vertex (a decay or secondary interaction). While in the first case the exact position of the vertex might seem a simple mathematical constraint, it is of some importance in the second case, since it determines the direction of the (possibly unseen) track connecting the two vertices.

In both cases the momentum vectors of all emerging charged tracks should be computed with the best possible precision together with their common covariance matrix, since they are the input for a subsequent kinematical fit. Here the LSM turns out to be the best method for the subsequent merging of information in the presence of additional constraints (Fig. 3.16). (The same holds for the final fit of kinematics by adding the momentum and energy conservation constraints.)

The method described below has been proposed and used successfully by the first generation experiments at the CERN-ISR (Metcalf, et al. 1973). It involves the *inversion of a matrix of the order* $3n$ ($n =$ number of tracks). Since the number of arithmetical operations for the inversion increases with the third power of the order, this method becomes prohibitive with a further increase in energy and the resulting higher multiplicities. Also, the events become more complex with an increasing need to eliminate tracks which do not belong to the primary interaction vertex. Therefore, a new algorithm for the computation of the estimates, their covariance and the χ^2 was developed (Metcalf et al., 1973; Billoir, et al. 1985; Frühwirth and Regler 1989), allowing the application of this method to very complex events also. In order to avoid unnecessary repetitions of the vertex fit, a *recursive method* is desirable, allowing a check to be made on the association of tracks to a common vertex. The input for the vertex fit consists of information about the tracks to be grouped together. Normally one considers the estimated track parameters at a reference surface as '*virtual measurements*'. The reference surface will in most cases be a plane, a cylinder (especially in storage ring experiments) or the beam tube.

320   *Track and vertex fitting*

The choice of the reference surface has a certain influence on the behaviour of the fit, since it is desirable that the virtual measurements, namely the track parameters, are to a good approximation *linear functions* of the vertex position and of the parameters determining the momentum vector at the vertex. In some cases the reference surface will coincide with a physical surface, e.g. the wall of a vacuum vessel or a vertex chamber. If multiple scattering in this wall is important, it can easily be taken into account, by augmenting the covariance matrix of the estimated track parameters (Equation (3.70b)).

If the position of the vertex is known *a priori* to some precision, as is the case for the interaction region of a storage ring, this knowledge can be

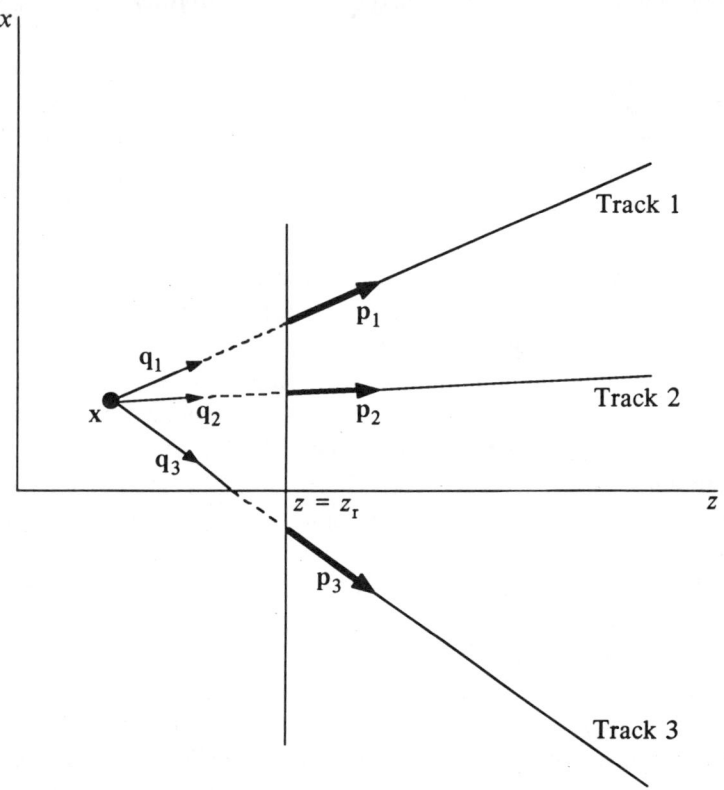

Fig. 3.16 On the right hand side, the track fitting for single tracks is demonstrated. For the vertex fit, the parameters $\mathbf{p}_i$, are considered as 'virtual measurements' (left hand side).

3.4 Association of tracks to vertices

considered as an *independent measurement* of the position, with its proper error matrix.

If a single track is poorly defined, its coordinates should be incorporated directly into the vertex fit, instead of a possibly biassed estimate of the track parameters. This should also be done if some *a priori* knowledge of the vertex position was used in a first individual track fit.

Some care is necessary when changing the reference surface, e.g. from a cylinder parallel to the z axis (with $R = $ const) to a plane at $z = $ const. Although the transformation may only concern the spatial components p_1 and p_2 of the parameter 5-vector, the *full dependence* on the old parameters *must be included* in the error propagation even if a component of the (virtual) measurement vector itself is not going to be propagated. That is, when changing from cylindrical coordinates with $R\Phi$, z at $R = $ const, to Cartesian coordinates x, y at $z = $ const, and when truncating higher orders in an expansion, the following arguments contribute by first order to the matrix of derivatives and therefore to the error propagation at a fixed point (R_H is the projected radius of the helix):

$p_{1,\text{new}} = x(R\Phi, z_{\text{old}}, \theta, \phi)$ at $z_{\text{ref}} = $ const

$\Delta x = - R \sin \Phi \Delta \Phi - \tan \theta \cos \phi \Delta z_{\text{old}}$ (with $R \sin \Phi = y$)

$p_{2,\text{new}} = y(R\Phi, z_{\text{old}}, \theta, \phi)$ at $z_{\text{ref}} = $ const

$\Delta y = - R \cos \Phi \Delta \Phi - \tan \theta \sin \phi \Delta z_{\text{old}}$ (with $R \cos \Phi = x$)

$p_{4,\text{new}} = \phi$

$\Delta \phi_{\text{new}} = \Delta \phi_{\text{old}} - (\tan \theta / R_H) \Delta z_{\text{old}}$

and p_3, p_5 remain unchanged. For the inverse transformation the corresponding derivative matrix is obtained by inverting the 3×3 non-trivial part of the derivative matrix. Note that the fixed position of the new reference surface (plane) is at $z_r = z_{\text{old}}$ but although $z_r = $ const, the error in z_{old} contributes to the errors in x and y via θ and ϕ terms in the matrix of derivatives, yielding dependence on θ and ϕ. The curvature, which is $\sim p_5 = 1/P$, contributes only to ϕ and the contribution to $\partial \mathbf{p}_{\text{new}}/\partial \mathbf{p}_{\text{old}}$ vanishes for $z_r \to z_{\text{old}}$.

If the direction were unknown, the matrix of derivatives would be *singular* for this special change of the reference surface, and so would the new covariance matrix. The transformation would not be a 'bijective' application, and the information about σ_z would be lost, while the correlation between x and y would be 1.

3.4.2 A fast global method for parameter estimation

As mentioned above, for the vertex fit the previously fitted $\mathbf{p}_i$ and their weight matrix $\mathbf{V}_i^{-1} = \mathbf{G}_i$ ($i = 1, \ldots, n$) are now regarded as 'virtual measurements' to find an estimate for the *vertex position* $\mathbf{x}$ and the three-dimensional *new track parameters* for direction and momentum, $\mathbf{q}_i$, at the vertex. Furthermore, we allow the inclusion of an independent measurement of $\mathbf{x}$ (e.g. for primary vertices by knowing the beam profile), yielding a 3-vector $\mathbf{v}$ with its 3×3 weight matrix $\mathbf{G}_0$.

The identity $\mathbf{v}(\mathbf{x}) = \mathbf{x}$ and a linear ansatz for $\mathbf{p}_i = \mathbf{p}_i(\mathbf{x}, \mathbf{q}_i)$ at an 'expansion point' $(\overset{0}{\mathbf{x}}, \overset{0}{\mathbf{q}}_i)$ yield

$$
\begin{pmatrix} \mathbf{v} \\ \mathbf{p}_1 \\ \cdot \\ \cdot \\ \cdot \\ \mathbf{p}_n \end{pmatrix} = \begin{pmatrix} \mathbf{1} & 0 & \cdot & \cdot & 0 \\ \mathbf{A}_1 & \mathbf{B}_1 & \cdot & & \cdot \\ \mathbf{A}_2 & 0 & \mathbf{B}_2 & 0 & \cdot \\ \cdot & & 0 & & \cdot \\ \cdot & & & & \cdot \\ \mathbf{A}_n & 0 & \cdot & \cdot & \mathbf{B}_n \end{pmatrix} \cdot \begin{pmatrix} \mathbf{x} - \overset{0}{\mathbf{x}} \\ \mathbf{q}_1 - \overset{0}{\mathbf{q}}_1 \\ \cdot \\ \cdot \\ \cdot \\ \mathbf{q}_n - \overset{0}{\mathbf{q}}_n \end{pmatrix} + \begin{pmatrix} \overset{0}{\mathbf{x}} \\ \mathbf{p}_1(\overset{0}{\mathbf{x}}, \overset{0}{\mathbf{q}}_1) \\ \cdot \\ \cdot \\ \cdot \\ \mathbf{p}_n(\overset{0}{\mathbf{x}}, \overset{0}{\mathbf{q}}_n) \end{pmatrix}
$$
(3.84a)

where $\mathbf{A}_i(\mathbf{x}, \mathbf{q}_i) = \partial \mathbf{p}_i / \partial \mathbf{x}$ at $(\overset{0}{\mathbf{x}}, \overset{0}{\mathbf{q}}_i)$ and $\mathbf{B}_i(\mathbf{x}, \mathbf{q}_i) = \partial \mathbf{p}_i / \partial \mathbf{q}_i$ at $(\overset{0}{\mathbf{x}}, \overset{0}{\mathbf{q}}_i)$ are 5×3 matrices composing the Jacobian matrix of derivatives.

Defining $\mathbf{p}'_i = \mathbf{p}_i + \mathbf{A}_i \overset{0}{\mathbf{x}} + \mathbf{B}_i \overset{0}{\mathbf{q}}_i - \mathbf{p}_i(\overset{0}{\mathbf{x}}, \overset{0}{\mathbf{q}})$, the linear ansatz for $\mathbf{p}'_i = \mathbf{p}'_i(\mathbf{x}, \mathbf{q}_i)$ becomes homogeneous:

$$
\begin{pmatrix} \mathbf{v} \\ \mathbf{p}'_i \\ \cdot \\ \cdot \\ \cdot \\ \mathbf{p}'_n \end{pmatrix} = \begin{pmatrix} \mathbf{1} & 0 & \cdot & 0 \\ \mathbf{A}_1 & \mathbf{B}_1 & \cdot & \cdot \\ \mathbf{A}_2 & 0 & \mathbf{B}_2 & \cdot \\ \cdot & \cdot & \cdot & \cdot \\ \cdot & & 0 & \cdot \\ \mathbf{A}_n & \cdot & \cdot & \mathbf{B}_n \end{pmatrix} \begin{pmatrix} \mathbf{x} \\ \mathbf{q}_1 \\ \cdot \\ \cdot \\ \cdot \\ \mathbf{q}_n \end{pmatrix}
$$
(3.84b)

Estimates for $\mathbf{x}$ and $\mathbf{q}$ can be determined by the LSM (see Equation (3.8)):

$$
\begin{pmatrix} \tilde{\mathbf{x}} \\ \tilde{\mathbf{q}}_1 \\ \cdot \\ \cdot \\ \cdot \\ \tilde{\mathbf{q}}_n \end{pmatrix} = \mathbf{M}^{-1} \mathbf{N} \cdot \begin{pmatrix} \mathbf{v} \\ \mathbf{p}'_1 \\ \cdot \\ \cdot \\ \cdot \\ \mathbf{p}'_n \end{pmatrix}
$$
(3.84c)

where $\mathbf{M}$ is a $(3n + 3) \times (3n + 3)$ matrix and $\mathbf{N}$ is a $(3n + 3) \times (5n + 3)$

matrix, both of which contain $\mathbf{A}_i$, $\mathbf{B}_i$ and $\mathbf{G}_i$. The covariance matrix of the fit is $\mathbf{M}^{-1}$, and the number of operations required for its calculation by inverting $\mathbf{M}$ is $\sim n^3$, causing a prohibitive CPU time if the straightforward method is applied when there are high multiplicities.

However, it has been shown (Metcalf, et al. 1973; Billoir, et al. 1985) that *faster algorithms* can be used. For the $(3n+3)$-vector of fitted parameters $(\tilde{\mathbf{x}}, \tilde{\mathbf{q}}_i)$, the $\mathbf{cov}(\tilde{\mathbf{x}})$ and the χ^2 of the fit, the number of operations required increases proportionally with n. The other elements of the covariance matrix can be calculated with $\sim n^2$ more operations. Comparing this with the straightforward method, there should be a break-even point for vertices about four tracks.

Moreover, the above formulae suggest an *iterative algorithm* to update the results of a vertex fit by the addition (or removal) of single tracks, yielding $(\tilde{\mathbf{x}}^*, \tilde{\mathbf{q}}_i^*, \tilde{\mathbf{q}}_{n+1}^*)$, the $\mathbf{cov}(\tilde{\mathbf{x}}^*)$ and a new χ^{2*} (Frühwirth 1988). Note that $\chi_+^2 \equiv \chi^{2*} - \chi^2$ can be used as a test criterion for the association of another track $(n+1)$ to a fitted vertex $\tilde{\mathbf{x}}$.

3.4.3 χ^2 *evaluation and track association strategy*

The separation of secondary vertices and the proper association of tracks are very important in the analysis of high-energy particle reactions. Even topologies to be identified include at least neutral 2- and 4-prong and charged 3- and 5-prong decays. However, in order to be able to select from many tens to hundreds of tracks those which belong to a particular secondary vertex, the primary vertex must also be reconstructed as precisely as possible with all the track fit information available. In this subsection, only decays inside the beam tube will be discussed. The treatment of 'kinks' (i.e. charged 1-prong decays) and of V^0 ('vees', i.e. neutral decays) outside the beam tube have recently been discussed elsewhere (Frühwirth 1988; Mitaroff, thesis in preparation). Thanks to the *block structure* of the matrices involved (see Equation 3.94a)), a vertex fit by the LSM may be regarded as a special case of the (nonlinear) *Kalman filter* with a variable dimension $(3n+3)$ of the state vector $(\vec{\mathbf{x}}, \vec{\mathbf{q}}_1, \ldots, \vec{\mathbf{q}}_n)$ (Frühwirth 1987).

Initially the state vector consists only of the prior information about the vertex position $\mathbf{x}_0$ and $\mathbf{C}_0 = \mathbf{cov}(\mathbf{x}_0)$. For each 5-vector $\mathbf{p}_k$ of fitted track parameters the state vector is augmented by the 3-vector $\mathbf{q}_k$ of momentum of track k at the vertex (Fig. 3.16). The system equation is particularly simple:

$$\mathbf{x}_k = \mathbf{x}_{k-1} \tag{3.85a}$$

Track and vertex fitting

The measurement equation, in the presence of a magnetic field, is nonlinear:

$$\mathbf{p}_k = \mathbf{h}_k(\mathbf{x}_k, \mathbf{q}_k) + \boldsymbol{\varepsilon}_k \tag{3.85b}$$

As usual, we linearize $\mathbf{h}_k$ at some point $(\mathbf{x}_{k,0}, \mathbf{q}_{k,0})$:

$$\mathbf{h}_k(\mathbf{x}_k, \mathbf{q}_k) = \mathbf{h}_k(\mathbf{x}_{k,0}, \mathbf{q}_{k,0}) + \mathbf{A}_k \cdot (\mathbf{x}_k - \mathbf{x}_{k,0})$$
$$+ \mathbf{B}_k \cdot (\mathbf{q}_k - \mathbf{q}_{k,0}) = \mathbf{c}_{k,0} + \mathbf{A}_k \mathbf{x}_k + \mathbf{B}_k \mathbf{q}_k \tag{3.85c}$$

Since there is usually no prior information about $\mathbf{q}_k$, we assign an infinite covariance matrix to the predicted vector $\mathbf{q}_k^{(k-1)}$:

$$\mathbf{D}_k^{(k-1)} = \operatorname{cov}(\mathbf{q}_k^{(k-1)}) = (1/\delta)\mathbf{1}, \delta \text{ small} \tag{3.85d}$$

Then the prediction equations have the following forms:

$$\left.\begin{array}{l} \mathbf{x}_k^{(k-1)} = \mathbf{x}_{k-1} \\ \mathbf{q}_k^{(k-1)} = \mathbf{q}_{k,0} \\ \mathbf{C}_k^{(k-1)} = \mathbf{C}_{k-1} \\ \mathbf{D}_k^{(k-1)} = (1/\delta)\mathbf{1} \end{array}\right\} \tag{3.86}$$

The filter equations in the weighted means formulation are:

$$\begin{pmatrix} \mathbf{x}_k \\ \mathbf{q}_k \end{pmatrix} = \begin{pmatrix} \mathbf{C}_k \mathbf{E}_k \\ \mathbf{E}_k^T \mathbf{D}_k \end{pmatrix} \begin{pmatrix} \mathbf{C}_{k-1}^{-1} \mathbf{x}_{k-1} + \mathbf{A}_k^T \mathbf{G}_k \cdot (\mathbf{p}_k - \mathbf{c}_{k,0}) \\ (\mathbf{D}_k^{(k-1)})^{-1} \mathbf{q}_k^{(k-1)} + \mathbf{B}_k^T \mathbf{G}_k \cdot (\mathbf{p}_k - \mathbf{c}_{k,0}) \end{pmatrix} \tag{3.87a}$$

with

$$\begin{pmatrix} \mathbf{C}_k \mathbf{E}_k \\ \mathbf{E}_k^T \mathbf{D}_k \end{pmatrix} = \begin{pmatrix} \mathbf{C}_{k-1}^{-1} + \mathbf{A}_k^T \mathbf{G}_k \mathbf{A}_k & \mathbf{A}_k^T \mathbf{G}_k \mathbf{B}_k \\ \mathbf{B}_k^T \mathbf{G}_k \mathbf{A}_k & \mathbf{B}_k^T \mathbf{G}_k \mathbf{B}_k \end{pmatrix} \tag{3.87b}$$

After some matrix algebra and taking the limit $\delta \to 0$ one obtains the following results

$$\left.\begin{array}{l} \mathbf{x}_k = \mathbf{C}_k [\mathbf{C}_{k-1}^{-1} \mathbf{x}_{k-1} + \mathbf{A}_k^T \mathbf{G}_k^B \cdot (\mathbf{p}_k - \mathbf{c}_{k,0})] \\ \mathbf{q}_k = \mathbf{W}_k \mathbf{B}_k^T \mathbf{G}_k \cdot [(\mathbf{p}_k - \mathbf{c}_{k,0}) - \mathbf{A}_k \mathbf{x}_k] \\ \mathbf{C}_k = (\mathbf{C}_{k-1}^{-1} + \mathbf{A}_k^T \mathbf{G}_k^B \mathbf{A}_k)^{-1} \\ \mathbf{D}_k = \mathbf{W}_k + \mathbf{W}_k \mathbf{B}_k^T \mathbf{G}_k \mathbf{A}_k \mathbf{C}_k \mathbf{A}_k^T \mathbf{G}_k \mathbf{B}_k \mathbf{W}_k \\ \mathbf{E}_k = \mathbf{W}_k \mathbf{B}_k^T \mathbf{G}_k \mathbf{A}_k \mathbf{C}_k \end{array}\right\} \tag{3.87c}$$

with

$$\mathbf{W}_k = (\mathbf{B}_k^T \mathbf{G}_k \mathbf{B}_k)^{-1}$$
$$\mathbf{G}_k^B = \mathbf{G}_k - \mathbf{G}_k \mathbf{B}_k \mathbf{W}_k \mathbf{B}_k^T \mathbf{G}_k$$
$$\operatorname{cov}(\mathbf{x}_k) = \mathbf{C}_k, \operatorname{cov}(\mathbf{q}_k) = \mathbf{D}_k, \operatorname{cov}(\mathbf{x}_k, \mathbf{q}_k) = \mathbf{E}_k$$

3.4 Association of tracks to vertices

The χ^2 increment is given by (see also Subsection 3.2.5)

$$\chi_+^2 = (\mathbf{p}_k - \mathbf{c}_{k,0} - \mathbf{A}_k\mathbf{x}_k - \mathbf{B}_k\mathbf{q}_k)^T \mathbf{G}_k \cdot (\mathbf{p}_k - \mathbf{c}_{k,0} - \mathbf{A}_k\mathbf{x}_k - \mathbf{B}_k\mathbf{q}_k)$$
$$+ (\mathbf{x}_k - \mathbf{x}_{k-1})^T \mathbf{C}_{k-1}^{-1} (\mathbf{x}_k - \mathbf{x}_{k-1})$$
$$\chi_k^2 = \chi_{k-1}^2 + \chi_+^2 \qquad (3.88)$$

If necessary, the linear expansion can now be repeated at the new point

$$\mathbf{x}_{k,0} = \mathbf{x}_k \qquad \mathbf{q}_{k,0} = \mathbf{q}_k$$

and the filter can be recomputed, until there is no significant change either in χ^2 or in the estimate (one more iteration should be sufficient).

Since there is no process noise, the smoother is extremely simple:

$$\left. \begin{aligned} \mathbf{x}_k^n &= \mathbf{x}_n \\ \mathbf{q}_k^n &= \mathbf{W}_k \mathbf{B}_k^T \mathbf{G}_k (\mathbf{p}_k - \mathbf{c}_{k,0} - \mathbf{A}_k \mathbf{x}_n) \\ \mathbf{C}_k^n &= \mathbf{C}_n \\ \mathbf{D}_k^n &= \mathbf{W}_k + \mathbf{W}_k \mathbf{B}_k^T \mathbf{G}_k \mathbf{A}_k \mathbf{C}_k^n \mathbf{A}_k^T \mathbf{G}_k \mathbf{B}_k \mathbf{W}_k \\ \mathbf{E}_k^n &= \mathbf{W}_k \mathbf{B}_k^T \mathbf{G}_k \mathbf{A}_k \mathbf{C}_k^n \end{aligned} \right\} \qquad (3.89)$$

If there is a significant change in the smoothed vertex position, it may be worthwhile recomputing the derivative matrices $\mathbf{A}_k$ and $\mathbf{B}_k$.

As mentioned above, only a small fraction of all tracks originate from secondary vertices (multioutlier problem); these tracks normally have a low momentum and therefore contribute with a relatively small weight to the vertex determination, so that the estimated position of the primary vertex has no noticeable bias. Again, the filtered or smoothed residuals can be used to decide whether or not a particular track really does belong to the primary vertex. The residuals and their covariance matrices have the following forms (Equations (3.36)–(3.39)):

$$\left. \begin{aligned} \mathbf{r}_k &= \mathbf{p}_k - \mathbf{c}_{k,0} - \mathbf{A}_k \mathbf{x}_k - \mathbf{B}_k \mathbf{q}_k \\ \mathbf{R}_k &= \mathbf{V}_k (\mathbf{G}_k^B - \mathbf{G}_k^B \mathbf{A}_k \mathbf{C}_k \mathbf{A}_k^T \mathbf{G}_k^B) \mathbf{V}_k \\ \mathbf{r}_k^n &= \mathbf{p}_k - \mathbf{c}_{k,0} - \mathbf{A}_k \mathbf{x}_n - \mathbf{B}_k \mathbf{q}_k^n \\ \mathbf{R}_k^n &= \mathbf{V}_k (\mathbf{G}_k^B - \mathbf{G}_k^B \mathbf{A}_k \mathbf{C}_n \mathbf{A}_k^T \mathbf{G}_k^B) \mathbf{V}_k \end{aligned} \right\} \qquad (3.90)$$

Since $\mathbf{R}_k$ and $\mathbf{R}_k^n$ are singular the filtered χ_F^2 and the smoothed $\leq_S^2$ have to be computed in the following way (see Equation (3.88)):

$$\chi_F^2 = \mathbf{r}_k^T \mathbf{G}_k \mathbf{r}_k + (\mathbf{x}_k - \mathbf{x}_{k-1})^T \mathbf{C}_{k-1}^{-1} (\mathbf{x}_k - \mathbf{x}_{k-1}) \qquad (3.91a)$$

$$\chi_S^2 = \mathbf{r}_k^{nT} \mathbf{G}_k \mathbf{r}_k^n + (\mathbf{x}_n - \mathbf{x}_k^{n*})^T (\mathbf{C}_k^{n*})^{-1} (\mathbf{x}_n - \mathbf{x}_k^{n*}) \qquad (3.91b)$$

where $\mathbf{x}_k^{n*}$ is the smoothed estimate x_n with the track $\mathbf{p}_k$ removed. It is obtained by the inverse Kalman filter.

$$\mathbf{C}_k^{n*} = (\mathbf{C}_n^{-1} - \mathbf{A}_k^T \mathbf{G}_k^B \mathbf{A}_k)^{-1}$$
$$\mathbf{x}_k^{n*} = \mathbf{C}_k^{n*} \cdot [\mathbf{C}_n^{-1} \mathbf{x}_n - \mathbf{A}_k^T \mathbf{G}_k^B \cdot (\mathbf{p}_k - \mathbf{c}_{k,0})] \quad (3.92)$$

If $\mathbf{p}_k$ belongs to the primary vertex, χ_F^2 and χ_S^2 are χ^2-distributed with two degrees of freedom. If $\mathbf{p}_k$ originates from a secondary vertex $\mathbf{z}$ not too far from the primary $\mathbf{x}$, one can write in linear approximation:

$$\mathbf{p}_k = \mathbf{h}_k(\mathbf{x}, \mathbf{\dot{q}}_k) + \mathbf{A}_k \cdot (\mathbf{z} - \mathbf{x}) + \boldsymbol{\varepsilon}_k$$

One may choose $\mathbf{z}$ in such a way that $\mathbf{d} = \mathbf{z} - \mathbf{x}$ is orthogonal to $\mathbf{\dot{q}}_k$. The impact parameter of track k is then given by $|\mathbf{d}|$ and the offset of $\mathbf{p}_k$ with respect to the primary vertex by

$$\langle \mathbf{p}_k - \mathbf{c}_{k,0} - \mathbf{A}_k \mathbf{x} - \mathbf{B}_k \mathbf{\dot{q}}_k \rangle = \mathbf{A}_k \mathbf{d}$$

One can now compute the noncentral parameters of χ_F^2 and χ_S^2:

$$\lambda_F = \mathbf{d}^T \mathbf{A}_k^T \cdot (\mathbf{G}_k^B - \mathbf{G}_k^B \mathbf{A}_k \mathbf{C}_k \mathbf{A}_k^T \mathbf{G}_k^B) \mathbf{A}_k \mathbf{d} \quad (3.93a)$$

$$\lambda_S = \mathbf{d}^T \mathbf{A}_k^T \cdot (\mathbf{G}_k^B - \mathbf{G}_k^B \mathbf{A}_k \mathbf{C}_n \mathbf{A}_k^T \mathbf{G}_k^B) \mathbf{A}_k \mathbf{d} \quad (3.93b)$$

With λ_F and λ_S one can in turn compute the power of the respective χ^2 test. The main advantage of this method is that it avoids an increase of computing time due to combinatorial effects. Once a primary vertex has been found with sufficient accuracy, the remaining tracks can be tested iteratively. However, the *power of the test* (Equation (3.30)) is variable during this procedure.

After having associated all the nonprimary tracks into secondary vertices, the event topology is fully determined. In the next step of the data analysis the kinematical constraint equations (three for momentum conservation and one for energy conservation) are imposed upon the geometrically fitted data, using mass assignments as provided by particle identifying detectors (e.g. Cherenkov counters). A good geometrical fit of the direction between the primary and any secondary vertex will be an advantage. (This is also important for '*lifetime measurements*'.) It is possible to readjust the geometrical vertex fit after the kinematics fit (Subsection 3.4.4). However, this update is very sensitive to ambiguities in the mass assignment (Forden and Saxon 1985). If the mass assignment is doubtful, only the three constraints from momentum conservation should be used.

3.4.4 Kinematical constraints

In Subsection 3.4.2 we discussed how the LSM can be used to add the vertex constraint to the information obtained from the individual trajectories, and in Subsection 3.4.3 the detection of secondary vertices has been discussed. In order to add the information obtained from the four equations of energy and momentum conservation, the problem of using the LSM with linearized constraints with the help of a 'Lagrangian multiplier' will be discussed in this subsection. This method is, of course, applicable to a large variety of problems.

Consider a system of n direct measurements (e.g. three virtual measurements for each particle coming from a secondary vertex), with some additional unknown parameters $p_1,\ldots,p_r$, e.g. momentum and mass (or, sometimes better, the square of the mass) of a decaying particle. Note that for those vertices which should undergo a subsequent kinematical fit the three Cartesian projections of the momentum P, $\mathbf{p}$, should be choosen as parameters $\mathbf{q}_i$. Since the errors are now smaller than for the individual track fit, the problem of the linearity of the model is less important. For a primary vertex, however, the old parameters should be kept.

If because of too large errors the parameter change is only done after the vertex fit, the matrix to be transformed is of dimension $3m \times 3m$ (m is the multiplicity). The measurement vector is then

$$c = \begin{pmatrix} q_1 \\ \vdots \\ q_m \end{pmatrix} = \begin{pmatrix} {}^t q_1 \\ \vdots \\ {}^t q_m \end{pmatrix} + \varepsilon$$

with dimension $n = 3m$.

The covariance matrix of ε (with $\mathbf{W} = \mathbf{V}^{-1}$) built up from the 3×3 matrices is (see beginning of Subsection 3.4.2 and Equation (3.89))

$$\text{cov}(q_i, q_j) = \delta_{ij}\mathbf{W}_j + \mathbf{W}_i \mathbf{B}_i^T \mathbf{G}_i \mathbf{A}_i \mathbf{C}_m \mathbf{A}_j^T \mathbf{G}_j \mathbf{B}_j \mathbf{W}_j$$
$$= \delta_{ij}\mathbf{W}_j - \mathbf{E}_i^m \mathbf{A}_j^T \mathbf{G}_j \mathbf{B}_j \mathbf{W}_j \tag{3.94}$$

and the four constraint equations are

$$\sum_{i=1}^m q_{x,i} = \sum_{i=1}^m q_{y,i} = \sum_{i=1}^m q_{z,i} = 0 \tag{3.95a}$$

$$\sum_{i=1}^m (m_i^2 + q_{x,i}^2 + q_{y,i}^2 + q_{z,i}^2)^{\frac{1}{2}} = 0 \tag{3.95b}$$

where $i = 1,\ldots,m$ (the multiplicity) and m_i is the mass of the ith particle. All *outgoing particle terms* in the sums *have opposite signs*, and *all particles,* measured or unmeasured, *must be included*. Momenta and masses are in GeV.

All the unknown quantities in Equations (3.95a, b) are contained in an additional parameter vector $\mathbf{p} = (p_1, \ldots, p_r)^T$. Note that for the masses only *discrete mass hypotheses* will be compared by two separate fit procedures for each set. The general form of the constraint equations is

$$g_j = (c_1, \ldots, c_n, p_1, \ldots, p_r) = 0, \quad j = 1, \ldots, k$$

If the problem is to be completely determined, it follows that $r \leq k$, and consistency requires $k \leq n + r$.

The χ^2 ansatz

$$M(c) = (\mathbf{c} - \boldsymbol{\alpha})^T \mathbf{W} \cdot (\mathbf{c} - \boldsymbol{\alpha}) = \boldsymbol{\varepsilon}^T \mathbf{W} \boldsymbol{\varepsilon}$$

should now be minimized together with the constraint equations $\mathbf{g}(\tilde{\boldsymbol{\alpha}}, \tilde{\mathbf{p}}) = \mathbf{0}$. This is done with the help of 'Lagrangian multipliers' $\boldsymbol{\mu}$

$$L = \boldsymbol{\varepsilon}^T \mathbf{W} \boldsymbol{\varepsilon} + 2\boldsymbol{\mu}^T \mathbf{g} \cdot (\mathbf{c} - \boldsymbol{\varepsilon}, \mathbf{p}) \tag{3.95c}$$

The Lagrangian method requires that the variation of δL with $\delta \varepsilon$ must vanish.

Now the constraint equation will be linearized (care is required if $\sigma(P)/P$ is not $\ll 1$)

$$\mathbf{g}(\boldsymbol{\alpha}, \mathbf{p}) = \mathbf{g}(\overset{0}{\boldsymbol{\alpha}}, \overset{0}{\mathbf{p}}) + \mathbf{A} \cdot (\boldsymbol{\alpha} - \overset{0}{\boldsymbol{\alpha}}) + \mathbf{B} \cdot (\mathbf{p} - \overset{0}{\mathbf{p}}) \tag{3.96a}$$

with

$$\mathbf{A} = \partial \mathbf{g}/\partial \boldsymbol{\alpha}|_{\boldsymbol{\alpha} = \overset{0}{\boldsymbol{\alpha}}, \, \mathbf{p} = \overset{0}{\mathbf{p}}}$$
$$\mathbf{B} = \partial \mathbf{g}/\partial \mathbf{p}|_{\boldsymbol{\alpha} = \overset{0}{\boldsymbol{\alpha}}, \, \mathbf{p} = \overset{0}{\mathbf{p}}}$$

and

$$\mathbf{g}(\boldsymbol{\alpha}, \mathbf{p}) = \mathbf{g}(\overset{0}{\boldsymbol{\alpha}}, \overset{0}{\mathbf{p}}) + \mathbf{A} \cdot (\boldsymbol{\alpha} - \mathbf{x} + \mathbf{x} - \overset{0}{\boldsymbol{\alpha}}) + \mathbf{B} \cdot (\mathbf{p} - \overset{0}{\mathbf{p}})$$
$$\equiv \text{const} - \mathbf{A} \cdot \boldsymbol{\varepsilon} + \mathbf{A} \cdot (\mathbf{x} - \overset{0}{\boldsymbol{\alpha}}) + \mathbf{B} \cdot \boldsymbol{\delta} \tag{3.96b}$$

It follows

$$L = \boldsymbol{\varepsilon}^T \mathbf{W} \boldsymbol{\varepsilon} + 2\boldsymbol{\mu}^T [\text{const} + \mathbf{A} \cdot (\mathbf{x} - \overset{0}{\boldsymbol{\alpha}}) - \mathbf{A} \cdot \boldsymbol{\varepsilon} + \mathbf{B} \cdot \boldsymbol{\delta}]$$
$$dL/d\varepsilon = 2\boldsymbol{\varepsilon}^T \mathbf{W} - 2\boldsymbol{\mu}^T \mathbf{A} = 0 \tag{3.97}$$
$$\mathbf{G} \boldsymbol{\varepsilon} = \mathbf{A}^T \boldsymbol{\mu}$$
$$\boldsymbol{\varepsilon} = \mathbf{V} \mathbf{A}^T \boldsymbol{\mu} \tag{3.98}$$

Substituting ε from Equation (3.98) in Equation (3.96b)

$$\text{const} - \mathbf{A} \mathbf{W}^{-1} \mathbf{A}^T \boldsymbol{\mu} + \mathbf{A} \cdot (\mathbf{x} - \overset{0}{\boldsymbol{\alpha}}) + \mathbf{B} \boldsymbol{\delta} = 0 \tag{3.99a}$$

$$\boldsymbol{\mu} = (\mathbf{A} \mathbf{V} \mathbf{A}^T)^{-1} \cdot [\text{const} + \mathbf{A} \cdot (\mathbf{x} - \overset{0}{\boldsymbol{\alpha}}) + \mathbf{B} \boldsymbol{\delta}] \tag{3.99b}$$

3.4 Association of tracks to vertices

$$\varepsilon = \mathbf{V}\mathbf{A}^T(\mathbf{A}\mathbf{V}\mathbf{T}^T)^{-1}[\text{const} + \mathbf{A}\cdot(\mathbf{x}-\overset{0}{\alpha}) + \mathbf{B}\delta] \quad (3.100)$$

From $\partial L/\partial \delta = 0$ it follows (with $(\mathbf{A}\mathbf{V}\mathbf{A}^T)^{-1} = \mathbf{W}'$)

$$2\mu^T\mathbf{B} = 0$$
$$\mathbf{B}^T\mu = 0$$
$$\mathbf{B}^T\mathbf{W}'\cdot[\text{const} + \mathbf{A}\cdot(\mathbf{x}-\overset{0}{\alpha})] + \mathbf{B}^T\mathbf{W}'\mathbf{B}\delta = 0$$
$$\mathbf{B}^T\mathbf{W}'\cdot[\text{const} + \mathbf{A}\cdot(\mathbf{x}-\overset{0}{\alpha})] + \mathbf{B}^T\mathbf{W}'\mathbf{B}\delta = 0$$
$$\delta = -(\mathbf{B}^T\mathbf{W}'\mathbf{B})^{-1}\mathbf{B}^T\mathbf{W}'\cdot[\text{const} + \mathbf{A}\cdot(\mathbf{x}-\overset{0}{\alpha})] \quad (3.101)$$

This requires, however, that $\mathbf{B}^T\mathbf{W}'\mathbf{B}$ is not singular, with the necessary condition $n \geq k$; otherwise the complete set of equations must be solved, which would also facilitate the evaluation of the covariance matrix between $\tilde{\alpha}$ and $\tilde{p}$.

Substituting δ in Equation (3.100) one obtains

$$\tilde{\varepsilon} = \mathbf{V}\mathbf{A}^T\mathbf{W}'[1 - \mathbf{B}(\mathbf{B}^T\mathbf{W}'\mathbf{B})^{-1}\mathbf{B}^T\mathbf{W}']\cdot[\text{const} + \mathbf{A}(\mathbf{x}-\overset{0}{\alpha})] \quad (3.102a)$$

and finally

$$\tilde{\alpha} = \mathbf{c} - \tilde{\varepsilon} \quad (3.102b)$$

$$\tilde{p} = \overset{0}{p} + \tilde{\delta} \quad (3.102c)$$

$$\chi^2 = \tilde{\varepsilon}^T\mathbf{W}\tilde{\varepsilon} \quad (3.102d)$$

If necessary a second iteration (or even a third due to the strong nonlinearity of the fourth constraint equation) can be performed by substituting $\overset{0}{\alpha} \leftarrow \tilde{\alpha}, \overset{0}{p} \leftarrow \tilde{p}$. Note that only during the first iteration $\mathbf{c} = \overset{0}{\alpha}$ holds, and therefore $\mathbf{A}\cdot(\mathbf{c}-\overset{0}{\alpha}) = 0$.

The covariance matrices are calculated by error propagation:

$$\left.\begin{aligned}\text{cov}(\tilde{\delta}) &= (\mathbf{B}^T\mathbf{W}'\mathbf{B})^{-1}\mathbf{B}^T\mathbf{W}'\mathbf{A}\mathbf{V}\mathbf{A}^T\mathbf{W}'\mathbf{B}(\mathbf{B}^T\mathbf{W}'\mathbf{B})^{-1} \\ &= (\mathbf{B}^T\mathbf{W}'\mathbf{B})^{-1}\end{aligned}\right\} \quad (3.103a)$$

$$\text{cov}(\tilde{p} - \overset{0}{p}) = \text{cov}(\tilde{\delta})$$
$$\tilde{\alpha} = \mathbf{c} - \tilde{\varepsilon} = \text{const}' + \{1 - \mathbf{V}\mathbf{A}^T\mathbf{W}'$$
$$\cdot[1 - \mathbf{B}(\mathbf{B}^T\mathbf{W}'\mathbf{B})^{-1}\mathbf{B}^T\mathbf{W}']\mathbf{A}\}\cdot\mathbf{c} \quad (3.103b)$$
$$\text{cov}(\tilde{\alpha} = \{1 - \mathbf{V}\mathbf{A}^T\mathbf{W}'\cdot[1 - \mathbf{B}(\mathbf{B}^T\mathbf{W}'\mathbf{B})^{-1}\mathbf{B}^T\mathbf{W}']\mathbf{A}\}$$
$$\mathbf{V}[\cdots]^T = \mathbf{V} - \mathbf{V}\mathbf{A}^T\mathbf{W}'\mathbf{A}\mathbf{V} + \mathbf{V}\mathbf{A}^T\mathbf{W}'\mathbf{B}$$
$$\cdot(\mathbf{B}^T\mathbf{W}'\mathbf{B})^{-1}\mathbf{B}^T\mathbf{W}'\mathbf{A}\mathbf{V}$$

The expectation value of the χ^2, $\langle\chi^2\rangle$, is $k - r$.

3.5 Final observations on track fitting

This chapter has described how to obtain the best geometrical resolution by a reasonably fast and flexible program, and how to test the hypothesis that the measurements grouped together represent a particle's track up to its final confidence level. A possible feedback of the geometrical fit to the trackfinding might be necessary in some cases, but this has not been discussed here, and the problem of outliers was mentioned only briefly. More robust tests than the χ^2 and more robust estimation procedures are under investigation (Frühwirth 1988); their application is essential if, in the presence of outliers, kinks are to be detected.

Application of the Kalman filter and smoother provides a new, flexible tool for many applications, and the application of the Kalman filter to the search for secondary vertices eliminates the combinatorial aspect of this problem. The Kalman filter also offers flexible possibilities for alignment, Fig. 3.17.

Throughout this chapter only the LSM has been discussed. It allows one to perform the fit on several levels: single track fit, the fit of vertices, and kinematics. Error propagation is trivial in the LSM, and in the ideal case of Gaussian errors in the coordinates, the errors in the estimated parameters are well known. This also holds for losses due to χ^2-cuts. However, it is essential *that all the program modules are carefully tuned* via the pull quantities and the χ^2, first for the single track fit procedure, and then for the vertex fit and kinematics, *step by step*, (pull quantities of single tracks, vertices, and finally for the kinematical fit).

Before analysing real data, the full algorithm should, in any case, be tested with the help of a Monte Carlo simulation, where the true values of the parameters to be estimated are also available, allowing a check on the reduced quantities $(\tilde{\mathbf{p}} - \hat{\mathbf{p}})/\sigma(\tilde{p})$. The exact errors in the fitted parameters which might be influenced by outliers and their removal procedure as well as by misalignment should also be carefully studied.

This chapter should be a guide for a programmer writing a geometry reconstruction program when he wants to adopt one of the described methods for his specific problem, or to develop a new, elegant, reliable, fast, and precise method himself; it should be mentioned that the manpower and computer load needed for program development and tuning are usually underestimated. Of course, when judging the speed of the method to be used for the geometrical reconstruction, one also has to estimate the data

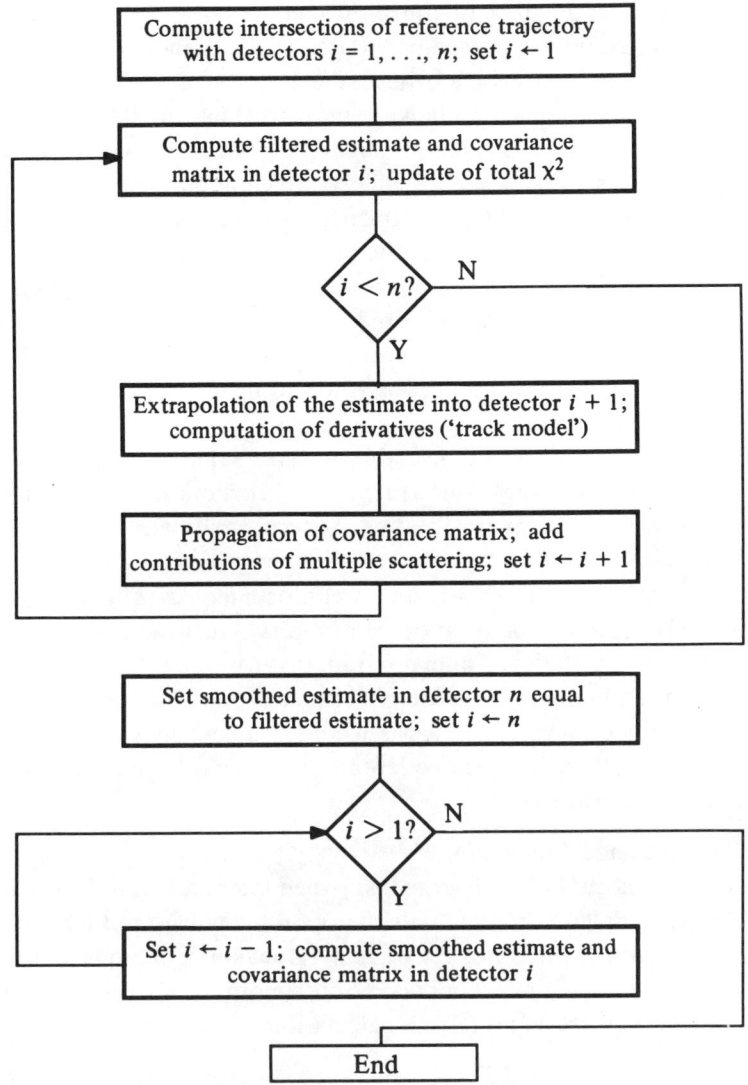

Fig. 3.17 Flow chart of the basic Kalman filter/smoother algorithm. Track extrapolation, error propagation and the matrix operations of the filter are clearly separated, in the spirit of modular programming. Thus it is easy to refine – and simplify – the track model and the evaluation of multiple scattering and energy loss, if needed.

reduction that might be obtained through on-line and off-line filtering and by the trackfinding procedures. Also, in a collaboration where several programmers are involved, both the lucidity of an algorithm and its modularity are aspects to be kept in mind (Metcalf 1986).

Sometimes in this chapter a careful judgement as to whether an algorithm is suitable for a special experimental purpose was given, and in certain cases this might also influence *the design of an experiment*. As this book is not a book of mathematics, not all methods are accompanied by a rigorous mathematical proof of consistency – as long as they work with the desired precision and have been applied successfully in the analysis of a perhaps quite unusual experiment. The experiments quoted should serve as examples to illustrate the different track fitting methods and should allow a comparison of them.

Of course it would be desirable to have some successful field-proven algorithms in a library. Unfortunately it turns out in practice that global track fitting programs are usually too specific to be used in several different experiments, and too complex to allow easy extraction of just parts of them. Structured, modular, and well-documented programming would not only facilitate the creation of a library, but would also facilitate the exchange of experience and program code inside a large collaboration. Also, the additional effort required to make some modules available to the community of particle physicists would be relatively small.

Although there is some overlapping, three main ingredients to the track fit can be considered quite separately:

The track model (including the derivative matrix).
Material effects such as energy loss (when interferes with the track model) and multiple scattering (with effects on the covariance matrix).
The matrix operations for the track fit (global fit, if multiple scattering can be neglected, Kalman filter and smoother otherwise, including the proper weight matrices) and for the calculation of the χ^2.

A well-documented library offering a variety of solutions would then encourage the programmer to choose the coordinate system, the field representation, and the track parameters such that they fit to one of the library modules, but are still adequate to his specific requirements.

It should be mentioned that track fitting is not a purely geometrical task, as trajectories cannot always be described by five geometrical parameters only. One has to know *a priori* what kind of particle has followed the trajectory in question. For a low-energy particle, knowledge of its mass is

3.5 Final observations on track fitting

needed when energy loss and multiple scattering have to be treated rigorously. In some cases, measurements which allow particle identification also have to be fitted together with coordinate measurements, e.g. TOF measurements. For coordinates measured by drift chambers, knowledge of the TOF of the particle yields a correction that might be necessary to apply to them.

Finally it should be stated that track and vertex fitting have successfully been applied in many experiments both for high- and low-energy particles. To facilitate this, however, the problems associated with track fitting should be kept in mind when planning the detector and the structure of the reconstruction program. The effort necessary for this is small compared with the possible loss of information, and with the total effort invested in an experiment.

4

Tools and concepts for statistical data analysis

In this chapter we discuss some practical problems concerning data analysis as it is currently practice in high-energy physics research. We show how the abstract world of mathematical formulae can be mapped onto practical computer and software systems *as they exist*, by formulating them as algorithms.

Chapters 2 and 3 both gave formulae in a rather classical description, as well as discussing how detectors are built and what their characteristics are. Here, we deal with a subject which is, in comparison, more technical. We are closer to Chapter 1, where we described triggering algorithms. The discussion is centred on the translation of the physics concepts ('selection of interesting reactions') into what can be implemented in electronics, firmware, or by programming (the 'algorithm').

In this chapter, we are again heavily dependent on the technical possibilities available: on the programming languages and computer architectures of today, on the hardware components and software products available now, and on the specific expertise existing in the relatively small high-energy physics community.

The manufacturers of computers and associated systems, and the producers of software concepts and products, seem to maintain, over many years, a breath-taking speed in developing novel and cheaper systems, resulting in a practically unlimited choice in combining hardware and software for a given application. This chapter consequently runs the risk of being overtaken faster than other parts of this book by the evolution of the computer environment.

Computers and associated software systems offer such an overwhelming number of possibilities, such scope for ingenuity and creativity, that for a novice, trying to choose amongst and to make use of existing systems, a clear irritation results initially. This irritation quickly gives way to an 'expert syndrome'; as soon as the novice has become an expert on some questions, he cannot but be seriously biassed by the specific influence and experience he has undergone. And he will find that state of bias, based on very local expertise, difficult to leave, because only a conscious effort will allow him to refind his virginity.

Whatever is said below is thus likely to appear incomplete and out of date within a short period of time. Should a book like this then refrain entirely from mentioning what the state of the art is today, and let the reader fend for himself? Given the choice between a text which is of limited use to the physicist who has to make his choices, because it addresses abstract problems only, and a text which risks sounding out of date after a few years, because it includes a discussion of components which are evolving rapidly, we quite consciously opt for the latter.

In looking back a decade or so, we can also, investing some optimism, discern a backbone of concepts and tools which have remained remarkably stable, even though such stability seems to be the exception. In the quicksand territory just outlined, any successful attempt to stabilize some aspects of computer-related systems will be, at least for the application-oriented user, a welcome reduction in the dimensionality of his problem: a piece of the puzzle solved independently. Our desire to contribute to such stabilization is the main motivation for embarking on this chapter. We realize that such effort cannot but navigate between the expert's *bias* and the nonexpert's *ignorance*. Nevertheless, in the complete absence of a metric for judging or even comparing systems, a serious effort to shed light on some of the existing components should at least help those who are obliged to make choices.

4.1 Data abstraction

This section deals with the process of bringing a problem to the computer, of 'programming' it. It has been mentioned above, that *formulae* for analysis and *algorithms* which implement such formulae as a computer program, are two different things. Formulae are usually expressed like: *Let x_i, y_i be a set of points measured in the plane of a photosensitive detector of a Ring Imaging Cherenkov detector* (RICH) *with depth L, due to Cherenkov radiation from a particle with velocity* β. *We fit a circle to these points by the least squares method, minimizing the sum of squared shortest distances of all points from the circle. Let us call the result of this fit a circle of radius R with error* ΔR. *The error on* $\gamma = 1/(1-\beta^2)^{\frac{1}{2}}$ *is then given by*

$\Delta\gamma = \gamma^3\beta^2 n\Delta R/(N_o \dot{L}^3)^{\frac{1}{2}}$, where N_0 is radiator's quality factor in photons radiated per unit length and energy interval.

Such a statement should sound perfectly meaningful to a physicist. Of course, it contains numerous implied approximations and abstractions: The detector in the formula is assumed planar, Cherenkov radiation all happens at the same distance L from this detector, and results in a circular distribution of points in that plane, an error on R can be estimated by assuming constant errors for all x and y coordinates, errors are Gaussian and can be propagated by linearization. We also have made reference to established knowledge: Physicists can be expected to *know* what a RICH is (see Section 2.6), or how the LSM works (details in Subsection 3.2.2, or in Blobel 1984b).

When translating such a formula into an executable algorithm, though, these abstractions and assumptions must now be further matched with the concepts that a computer, or at least a programming language can understand (there are computer details finer yet, usually hidden from the application programmer who chooses his high-level programming language close to mathematical formulations). In accomplishing this step from formula to algorithm, multiple addition abstraction are necessary.

Let us assume we want to write the above algorithm in FORTRAN 77 (the choice has little influence on what follows). We cannot then escape formulations like this: *The coordinate pairs x, y are stored in single-subscript arrays with the names XRICH and YRICH, and their number is given as the variable NXY. These variables are found in a COMMON block named RICDAT. This COMMON also contains the distance L in variable RICDEP, and the definition of the position of the plane in space in the variables XPLAN(4), YPLAN(4), ZPLAN(4), which give the coordinate triplets for each of the four corner points of the plane.*

It is readily seen, that much more detail of a nonphysical nature enters an algorithm, than is necessary in a formula. However, we have left even more detail unmentioned. In our simplified example, we have assumed, sheepishly, that the plane is bounded by exactly four edges, and hence have to decide if consistency checks on the four points XPLAN, YPLAN, ZPLAN still have to be made ('are the points in a plane?'). We have to relate the coordinate pairs XRICH, YRICH to some origin and orientation in the plane. We have to define criteria for how far we allow our points to deviate from a circle ('what action does the program take if the χ^2 of the fit is larger than a maximum value?'). We also have to define what we should do with

4.1 Data abstraction

photons which fall outside our plane bounds ('will they be detected in some other chamber or lost?')

It is important to realize that *all* possibilities, *even the most irrelevant and trivial ones* have to be spelt out to the computer by the program: *Computers simply have no a priori knowledge whatsoever.* At best we can rely on previously programmed algorithms, available in a local program library. But the need to foresee all possible conditions and to deal acceptably with them is our own, in every detail.

All of this is no more than a very modest beginning for a detector analysis code, of course. Our trivial example algorithm will normally be only a small part of such a code. Real detectors do not just have a single RICH, a RICH detector does not consist of a single plane, and detector planes are not typically bounded by four edges. Also our technical assumptions for data storage are naive. The measured points, for example, are not usually found in isolation in a handy COMMON. If we were to arrange for that, we would quickly find ourselves with hundreds if not thousands of such simple COMMON's and multiple 'interface' routines transcribing our data between them, in order to have simple COMMON definitions for all the subtasks we have to solve. More typically, the few measurements, for which we formulated our algorithm, are a subset of many measurements. Other algorithms make use of these and the entire set of measurements and the correlations between them have to be accessible. Such data have to be organized with great care in order for programs to access them as simply as possible (more about data access in Section 4.5).

We also have to decide, in our example, which least squares package we want to use, or if we want to code the fitting procedure, which converts measured coordinates into a circle radius, ourselves. If we use a package, will it fit our problem without modifications? How do we evaluate its reliability for our case? Is it also available in the program library of all of the collaborating institutes which will want to run the same code?

We have thus discovered the problems of *organization* and *size*, both for data and programs. As programs and data reach a certain size, efficient ways are needed both to define and document their details, and to communicate these between those who collaborate in the writing, testing, running, and maintenance of a program. We should be conscious that this task usually takes a much larger fraction of the design and development time than the definition of the formulae and their translation into an algorithmic language. Contrary to what intuition tells us, any serious program design *begins* with the organizational aspects, by defining program and data structures at the highest level. Details of algorithms are the last fine detail which are brought into the program. More about this 'top-down' approach in Chapter 5. The relative domination of organizational over algorithmic aspects in a program is analogous to the visual aspect of programs, in which the *algorithmic* statements occupy only a small fraction of the space that is taken by data and program *organization*, and declarations to the computer, interfaces, validity checking, exception handling, and many more.

As large programs are written as much with human communication in mind as for simply submitting an algorithm to a machine, the abstract algorithm has to be embedded in conventions that make a program easy to understand (this is necessary and vital even for the person who develops it), easy to communicate among coworkers, and easy to change as the detector evolves ('program maintenance'). For these reasons the 'engineering of software' has been a subject of intensive work on behalf of many computer scientists, of industry, and also of some large scientific computer users. The techniques emerging in this field have also been introduced in high-energy physics, we will discuss them briefly in Chapter 5, but we concentrate for now only on data abstraction.

We should mention here that one very basic understanding is of prime importance: A suitably structured *program* will consist of modules that are *functionally* as independent as possible. Thus, a certain detector part will have the corresponding code in a program module. At a lower level, a routine will typically implement the transformation of a specific detector's digitizings from the least count like the inverse clock frequency of an analogue-to-digital converter, to a physical unit like the position in millimetres. Suitably structured *data* will again consist of *conceptual* entities such as the description of part of a detector, an association between digitizings into a candidate track, or the jets found in a calorimeter. We will give technical meaning to 'structuring' in Subsection 4.5.3. We assume here that the programming language allows the structuring of data, or that one of the widely used extensions of FORTRAN is used, as described in Subsection 4.5.3.

These criteria for structuring programs and data have the advantage of being close to the physicist's thinking, hence are more natural to conceive and to communicate to other physicists than any other structure. Being compatible with each other, they allow the visualization of a program part as a 'transforming operator' on well-defined parts of the data structure. Such structures will correspond to the most economic way of writing a program, if we define good economy as a short time spent by the physicist, who has to write such an application. Such economy may well, in fact *is likely to*, result in programs that are not well adapted to specific computer architectures (*viz.* the emerging vector machines), and that do not hide implementation details well, as is ideally required from a program (*viz* many scattered references to the same data).

When discussing the *statistical* analysis of experimental data, the 'final event analysis', we are beyond the multiple paths that lead from the detector digitizings to physics information. We are then concerned only with the aspect of abstracting (condensing) the ultimately relevant information for each of our events into a few numbers, so that we can perform our analysis on these alone. This task is apparently innocuous; an interaction is described by an interaction point, by emerging tracks or possibly jets whose hadronization details may or may not be given. For a proper description, we need coordinates for points, and describe each track or jet by a momentum 4-vector. We also assume that all events are uncorrelated, so that the order in which they are stored is irrelevant. So what more is needed?

True enough, this is the idealized description which experiments are ultimately aiming at. In practice, it is rarely sufficient. The multitude of digitizings produced in a detector's inner guts is riddled with a variety of technical details the influence of which is not foreseeable at the outset, but which potentially have the effect that neither the sample content nor the information content of each event in the sample, as simplified above, may be sufficient for any physics conclusions to be drawn. These details have to be brought to light, one by one, and their influence has to be reliably estimated and corrected for. If they blur or mask the desired result in too obvious a way, remedies will have to be introduced that may implicitly alter the definition of some variables, or require entirely new algorithms to be written. Each such updated version will have to be introduced whilst fully maintaining the ability to reanalyze and reproduce any result previously achieved. It must be understood that high-energy physics experimentation continuously revolves around this constantly changing situation. It is thus absolute suicide not to plan systems such that they can cope with change.

4.2 Data selection

One of the most frequent problems in preparing a data sample for final analysis is that of deciding which events to include and which to reject. Equally important is the question of how to account for the errors that will result from inevitable wrong decisions. Except in the most simple cases, this is a process that needs iterations, test runs, and extended simulation. Very often results obtained after many successive selection filters indicate the need to go back and verify that an observed effect is indeed inherent in the physics or the detector, and not one that was introduced artificially by the selection criterion.

We will refrain here from discussing, in general, the ways of deciding between different hypotheses. Textbooks of statistics deal competently with the subject. In some situations, philosophical arguments cannot be avoided e.g. the Bayesian argument dealing with the use of *a priori* knowledge. Eadie, *et al.* (1971) and others have said what can be said, in general, about decision making in high-energy physics applications.

It may seem appropriate and of practical use, however, to attempt to give a few rules of thumb, which would at least be applicable to the simpler cases. Let us therefore define a restricted decision problem in which we are interested: Let each event be characterized by a fixed number n of elements (an n-vector or n-tuple). Let the events be drawn randomly from a mixture of two distributions. Call the members of these distributions *good* and *background* events. They should form different, ideally separate, clusters in n-space (the feature space of Subsection 2.1.5). We can make assumptions about the shape of the distributions of the two classes, and about the mixing ratio. We can also assume that we can tune the performance of our algorithm on representative *training samples* of good and background events.

We seek a classification algorithm expressed as one or several test statistics, i.e. single-valued functions of the n elements. In order to be useful, a test statistic must separate the two samples into two clusters which are as distinct as possible. To each test statistic we can therefore associate a cut parameter which divides that projection of the feature space into a 'good' and a 'background' fraction. These are called the 'acceptance region' and the 'rejection region' respectively (see also Subsection 3.2.4).

In other words, our selection mechanism is meant to operate like this: We form the test statistic(s) from a given event, and obtain a value which is a random element from one of the distributions, good or background. This element is now compared with a cut parameter. Depending on the sign of the difference between the cut parameter and the function value, the event is classified as a *good* or *background* event. A *correct decision* will then classify a good event as good, a background event as background. A false decision will classify a background event as good (*contamination*), or a good event as

4.2 Data selection

background (*loss*). Assuming sufficiently different distributions for good and background elements, the 'significance of the test' (the probability of rejecting good events) will be small, and the 'power of the test' (the probability of rejecting background events) will be large, i.e. the procedure will classify individual elements correctly most of the time (cf Subsection 3.2.4). Note that various terminologies are used: significance is also denoted by 'cost', and power by 'purity'.

When there are several test statistics, the problem is multidimensional and a decision can be formulated as a Boolean function of several different tests (e.g. a good event has to have a χ^2 less than a certain value, *and* a minimum number of points in a given tracking chamber).

We call the set of test statistics with associated cut parameters, combined with the prescription of how to connect them logically, if this applies, the *decision function* for a set of measured events. To any numerical choice of the cut parameter(s), there will be a fraction of correct and false decisions for both good and background events. In other words, a certain loss of good events will correspond to a given contamination (see also the discussion of tests on the goodness of fit, in Subsection 3.2.4).

For a single test statistic and cut parameter, contamination versus losses can be plotted as a curve in a two-dimensional Cartesian coordinate system. (for an illustration, see examples and Fig. 4.1(*d*)). Different points on the curve correspond to different values of the cut parameter. Due to the integral character of both loss and contamination, they will both increase/decrease monotonically with the cut parameter. Let us call this curve the 'decision quality diagram'.

If the decision function includes several test statistics and cut parameters, a higher-dimensional surface results from varying the cut parameters. Each additional test statistic corresponds to an additional axis. The decision quality diagram can then no longer be shown as a curve in two dimensions. Only its projections can readily be visualized, e.g. they can obtained by fixing some of the cut parameters.

An *optimal* decision function will use algorithms that minimize the losses for a given contamination. The optimal decision function will have the property that *no other decision function will have smaller losses for the same background*. In the language of statistics (Kendall and Start 1967) this would be a 'uniformly most powerful test'. In the decision quality diagram, the optimal decision function results in the surface of closest possible approach to the point where both losses and contamination are zero. The cut parameter(s) can then be chosen to define a point on the curve or surface in the region of that minimal distance. As there is usually a wide choice of possible decision functions, one might be tempted to give a recipe for constructing the optimal decision function based on mathematical principles. This seems a possible goal as we have assumed that we know the statistical properties of our good and background event samples. Even more modestly, it might seem possible (and useful) to give a rule such as *if criteria of distinction exist in several independent dimensions, use them independently rather than combining them into a single-parameter criterion.*

Unfortunately, in general even such a simple rule cannot be formulated. It is comparatively easy to find background shapes which favour, for the same sample of good events, different decision functions. And it is typically the assumption about the shape of background which is the least robust, subject to numerous hypotheses. The finding of an optimal decision function invariably needs a heuristic approach, one of trial and error. We shall now demonstrate this difficulty by a fictitious analytical example. We also give an example taken from a real experiment. Further examples can be found in the case studies in Subsection 2.5.6.

A simple fictitious example:

In order to show the problem of optimizing decision functions in more detail, we will concentrate on a very simple case. Let each event be characterized by two elements. For the sample of good events, let these be random elements drawn from two independent χ^2 distributions, each with three degrees of freedom (see Subsection 3.2.5 for an explanation). Let the corresponding elements in the sample of background events, instead, be random variables from two identical flat distributions, extending (for normalization) from 0 to 15 (any background beyond that value is considered to be without impact on our study). Fig. 4.1(*a*) shows these distributions for a single variable.

We now limit our study to comparing two decision functions, hoping that one of them will be uniformly superior, i.e. provide a better recipe for separating events in a mixed sample into good and background events, whatever contamination or loss we are willing to tolerate.

(1) We use the two elements as independent test statistics, with the two cut parameters assumed to be equal. An event is considered good if both its elements are below the cut parameter; it is considered background otherwise. This reflects the fact that the χ^2 distributions peak at low values, whereas the background distributions are flat.

4.2 Data selection

(2) We construct a single test statistic by forming the sum of the two original test statistics, and then use a single cut parameter.

The first decision function uses two variables separately, and hence is two-dimensional. The single cut parameter used on both statistics translates into the prescription that good events have to have both χ^2s below the single cut parameter. This corresponds to cutting out a square in the plane (2-space) of the two elements, in which the probability density is a two-dimensional surface. This surface is peaked towards the origin for the good events (but with a zero at the origin), and is entirely flat for the background event sample. Fig. 4.1(b) shows the probability density integrated over a square with a given side length corresponding to the cut parameter, and normalized to the integral over the entire plane.

In the second decision function, we have reduced the problem to a one-dimensional case. The test statistic will be a χ^2 with six degrees of freedom for good events, and will have a linearly rising and falling probability density for background as shown in Fig. 4.1(c). Here again, low values are more likely to be associated to good events, and the cut parameter is used to separate good from bad.

The comparison of the two different decision quality diagrams corresponding to cases (1) and (2), is shown in Fig. 4.1(d). It clearly shows that it is impossible to favour one of the two approaches over the other systematically, i.e. neither is uniformly superior. For this to occur, the superior decision function should show a systematically smaller contamination for any given loss in the decision quality diagram. The two decision quality diagrams actually obtained intersect and favour decision function (1) or (2) depending on which loss/contamination fractions we are willing to tolerate.

It is understood that the two approaches are not very different qualitatively in this particularly simple example, and our interest is somewhat academic. The example is, however, a useful demonstration of a *genuine difficulty*.

An example from an experiment:

The experiment UA1 at the CERN SPS p$\bar{\text{p}}$ collider isolated the reaction p$\bar{\text{p}} \to W^\pm + X$, $W^\pm \to \tau + \nu$, by using a test statistic optimized and invented for the purpose, called the 'τ-likelihood'. The signal of τ-decays of the charged intermediate vector boson $W^\pm$ is seen with a background of Quantum ChromoDynamics (QCD) jets, and for the optimal separation of the two samples some specific properties of τ-decays have been used. τs appear as

344 Statistical data analysis

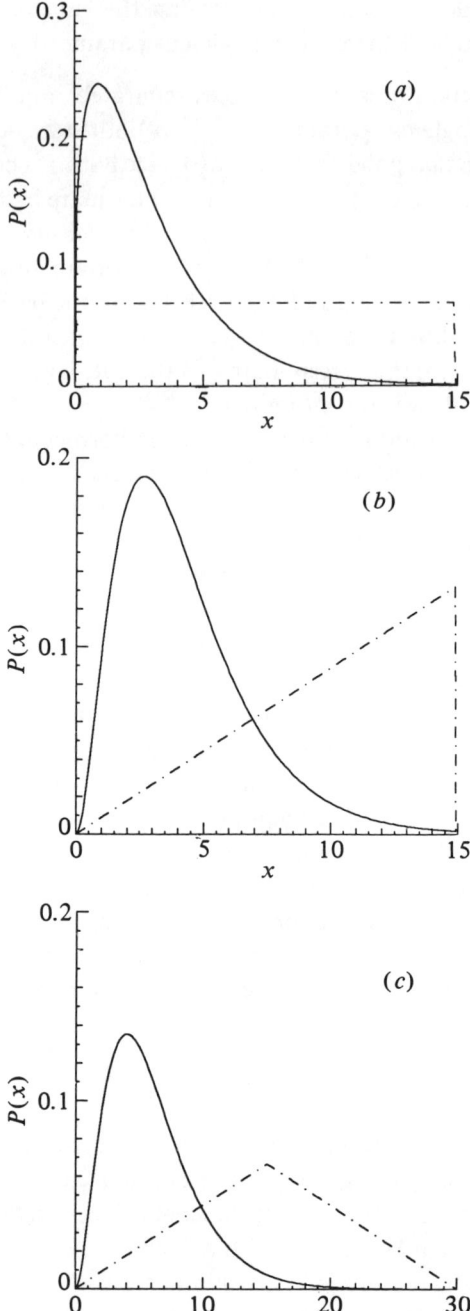

4.2 Data selection

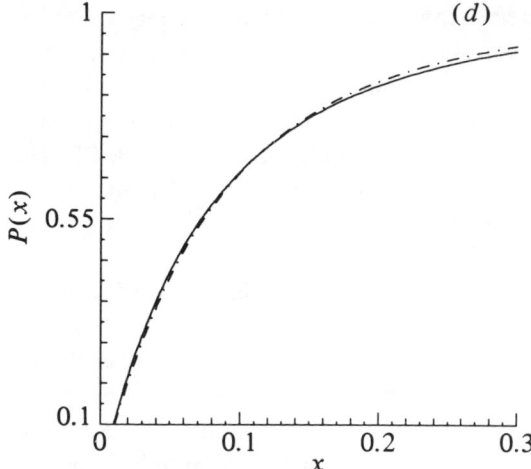

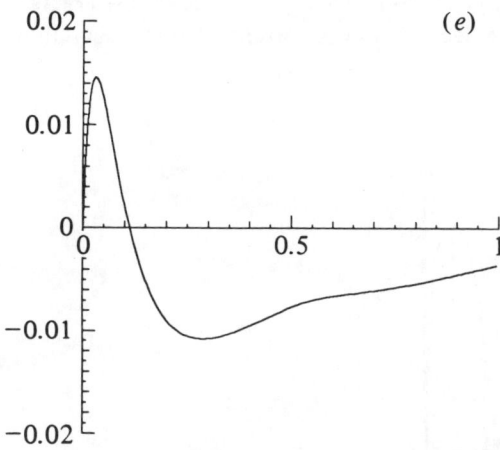

Fig. 4.1 An example of constructing decision quality diagrams. (*a*) A single χ^2 distribution (three degree of freedom) compared with a flat background of the same unit integral. (*b*) Two independent χ^2 distributions, and two background curves as in (*a*), used as the probability density in two dimensions with the same cut parameter on both axes. The curve shown gives the probability density associated to a given cut parameter, i.e. over an incremental side length of a square in the plane. (*c*) A single χ^2 distribution (six degrees of freedom) compared with a background obtained by summing two independent random elements from the flat background of (*a*). (*d*) The 'decision quality diagrams' explained in the text, for the distributions of (*b*) and (*c*). The curves intersect, as is shown more clearly in (*e*) which shows the difference between the two decision quality diagrams shown in (*d*). The fact that the curve passes through zero indicates that a unique statement as to the superiority of one of the two tests is impossible.

very collimated jets due to their decay into few hadrons, whereas QCD jets are due to parton hadronization, and may radiate gluons. τs are associated, like some rare QCD events, with a large unobserved energy vector, inferred by forming the vector sum of all energies observed in the hermetically closed calorimeter (see Subsection 2.5.6). Both τ-decays and QCD jets appear on top of an event-internal background of tracks due to the remnants from the two spectator quarks and antiquarks (that do not participate in the hard interaction).

After various cuts (see the full description in Albajar, et al. 1987), three *distinct* relevant parameters (test statistics) are evaluated: (*a*) the ratio F of the jet energies, contained in a narrow and a wider cone around the jet direction, (*b*) an angular separation r between the leading track and the jet direction, (*c*) a multiplicity n of charged particles with high transverse momentum in a narrow cone. All of F, r, and n are expected to be larger for the signal (τ-decays) than for the background (QCD jets). A *single combined test statistic L* is therefore defined by $L = \log(nrF)$ (the 'τ-likelihood'). The probability densities for L as measured in a sample of known QCD jets, as Monte Carlo simulated for τ-decays, and as observed in the candidate

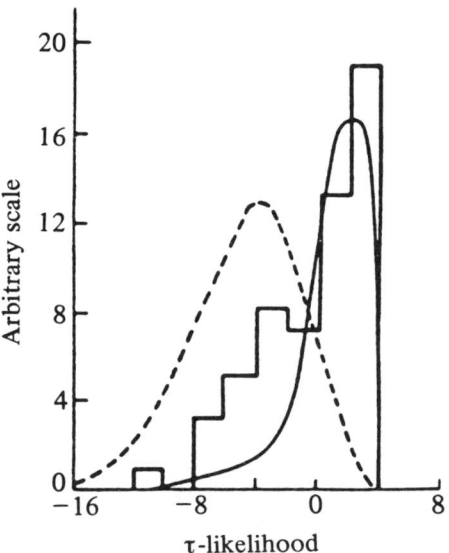

Fig. 4.2 The test statistic 'τ-likelihood' explained in the text, plotted for UA1 events (histogram), for τ-decays from Monte Carlo (solid curve), and from observed jet events (dashed curve), from Albajar, *et al.* (1987). All curves are normalized to the same number of events.

events, are shown in Fig. 4.2. A cut parameter $L = 0$ results in a loss of signal of 22%, with a contamination of 11% (of QCD events!).

Practical conclusion

What of practical help, then, can be said in general? Little more than the following:

Hypothesis deciding deserves careful study, making models for both good and background events, and trying different decision criteria on *training samples* of events, constructed typically by a Monte Carlo program with the statistical properties of the samples under study.

The more test statistics that can be found that are not interdependent and in which probability densities are significantly different, the easier it will be to find a satisfactory cut parameter(s) and hence an acceptable decision function.

In a systematic study, the graphical presentation of decision functions in a decision quality diagram may be helpful, in particular to show the sensitivity to background assumptions.

If test statistics closely relate to physical or detector properties, it may be an advantage to use them separately in order to identify 'outliers'. Outliers are elements which fail to follow the statistical properties assumed in the model, either because of a truly exceptional low probability, or, more likely, because of some gross errors, e.g. δ-rays in a chamber, readout or transmission errors, wrong association in the pattern recognition etc.

4.3 Data accumulation, projection, and presentation

We now assume that we have reduced our data sufficiently to be left with a random sample of events, each characterized by n elements which represent the detector- and physics-dependent variables of interest in further analysis. In other words, we have used successive steps of *abstraction* (to represent each event by few variables) and *selection* (to define the event sample). Whatever information has been lost in these steps, has to be carefully and often independently evaluated, and the sample may have to be corrected for the losses. The corrections must also be as independent as possible with respect to the cut parameters ('robust correction'). Typically, the selection and abstraction process is iterative and tentative stepwise, i.e. samples are formed under certain assumptions, studied, and the result verified by changing some of the assumptions, Consequently, final event samples as stored on tape or disk are often not purified nor abstracted to a fixed number of variables for each event; in practice, gross event samples are often kept together, maintaining highly redundant data for each event, the selection and abstraction processes being left to what is called the 'final event analysis' or, in jargon, DST (for 'Data Summary Tape') analysis.

In this final event analysis, then, the sample will be subjected to the usual statistical methods. In order to extract the physics content of quantum-mechanical nature, most of the analysis will concentrate on projecting the multidimensional space spanned by the event variables onto lower-dimensional spaces, and usually one will use probability densities in one or two dimensions, i.e. *accumulation* and *projection*. Accumulated and projected representations of data stand for observations of probability densities. Typically, they are shown as graphical presentations, i.e. as histograms in one or two dimensions (here we somewhat generalize the notion of a histogram, which is usually restricted to the one-dimensional projection). The contents n_i of bin i of a histogram allow the deduction of the probability frequency $P(x)$, from:

$$\langle n_i \rangle = \sum n_i \int P(x)\,dx \quad \text{(one-dimensional)}$$

or

$$\langle n_i \rangle = \sum n_i \iint P(x,y)\,dx dy \quad \text{(two-dimensional)}$$

with the sums extending over all bins, and the integrals over the limits of bin i in x and y). Brandt (1970) or other statistics textbooks give more details. In these projections, comparisons with theoretically predicted or independently derived experimental probability densities become possible, unexpected structure may be observed, correlation may be seen and may lead to regression analysis, etc.

In what follows, we raise a few practical problems concerning projecting and presenting data, whose solution is not always apparent from the more general textbooks; we do this under a few unrelated headings.

Binning

In order to obtain probability densities, interesting data which represent typically continuous variables are usually discretized, i.e. all elements between given limits are grouped into 'bins'. The frequency of events in a given bin is a measure of the average probability density in the region spanned by the bin. Any resolution contained in the original measurements

4.3 Data accumulation, projection, and presentation

finer than the bin size will be lost; the loss is more severe the smaller the experimental resolution is compared with the bin size. The choice of bin size, therefore, is of prime importance. Bins should be chosen such that the bin content should not be less than some statistically significant number (no rule of thumb exists!). In other words, there should not be too many bins. Too few bins, on the other hand, result in loss of information about the probability distribution.

If this dilemma presents a serious problem, the computational advantage of equal-size bins may be abandoned. Statistically optimal binning has been shown (Eadie, *et al.* 1971) to correspond to bins of equal probability, hence not to equal-size bins, in general. For small samples, binning may be outright impossible. The use of unbinned data may instead be indicated, and methods other than standard ones (often based on the laws of large numbers) should be considered. James (1981) gives some specific hints for small samples. A method that works reliably on unbinned data, without being resticted to this application, is the MLM (see Subsection 3.2.5).

Note that a compromise between the data reduction inherent in binning and the necessity of keeping analysis options open, is frequently found by *truncating* numerical data to some acceptable level of precision without accumulating them statistically into bins. This results in creating a small set of (truncated) numbers, for each event (often called an 'n-tuple'). All correlations between variables, lost when histogramming (i.e. projecting onto a lower-dimensional surface), are preserved, and later statistical analysis for any bin size larger than the truncation precision is possible (with the caveat, though, that periodic effects due to interferences between truncation precision and the chosen bin size can cause unpleasant surprises). In fact, n-tuples are more often than not used without truncation, which is only useful for reducing the space occupied by large data sets.

Error analysis

A general classification of errors (see also Subsection 3.3.2) is into the following four groups: *random errors* (those caused by random processes, decreasing by a factor $f^{\frac{1}{2}}$ if the sample size is increased by a factor f), *truncation and rounding errors* (arising from digital signal processing, and decreasing with sample size only if kept carefully uncorrelated), *systematic errors* (caused by incomplete knowledge or inadequate correction, and causing the same bias independent of sample size), and *gross errors* (due to wrong assumptions such as including measurements in the wrong sample).

Most error estimates are based on random errors. Great care therefore has to be taken to estimate the influence of non-random errors correctly, or to ensure that their influence is negligibly small. In judging the influence of errors on statistical estimators derived from a sample, the most general and computer-adapted method may be the 'bootstrap' error estimate of Efron (1982): New artificial samples, of the same size as the original sample, are formed by randomly drawing the sample members from the original sample (hence possibly introducing some members several times). The estimator in question will then show a probability density function of its own, which allows the derivation of confidence limits.

Presentation:

This may seem a purely technical or aesthetical issue, but the choice of presentation is *not* only that, but may well help you recognize what otherwise would escape your eye. Take the example of Fig. 4.3: data have been generated from the hypothetical function $z = (1 + \alpha \sin n\phi)(1 + \sin r)$, with a small coefficient α, $r^2 = x^2 + y^2$, and x, y proportional to $\cos \phi, \sin \phi$. The very same data are represented as two-dimensional probability densities in three different ways. Only one of them shows clearly the fact that along the rim, extremes (maxima and minima) do exist.

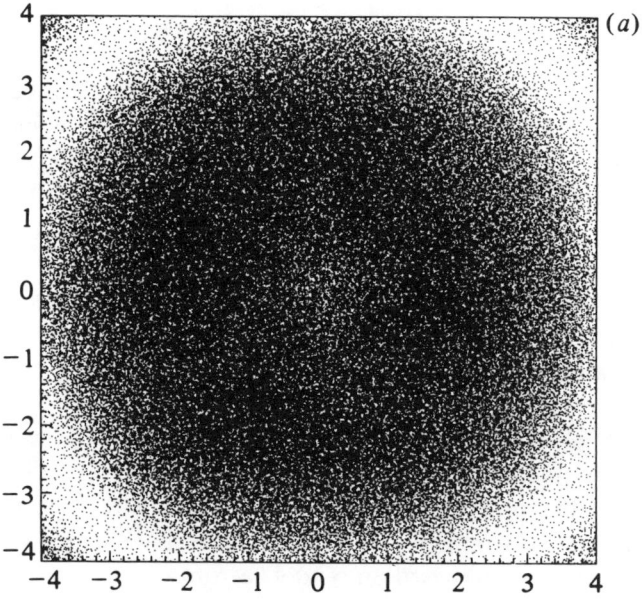

4.3 Data accumulation, projection, and presentation

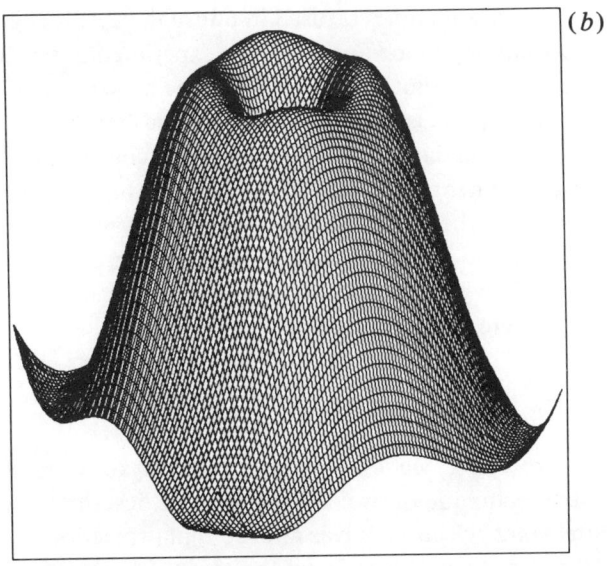

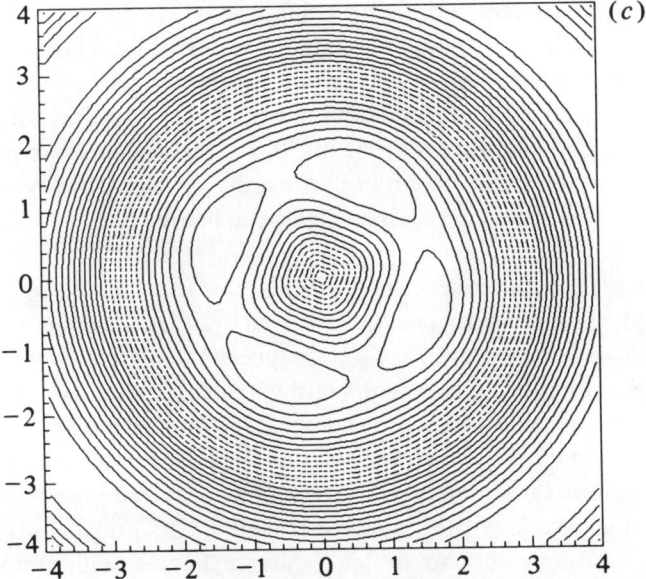

Fig. 4.3 Three different ways of presenting the same data: The probability density function $P(x, y) = (1 + 0.002 \sin 4\phi)(1 + \sin r)$ is shown as (*a*) a scatter plot, (*b*) with interpolated surfaces, (*c*) as a contour plot (lines of equal probability density). The extremes along the edge cannot be seen in the scatter plot, and the distortion of contour lines is only visible in the contour plot.

Histogramming can sometimes be used in unusual ways: Ideograms are histograms with a binning much finer than the experimental resolution, in which the measurement resolution is folded into the presentation. In other words, every measured point is entered as a (discretized) probability density, normally given by a Gaussian curve with an integral corresponding to the weight of the measurement. All such probability density curves are then added on top of each other to obtain the overall probability density. An ideogram will have a smoother appearance than a histogram, because data have been smeared. The width of any narrow structure in an ideogram will appear artificially increased, however.

4.4 Multidimensional analysis

In the preceding sections, aspects of analysis were discussed which rely on the technique of projecting the event description onto one or two axes, after some selection. It was implied that projections are chosen which reveal the statistical properties of the class of events selected.

Such projections result, of course, in a loss of information. Every reaction of a given type is characterized by a number of parameters which is multiplicity-dependent, and only the simplest reactions like elastic scattering can be represented without information loss in a single histogram. All variables are, in addition, potentially influenced by distortions or other detector-dependent calibration problems, or biassed by selection inefficiencies, therefore such a reduction in dimensionality nearly always has to be considered a possible source of problems, and hence must be avoided as long as possible.

It seems a logical consequence of these thoughts that techniques should be developed which consider *simultaneously* all or at least a larger number of variables for a sample of events of a given type. If the sample is sufficiently large, statistical methods in several dimensions exist which can be expected to be superior to any analysis in projection, i.e. methods applicable for one- or two-dimensional histograms. Such multidimensional methods have, of course, been in use for a long time; the technique of extracting amplitudes with different spins and parities from experimental data known as 'Partial Wave Analysis,' is one of these. In this technique, limited assumptions are made about the quantum numbers of the contributing waves, and the higher contributing spin amplitudes are either ignored or calculated from some model (for a better introduction and a bibliography, Litchfield (1984) should be consulted).

4.4 Multidimensional analysis

Such analysis methods using *a priori* knowledge, or at least a model about reactions, are multidimensional, but do not correspond to what is to be touched upon in this section. We refer here to the attempts, popular in the 1970s and somewhat more dormant today, to extract information in several dimensions simultaneously *without* any preconceived idea about the physical process. Mostly, this amounts to looking for *structure* in event distributions without making assumptions as to what might be at the origin of such structure. The natural consequence of detecting structure would, of course, be to *classify* events into categories, belonging to different 'clusters' of which the structure is made.

In order to define structure and clusters, great care has to be taken in choosing the *variables* which are meant to reveal the structure, and in defining a *metric* which corresponds to the space spanned by these variables. By metric, we mean the recipe which tells us how the variables are interconnected and which allows us to define a measure of *closeness* or *similarity*. In Euclidean space, two points are said to be close if their distance is small, and distance is defined by $(\Delta x^2 + \Delta y^2 + \Delta z^2)^{\frac{1}{2}}$. In a space defined by variables such as the scattering angle (or rapidity), the azimuth, the energy ratio, how does one define closeness? Without this very central notion of a metric, there is no defined procedure for finding clusters.

Various attempts have been made in the last 20 years to define general clustering techniques that would produce good physics results in the context of high-energy physics data analysis. Projection pursuit techniques were tried in order to enhance signals in one- or two-dimensional projections optimally (Friedman and Tukey 1974). Hierarchical clustering methods were introduced to give essentially *all possible* clusterings and then to allow manipulation of the various cuts in order to extract the most 'significant' (by some definition) clusters.

It is not unfair to say that the pursuit of such techniques has not, so far, been able to enrich high-energy physics with result that would not have been accessible otherwise. For the results these efforts have produced, there is a good overview with a bibliography in Naumann and Schiller (1984). The hope that interactive techniques with modern graphics might be helpful, has also not quite been borne out so far; it is difficult to display even three dimensions on a screen (see also Subsection 4.5.1). Attempts to 'visualize' even further dimensions were reported by Friedman, McDonald and Stuetzle (1987), and were not found to yield easily achieved superior results. It must be assumed that methods that exceed human perceptual abilities, although conceivable on a computer, present serious implementation problems – hardly a surprising result. That does not preclude future breakthroughs in this area, and the potential available in modern equipment is certainly enormous (e.g. Farrell (1987)).

4.5 Technical aids for high-energy physics data analysis

This section touches most closely upon the problems discussed at the beginning of Chapter 4, problems of discussing techniques which are in constant and sometimes rapid evolution. We are fully aware that our discussion of technical tools will became out of date comparatively rapidly, and that the viewpoint from which we present it must be one influenced by our limited experience. No author escapes this dilemma unless helped by the distance that only time can provide. Consequently, there exist only few textbooks, and practically none that are up to date. It is precisely for this reason, that we have thought it worth our effort to discuss the practical questions connected with graphics, data presentation, and data access in this book. We do this without feeling constrained to achieve completeness or to give solid theoretical foundations, constraints which haunt the authors of a textbook on these specific subjects.

4.5.1 Basic graphical communication with computers

The semantic meaning of the general adjective 'graphic' carries the connotation vivid, life-like, or more generally, clear. We know that human memory is largely pictorial, and nearly all the information surrounding us is perceived not in numerical but in pictorial, in graphical, form. Even some abstract mathematical problems, especially in topology, are solvable only when mapped onto graphical representations (graph theory).

It seemed natural to expect computers to become sufficiently 'human' to cover this domain in their interaction with man. In the 1950s and early 60s the field of computing was in its infancy and resembled a playground for mathematics-oriented specialists rather than the present industry of overwhelming economic and social impact. In those early days, the importance of graphics was already fully recognized, although the tools were not available.

Graphics then was mostly understood as a means of presenting data to the human, i.e. as a passive form of output from computers. Later, the role of interaction, i.e. of some form of dialogue between computer and human, in graphic or mixed graphic/text mode, started to emerge as a means of easing

4.5 Technical aids

the burden of communication for the human. This was fully understood to be at the expense of some clever programming which makes the computer 'converse' with the end user sitting in front of a graphics terminal. Those who decried this as 'computer inefficiency' have by now been silenced by the obvious gains in human resources which such interactive programs were able to demonstrate. These gains outweigh any losses in computer power by large factors. Simultaneously, the Artificial Intelligence community began its efforts to enable computers to not only 'read' graphical information by digitizing it into small black/white cells, but also to 'understand' pictorial content at least at some limited level.

Each of these efforts in turn met its own share of difficulties. Passive graphical output was solved relatively quickly, albeit with devices that were either very crude, not to say inadequate, or very expensive and thus mostly accessible only to commercially and politically important applications, such as training airline pilots, space research, or defence. For physicists with their tight budgets, plotters driven from the computer centre were the rule, and using them, typically through a procedure involving several operator interventions, was not an easy job for the casual user. Interactivity with graphics remained the domain of the very few who could afford the necessary gadgetry.

Finally, the hope that computers might one day truly 'understand' what they are given to 'see', has been more or less shattered, or at least has not yet been borne out. To establish a proper description of how visual stimuli in our brain cause us to come to our routine conclusions about our environment, to perceive objects or their motions, is a fascinating and challenging field of research, but one that deals mostly with unsolved problems. Much more progress has been made in understanding the sensory aspects of seeing and the brain's physiological aspects, than in describing or even modelling 'human data processing'. Cognitive processes are thought by many to be fundamentally nonalgorithmic (Dreyfus 1979).

All this is to introduce a statement which is true despite the extraordinary progress that graphics has made in the last few years, and which applies to some extent even to the simple graphical output tasks that are performed perfectly daily: *Graphical communication is not natural for a computer. If graphics has to be programmed at low level, specialization and hence a substantial effort is required.* Unfortunately, it is only the low level of graphics that has converged to a somewhat standardized applications interface, whereas the easy *human* access to *computer* graphics is confined to packages linked to a specific hardware and application environment.

We will now examine to what extent the problems in the area of graphical man–machine communication have been solved, and how the more common high-energy physics applications can take advantage of the solutions.

4.5.1.1 General problem areas in graphics programming

Graphics communication is nontrivial

When programming graphics applications directly, much of the mismatch between the computer and human concepts will come to the surface. The basic programming of graphics devices, whatever the application may be, remains an area of programming requiring special expertise. Despite undeniable successes in standardizing hardware and software, graphics programming is still full of nitty-gritty details and nasty pitfalls. These make the field unattractive to anyone who has primarily only his own application in mind and has no desire to become an 'expert'.

As in other neighbouring areas which are rapidly evolving, such as support software in detector design or software management, there is not as yet a body of knowledge or a collection of high-quality standard books (despite the very good Foley and Van Dam (1982), Enderle, *et al.* (1986), and Ten Hagen (1986), which one can rely upon or refer to. Implementations of graphics support products differ widely between computer centres, and the portability of application programs including graphics input or output is an unfulfilled dream.

Interaction comes closely linked to graphics

Substantially improved graphical devices have been marketed, in large numbers and at low cost, in recent years. They have made marked improvements in presenting graphical output from programs, on a screen or a hardcopy device. At the same time it has been recognized that modern screens allow a complete change in the dialogue between a computer user and a program. Interaction between man and machine benefits in an essential way from the cheap availability of graphical interfaces. Many now popular small systems owe their success to the ease of use of the software that is delivered with a system, and graphical interaction is a vital part of it.

The most prominent example of clever interaction is the program *menu* interface. Instead of expecting the user of a program to study a manual and then express his intentions as a list of typed commands, a menu-driven program will offer a small range of choices, each of which will either result in executing a command or offer further, more detailed choices at a lower level. Most successfully, the choices are offered in an easy-to-understand form, often using graphical symbols instead of lengthy text. Well-designed graphical menus thus are a documentation, a tutorial, and a learning tool all in one, but only well-designed ones.

Take text processing as another example. A modern text processing system again converses with a user through graphical menus, and a subset of actions allows the user to compose, as bit patterns, his own character set, including for instance, mathematical symbols or special character fonts. The input will have to pass through the bottleneck of the keyboard, but where graphical input is needed, e.g. in defining patterns, the screen and mouse can be heavily involved. The graphical dialogue allows the user to follow the composed text including special symbols *as soon as he types* (so-called WYSIWYG system, for 'What You See Is What You Get'). Such interaction is by far superior (and hence more successful) than a system in which a rigid interface of typed commands is used to compose an input file, which after editing is processed in a separate program producing output for a plotter or special printer.

Graphics software often does not match the hardware

The many breakthroughs in recent years in graphics and associated computer hardware and software have entirely changed the availability of graphics devices. Modern screens, usable for interactive applications, are virtually exclusively bit-addressable cathode ray tubes. The screen elements are refreshed at regular intervals from a 'bit map', a digital memory representing a matrix of picture cells ('pixels'). To each cell is associated one of a limited number of grey scale or colour values, in the simplest case a single bit deciding between 'black' and 'white'. From such cells, pictures are generated with shades of colour or grey, much as in the raster techniques that newspapers have been using for many years. A large number of pixels, e.g. of the order of a thousand along each axis, allows a fine resolution to be achieved. Pushed further, it would become comparable to the granularity of silver atoms as used in photographic emulsions, which also consist of a finite number of (irregular) pixels.

Cheap memories (from which refresh displays work) and microprocessors allowed the development of modern software which communicates directly in text and graphical form with a user. Such systems solve *popular* problems *locally*. They solve problems of interest to the public at large, the customers: integrated text processing and drawing tools, games, etc. They solve them optimally on a given hardware device, because that is what makes the customers spend their money. The commercial success of such systems has resulted in the widespread acceptance of computers by the public, mostly in the form of Personal Computers (PCs), and high production numbers have brought down prices even more. Graphics screens thus have evolved from expensive devices for specialists to routine tools in daily use. Good devices are now found as adults' and children's toys in many homes.

Display screens are, of course, backed up by passive output 'hardcopy' devices, and most of these, again, are bit-addressed. However, there is not normally a match between the bit map of a screen (typically $< 10^6$ pixels) and a good hardcopy device (which often has a much finer resolution). A general graphics interface, therefore, needs a description of images, at a high level, from which both representations can be derived.

General application interfaces, which physicists can resort to, are invariably vector-oriented, i.e. all lines are drawn as connections between two points. Curved lines are represented as many small straight vectors. This interface is nearer to human thinking, and also corresponds to early graphics devices which all were of the plotter type, obeying instructions like 'pen down', 'move to point x, y, 'pen up' etc. This vector-oriented interface is also the one most commonly used to preserve full precision for different hardware, e.g. the hardcopy devices mentioned above.

Modern hardware concepts in the area of graphics do not match the most natural thought concepts. A good fraction of the interfacing problems between an application programmer's thinking and graphics hardware has its origin in this mismatch. The most acute problems arise not in the purely passive graphics output, but when interaction is involved, i.e when a mixture of locator and text input on the screen has to be mapped onto the vectorial structures of an application program.

True, the class of people that have become graphics experts has widened, the necessary degree of specialization has lessened, and the serious efforts to standardized the application interface seem to be slowly converging. Such standards help to introduce a general terminology, and provide some level of portability even for graphics programs. They ensure that diverse equipment can be addressed in a high-level and long-lived form. In turn, standards usually have to settle on a level of functionality below what can be achieved for any specific system: in order to be portable, only that subset of functions common to all devices can be supported. Also, the adherence to standards typically results in bulky and not necessarily efficient graphics packages which do not take advantage of intelligent local solutions.

The merit of standards in making graphics applications accessible to not-so-expert programmers is undeniable. Nevertheless, the difference in concept between programming ('vectorial') and implementation ('bit map') is at the origin of many programming pitfalls, particularly when human interaction on a graphics screen is required, i.e. the hardware's bit map concepts have to be translated to the application's vector concepts.

Colour provides beauty, but is difficult to use intelligently

The addition of the colour dimension to the graphics interface is not a technical problem for either displays or hardcopy devices. Colour also causes little if any additional nuisance in programming, it simply gives more choice in selecting the line or fill area style. Colour does, however, find surprisingly narrow limits of applications simply because there is limited understanding of its intelligent use, i.e. of how to enhance information rather than just provide beauty. Some of the perceptual limitations and possibilities have been discussed by Galitz (1985), Friedman, McDonald, and Stuetzle (1987), and Farrell (1987). Some practical hints on how to combine colours efficiently and with the desired effect ('colorimetry') are given in Hopgood, Hubbold, and Duce (1987). Colour further experiences manifest problems in entering our daily life. Presently, among the thousands of paper copying devices which we use, only a tiny fraction will transmit colour information. These observations, of course, may quickly become out of date if the market goes through one of those sweeping changes that industry has so often effected.

High-quality graphics remains confined to special applications

Not only are there cheap and generally available graphics devices on the market today but also the art of presenting pictures with a high degree of naturalness has made substantial progress. Three-dimensional hardware, colour, and software support for the removal of hidden lines and for rendering surfaces with shading and highlighting do exist, and this 'solid modelling' may be complemented by 'animation', i.e. the ability to rotate three-dimensional pictures dynamically. These efforts have resulted in remarkable tools for engineers, architects, and filmmakers. We will confine the concepts of such high-level graphics to the glossary in Subsection 4.5.1.3. The ultimate possibilities of these devices and of imaging techniques, in general, are as yet largely unexplored (Farrell 1987). It should be noted that high-level graphics is a computer-intensive application, and hence often done on large computers with passive output devices. If combined with interaction, they necessitate systems of high CPU capacity and high-quality interactive graphics, hence are of very considerable cost.

Our concern here is the casual physicist programmer who is confronted with some aspect of analysing the data of a high-energy physics experiment, and possibly also with the problem of visualizing his detector. For the physicist, high-end graphics tools do not exist off the shelf, and their price does not match a typical physicist's budget. At least, it is generally accepted that the potential gain by introducing such devices and the associated software in a physics experiment do not warrant the substantial expenditure in money and manpower. Excellent human interfacing including high-end graphics is, of course, being used in some experiments; but the examples are *local and atypical*, and allow no generalization.

4.5.1.2 Graphics in high-energy physics

Let us come back to the historical remarks in our introduction to Subsection 4.5.1. High-energy physics has participated in many of the learning processes of the past. There were early initiatives to encode numerically, and then analyse in the computer, the (pictorial) analogue output of the then most general and popular high-energy physics detector, the bubble chamber picture recorded on film. This approach experienced the same difficulties as pattern recognition projects in other parts of science. The numerical description of pictures in terms of pixels could be solved, albeit at a price. To turn that cellular description, however, into a much condensed description in terms of track images and vertices, turned out to be a serious problem. The process could never be more than partially automated, despite major investments of high-level human effort.

The successful systems in bubble chamber picture data analysis were characterized by using pixel-like digitizings, by track recognizing algorithms that typically worked at some 95% level, and by passive graphical output methods for showing events or event parts to a human operator on a computer screen together with the program's understanding of the information. Such systems acknowledge the human's superiority in deciding the more complex recognition issues. True interaction in the modern sense was not sufficiently supported by the hardware then available. Successful semiautomatic systems with operator guidance thus combined the speed of computers for the multitude of trivial low-level tasks, with the specific skills of human operators for tasks of context. Conceptually similar

4.5 Technical aids

solutions are today commonplace for the translation of natural languages or in the computer-aided (aided, not computerized!) design of mechanical or electronic components.

Since bubble chamber times, detectors have gone through several stages of evolution. Today, nearly all data acquisition is done in real time. The local and heuristic methods used in recognizing patterns as recorded by these detectors work at a similar 95% level. Their limits have been understood and accepted. Human operators are not generally required. In Chapter 2, the status of pattern recognition with detector-recorded data has been presented. It is not today a source of major headaches. The trend to fine-grained detectors, coupled with higher event rates, may, of course, necessitate some rethinking in this area in the years to come.

In high-energy physics, three main areas of applications seem to demand custom-made programs with a graphical interface:

The *statistical analysis* of physics experiments. The presentation of statistical data in the final event analysis of physics data by interactive graphics is a formidable tool for extracting physics from an experiment. Graphical presentation of data becomes a particularly powerful means of communication when associated with interaction. Both are today commonly accessible through inexpensive graphics workstations or PCs.

We have mentioned before that graphics has found its way to the general public, where it is surrounded by clever application programs that give the user a limited number of choices in a specific application area. In other words, the interactivity is limited to the choices foreseen by the application program. If the programmed possibilities suffice, the user is freed from any necessity of programming either devices or interaction. A good example for such applications are the popular drawing tools available on PCs, or, at a high level, Computer-Aided Design (CAD) systems for mechanical engineering and electronics design. In such systems, the program controlling the graphics and dialogue is hidden from the user.

Analogous systems, adequate for the task of physics analysis, are not found on the general market, but a number of physicist-created packages do exist. We discuss these in Subsection 4.5.3.

The *display of events* in a detector. This is an application that eases substantially the design and optimization of detectors with the help of Monte Carlo events. In a detector that exclusively records nonvisible data, event display is also a necessary aid in the monitoring of detector performance. Furthermore, looking at events is of substantial help in the development of analysis software at the detector level. And finally, the possibility of reverting to visual single-event analysis involving a physicist as a human ultimate decision maker for some exotic cases is also used at times.

No general system exists that can economically handle this application. Experiments usually struggle with the best hardware they can find on the market, programming it at the basic level.

Visualization. A third type of graphics application we can only mention in passing: the fascinating task of using graphical methods to represent the normally invisible. One may think of the well-known fractal functions in mathematics, or applications in chemistry for hypothetical molecules, in astrophysics of models of intergalactic interactions, in the aircraft or car industry for simulating deficient product behaviour, or in plasma physics to peek into calculated states that can only exist for extremely short time intervals. Examples also exist in high-energy physics, e.g. to render the short range forces between particles visible, or to visualize the dynamical evolution of lattice calculations.

These potential applications of high-end graphics are not, at present, under very active development, but theory departments are likely to discover the possibilities of high-level graphics some time in the future. Note again that computer-intensive graphics has a limited potential as interactive application, because response times are long.

4.5.1.3 *Graphics notions: a glossary*

We end this rather cursory discussion of graphics with a general glossary of terms and notions concerning graphics devices and interfaces. It is mostly intended to be a guide for a newcomer to the graphics terminology. It addresses the questions of graphics programming at the basic level, and includes some of the notions of modern high-level graphics, despite their limited applications in high-energy physics, and the fact that they are not part of generally available software packages. We obviously recommend to the interested some more serious reading; there exist introductory texts at various levels of complexity and up-to-dateness, e.g. the very comprehensive Foley and Van Dam (1982), the introductory Scott (1982), the very useful if somewhat ageing IEEE tutorials Booth (1979) and Freeman (1980), the Eurographics Seminars series (e.g. Ten Hagen (1986); Enderle, *et al.* (1986)), and the very short but readily available

4.5 Technical aids

(for the high-energy physics community) Myers (1983) and Hopgood and Duce (1986).

Area filling. An operation (a graphics primitive) that allows a two-dimensional area to be filled with a defined pattern. The area's boundaries are defined by a closed polygon. Area filling gives areas the appearance of surfaces, or differentiates areas visually by highlighting them through hashing or colouring, thus achieving better clarity. Area filling is particularly easy to implement efficiently on displays backed up by a bit map. On most devices of this type, raster operations exist that allow the transfer of rectangular parts of the bit map. Area filling of polygons defined in three dimensions is possible, but defined only after projecting onto two dimensions.

Aspect ratio: Coordinate systems used in graphics have to be mapped onto each other by rotations and translations. Window and viewport sizes can result in transformations that not only change the overall scale of objects, but also may scale the x and y axes by different factors, thus changing the aspect of the object. Desired changes in aspect ratio can be brought about by changing the viewport definition. Mathematically, changing the aspect ratio corresponds to using a transformation matrix which is not orthonormal.

Attributes: In connection with graphics primitives, attributes describe the rendering style: font for text, line style, surface texture, also width or size, colour, and similar properties. In a wider sense, transformations between coordinate systems, viewing parameters, or area filling, can also be classified as attributes.

Bit map: The name given to the memory containing the picture information in terms of pixels, for a raster display. The term bit map display is used for a raster display that is refreshed periodically from a bit map memory.

Clipping: The operation that reduces a picture to the content that can be shown in a given viewport, cutting off any part of the picture that falls outside. This well-defined task is tedious, particularly in the viewing of three-dimensional objects, and should not be left to an application program. At present, some graphics packages, particularly early implementations which drive plotters and storage tubes, do not perform clipping as part of the standard user interface. Such packages will produce nonsense pictures or even crash if given coordinates, outside the specified range.

Core graphics system: A graphics package proposed (by the ACM

SIGGRAPH committee) as standard to ANSI in 1977, seen somewhat as a predecessor of and in competition with (the accepted ISO standard) GKS, the 'Graphics Kernel System'. Implementations of Core systems with FORTRAN binding are widely spread, primarily in the USA. See also the discussion of graphics standards later in this glossary.

Cursor: The visible mark on the display screen moved under control of a mouse, thumbwheel, track-ball, keyboard (up-down-left-right keys), or any other similar device. It shows where current locator input is positioned. Usually implemented as a blinking rectangle or arrow, or as two lines intersecting at right angle.

Device coordinates: One of the important coordinate systems in graphics, device coordinates refer to a specific device, and are typically given in pixel addresses and ranges (e.g. from 0 to 1023 in a given coordinate). Preferably these should never to be introduced in an application program. *Normalized* device coordinates go from 0 to 1 in each coordinate, and thus offer greater portability possibilities to other devices.

Device driver: The code which interfaces a graphics package to a specific hardware device, e.g. the GKS package to a Pericom Monterey display terminal.

Display list: The list of graphics objects to be presented as a picture or part thereof. The display list describes graphical objects in terms of primitives, attributes, and viewing parameters. Display lists are the only information for plotting and storage tube devices, which obey the list's directives sequentially. They are also close to a programmer's thinking, and to the graphics packages' interface. Their counterpart is a graphics description in terms of pixels. Display lists on external media are largely equivalent to metafiles (see below).

DVST (Direct View Storage Tube): See storage tube.

Event: In the context of graphics, events are interrupts caused by the (interactive) user on one of the available input devices (locator, valuator, keyboard). An event-driven program is one which relies on the user to trigger its actions. A good interactive interface carefully enables different inputs at any given moment, including the ability simply to interrupt any computation that may be going on in response to a previous event.

Facets on surfaces: In shading curved three-dimensional surfaces to render them more natural, the fact that they originate from wire-frame representations corresponding to discrete plane surfaces, results in unnatural facets (discontinuities along the plane surface boundaries).

4.5 Technical aids

Facets are removed by a shade smoothing algorithm in high-level graphics software.

Flicker: A disturbing property associated with displays of the vector type. If the material to be displayed takes the computer more time to refresh than the typical phosphor storage time or the human eye's latency (about 0.04 s), a continuous on/off of information results which makes looking at the display tiresome. Vector displays are therefore rated by their capacity to display so many vectors without flicker.

GKS (Graphics Kernel System): A graphics standard accepted by ISO in its two-dimensional version in 1984. GKS does not define a single standard implementation, but allows for different levels of both input and output, resulting in a matrix of possible implementations depending on the degree of complication desired and devices available. Several implementations for the FORTRAN 77 (and Pascal, ADA or C) binding exist. A three-dimensional extension without relevant new features beyond the added third coordinate axis is in the process of being standardized, and implementations exist. For more information, see Enderle, Kansy, and Pfaff (1984) or Hopgood, *et al.* (1983).

Graphics packages/graphics standards: A graphics package is the software which implements a graphics interface at the application level for a given set of input and output devices. The history of graphics standardization efforts spans many years, but recently have there been signs that practicable standards with international acceptance are emerging (see the glossary entries for the standard GKS and the *de facto* standard Core). For three-dimensional graphics, in particular with high-level facilities, standardization efforts are being made by the bodies taking care of such questions: SIGGRAPH, ANSI and in particular ISO. The standards develop through a series of drafts and proposals, until they are finally accepted. Presently, GKS-2D is an accepted international standard (ISO 7942), Core (2D) is widely used although not an official standard, GKS-3D is a draft international standard (ISO/DIS 8805), and PHIGS (3D) is a draft proposal (ISO/DP 9592) (status 1988).

Hidden edge/hidden surface removal: When three-dimensional objects are brought onto the screen, this may be done in the form of a wire-frame picture. Even for objects with modest complexity such as a small part of a detector in a high-energy physics experiment or the example shown in Fig. 4.4, the resulting view is not easily interpreted, because the notion of surface is totally absent from the picture. The human eye is thus confused and has difficulty in establishing the context-conditioned interpretation of lines as boundaries (edges) of surfaces. The removal of lines and surfaces which are physically hidden by surfaces closer to the viewer, requires a definition of surfaces and non-trivial and time consuming algorithms (Foley and Van Dam 1982; Booth 1979, which contains a list of 111 publications on this subject). If the viewing order of surfaces is known and a bit map display used, area fill commands given in the

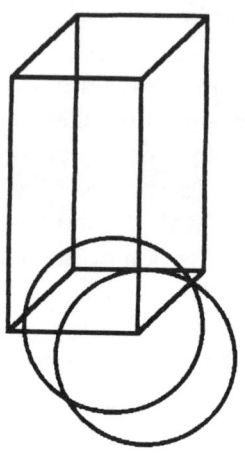

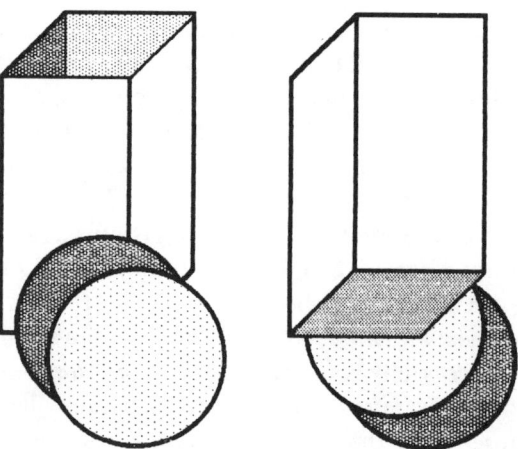

Fig. 4.4 A wire-frame drawing and a demonstration of its ambiguity.

4.5 Technical aids

correct order (from far to close) will result in partial and possibly repeated overwriting of pixels. This will result in an often acceptable poor-man's surface representation.

Hierarchy: Data to be displayed on a screen frequently have some structural hierarchy which can be usefully preserved in the display list. This allows the assigning of attributes not only to individual objects, but to groups of associated objects. The existence of hierarchy makes colour changes or coordinate transformation operations for picture parts much simpler to program. Hierarchy is also needed to define 'pick objects' i.e. image parts that can be selected by a 'locator device'. Current standard graphics packages do not allow for true hierarchy in data. GKS supports only a single level (linear structure), whose elements are called segments. The possible future high-level standard PHIGS features extensive hierarchical possibilities and the inheritance of attributes. Note that data hierarchy for graphics may, but often does not, coincide with the data hierarchies necessary for data processing, as implemented in data structures.

Highlighting: To make projections of three-dimensional surfaces appear closer to nature, simple shading is not considered to be sufficient. Impressive visual effects are obtained by assuming a bright light source at some fixed point with respect to the synthetic camera, and brightening surfaces where this light is reflected. Highlighting usually consists of light reflection, which may be diffuse or specular depending on the impression of texture that it is desired to give to the surface. Highlighting needs ray tracing for very small surface elements, and due to the large amount of computing necessary is restricted to very high performance systems.

Icon: A little pictorial object appearing on the screen and having a suggestive value as an object that can be picked by cursor in order to induce a program response. Icons are often used for menu implementation.

Inheritance (of attributes): The name given to the propagation of style or viewing attributes through a hierarchy of graphical objects in a display list.

Jaggies: See staircase effect

Joystick: An input device associated with graphics stations, of the locator type. It consists of a vertical stick like a car's gear shift, which controls the movements of a cursor in two directions by being moved out of the vertical in any direction; largely replaced by the track-ball or mouse today.

Light pen: A pencil-like input device associated with graphics devices. It allows points to be located by moving the pen directly to a point on the display screen.

Logical device: A notion used in programming input (sometimes also output) devices on graphics stations. It allows some independence from the implementation of an input device. In GKS, six classes of logical input device are defined: locator ($x-y$ position), pick (identifying a displayed object), choice (select from alternatives), valuator (numerical value), string (character string), and stroke (sequence of $x-y$ positions). Examples: any of mouse, joystick, or light pen can be programmed as a general locator or stroke device, and hence also used for pick or choice; any of a screen-implemented dial, a physical knob, or a numerical keyboard input can be programmed as a valuator device.

Locator: Any device that allows the indication of the coordinates of a point with respect to a given viewport.

Menu: A choice of items proposed on the screen, usually in written form, from which the user picks one in order to obtain a program response. An inexperienced user can be guided by structuring the menu into a hierarchy of questions. Many user picks in a menu will thus result in a next-level menu as computer response. Menus may be implemented in different ways depending on the device. Modern workstations work with a mixture of 'pull-down' and 'pop-up' menus using graphics. Text-oriented systems use full-screen implementations with few input fields, some implementations make heavy use of graphical symbols (icons). Whilst it is a pleasure to work with a well-designed menu as a novice *user*, it is by no means trivial for the *writer* to structure the guiding information for this type of human interface. The consistent choice of keywords, the ability to express the essentials in few words, and the structuring into few levels are more difficult than most amateurs realize, and an ill-conceived menu is more frustrating to use than a command language with an associated manual.

Metafile: The collection of all calls to a graphics package, with calling parameters added, constitutes a full record of what has been put onto the screen in an interactive (or other) session. Recording these in some format will therefore also allow the same or a similar picture to be brought onto different output devices (similar because the same facilities may not exist on different devices, e.g. black/white devices cannot give colour). A file corresponding to such a record for one or several pictures, is called a

4.5 Technical aids

metafile. Deficiencies in the definition and standardization of metafiles have led to the definition of a simple fully portable (and editable) picture description for use in high-energy physics (Bock, *et al.* 1987a).

Mouse: An input device of the locator type, popular in connection with graphical menu presentation. A mouse implements the cursor movement by moving a small and lightweight device on a plane horizontal surface like a desk, and achieves high speed *and* precise locating in a natural and quick way. Frequent switching between mouse and keyboard, however, is not a very efficient input form. It should be avoided unless such mixed input is necessary.

PHIGS (Programmer's Hierarchical Interactive Graphics Standard): A proposed standard, now under discussion and not yet widely implemented. PHIGS attempts to provide a standard interface for high-level graphics applications, with special emphasis on the support of many geometrically related objects and on the dynamical use of the display (i.e. frequent changes of transformation matrices). PHIGS does *not* contain the notions necessary for solid surfaces, be they plane or curved, and hence there is no hidden edge removal, shading, or highlighting. For details see Shuey, *et al.* (1986), or Myers (1988).

Picking: If objects are shown on a screen, a locator device may be used to pick any one of them. Modern graphics packages contain the algorithms necessary to pick objects, leaving to the application program the task of defining which objects are enabled for picking. The picked object in the simplest case is a primitive, GKS will allow a segment to be picked. Pictures based on a multilevel hierarchy of graphics objects can, of course, pick objects at different levels. Note that picking objects in three dimensions is, computationally, far from trivial.

Pixel: Contracted from 'picture cell'; pixel designates the smallest addressable unit on a raster display. A pixel can be set at a black/white or colour value.

Polygons and Polylines: These are graphics primitives allowing a succession of straight lines, i.e. a polygon to be drawn. The polygon is defined by an ordered list of point coordinates (two- or three-dimensional), given as a single set of 2- (or 3-) vectors. A smoothly curved line is obtained by giving a sufficient number of points with slowly varying gradient. This causes the notorious problem with blow-ups: the small discontinuities in the first derivative become visible when zoomed in. The difference between (closed) polygons and (open) polylines is sometimes made to indicate the possibility of area filling for the former, where an additional line is expected to exist between the first and last points given, thus defining a closed boundary.

Polymarkers: A graphics primitive which allows a succession of identical symbols (markers) to be drawn in a picture, indicating for each of them a set of two or three coordinates.

Primitives: The most basic drawing information given by an application program to a graphics package. Graphics primitives include polylines or polygons (possibly with area fill), polymarkers, and text strings, and also the setting of line style, text font, etc.

Raster display: A display driven in raster mode generates a picture by deflecting the electron beam in a regular line-by-line pattern as in a television set. The picture is drawn by controlling the beam's intensity, in the simplest case by switching it between 'on' and 'off'. Colour is usually obtained by separately controlling three beams exciting three different phosphor layers. Thus a picture is made of pixels which have a grey or colour value. Raster displays are always refresh displays and thus call for a memory corresponding to all the pixels to be displayed (the bit map). Inversely, the simple deflection pattern and control make raster displays cheaper than vector displays. Since the refreshing operation in a raster display is independent of the information displayed, raster displays have no problem of flicker.

Rasterops (raster operations): A raster display works from a pixel memory, where each cell on the screen is represented as one bit (Black/White) or several bits (grey value, colour) of information. This bit map representation allows information to be moved around simply by copying parts of this memory. A full bit map may be decomposed into several rectangular windows, and these windows may be copied over other parts of the memory, giving the viewer the impression of moving windows on the screen. The special instructions to copy, displace, overwrite parts of the pixel memory are called rasterops.

Ray tracing: In the context of graphics, ray tracing is used to make solid surfaces more natural. The operation is connected with high-quality graphics. It requires substantial computer capacity. The operation, similar in algorithmic implementation to hidden-surface removal, consists of finding the light ray(s) from the hypothetical light source(s) to the synthetic camera, thus allowing shading and highlighting effects.

Refresh display: A refresh display operates from a picture memory, and regenerates the picture at a rate of typically 30 times per second, to

4.5 Technical aids

counterbalance the decay time of the display screen's phosphor (typically milliseconds). Refresh displays may be of the vector type, in which case They work from a display list, or of the raster type, which work from a pixel memory (bit map). Their counterparts are storage displays which need no refreshing, but which therefore also do not allow dynamic movement on the screen as needed for rotation or translation movements or for animation.

Segment: A modest structuring possibility given by graphics packages (e.g. by GKS) to allow the grouping of graphics primitives into objects on a (single) higher level, which can subsequently be separately addressed as entities. Segments may constitute a primitive hierarchy, but are eminently useful in global attribute changing, selective erase, and in picking operations.

Solid modelling: The set of procedures which uses a high-level display with a graphics interface for showing solid objects with a highly natural aspect. All advanced techniques such as shading, highlighting, hidden-edge and surface removal, removal of facets, colour, and often dynamic motion are part of the solid modelling environment.

Staircase effect: The discretization into pixels which a raster display implies, has the effect of decomposing a line boundary which is neither horizontal nor vertical, into small rectangular steps, sometimes called jaggies. With fine resolution, this is not normally visible, but lines close to horizontal or vertical direction will show the effect even then. Mathematical methods are sometimes used to make the effect less visible on high-quality displays, e.g. by introducing different grey values for pixels near the edge.

Storage tube (DVST or Direct View Storage Tube): A vector display storing the picture pattern to be displayed as charges on a nonconducting plane in front of the phosphorous screen, and using this charge distribution as a picture store rather than a digital memory (as in refresh displays). The technique is based on using electron beams with different intensities; a high-intensity beam stores the charge pattern corresponding to the picture (in fact, it erases prestored negative charges); a lower-energy beam floods the entire screen permanently and is repelled by the stored charges, thus creating and refreshing the image on the phosphor; a medium-energy beam may be used in addition to impinge on the phosphor without altering the charge pattern (write-through). Storage tubes allow no dynamic movement of objects on the screen, except in the write-through mode (usually very limited). They also have no selective erase possibility, the entire screen has to be flushed and rewritten. Storage tubes were, however, for many years the workhorses of good resolution vector graphics.

Structure of graphics information: See hierarchy or segments.

Synthetic camera: The concept of a viewing point in space from which a three-dimensional object is to be seen through a window. If this viewing point is at infinite distance, a parallel projection results. At a finite distance, a perspective view will be obtained. Usually, a camera definition is accompanied by a window definition, which translates into a three-dimensional clipping volume.

Tablet: An input device of the locator type which allows the reading of coordinates from a flat surface on which a pencil-like stylus or hand-held platter usually equipped with one or more control buttons is moved. The tablet operation is not unlike that of a light pen, but coordinate reading is from a specially prepared horizontal surface. Like light pens, tablets do not show the stylus movement as movement of the screen cursor (as does a mouse). Tablets are ideal input media for digitizing simple drawings, and have also been used for three-dimensional input using two pointing devices.

Touch screen (touch panel): A crude locator device implemented as a touch-sensitive screen on which pictures and/or text can be displayed under program control. Usually used as a picking device, e.g. to make choices in a menu using a finger.

Transformation matrix: The relations between the different coordinate systems necessary in graphics are given by rotations, translations, and possibly changes of aspect ratios. Modern graphics devices perform some of these transformations in special hardware, hence very efficiently, thus allowing high-speed dynamical movements on the screen. It should be noted that translations coupled with rotations are sometimes implemented as single matrix multiplication. To do this, a fourth column (when discussing three-dimensional vectors) is added to the rotation matrix which operates on an additional vector element containing the value 1.0.

Valuator: An input to a graphics package from a dial, knob, or any similar implementation (often using a screen display and a locator device), which results in a numerical value, typically normalized to be between 0 and 1.

Vector display: A display which executes display instructions from a list of vector commands much like a plotter: every vector is given by its start and end-point coordinates, between which a line is drawn. Text is typically decomposed into small vector strokes. Vector displays have hardware that controls the electron beam in exactly this way. The beam is guided to the start point position, turned on, and then continuously moved to the end-point position by linearly changing the deflection currents.

Viewing parameters: The set of parameters which together allow the derivation of all transformations from the world coordinate system to the screen, including clipping. This comprises only few indications such as viewport and window in world coordinates for two dimensions, but may become very complicated when used for perspective viewing in three dimensions (synthetic camera, pyramid-shaped window, light sources).

Viewport: A rectangular portion of a screen, into which an image is to be directed. Changing the size of the viewport will scale the image.

Window: In graphics language, a window is defined as a subspace in the world coordinate system, which is to be mapped onto the screen or, more generally, onto a viewport. Note that some confusion reigns in terminology, because 'window' is also used to designate a region on the screen through which information (not necessarily of graphical nature) is viewed. This is standard terminology for modern workstations, and such screen windows then differ from viewports in that a clipping operation is performed, i.e. reducing the screen window size reduces the amount of information visible, but implies no scaling. The term window management refers to the control of such screen windows. Window management is notably absent from all existing or proposed graphics standards.

World coordinates: The application writer's most cherished coordinate system, whose boundaries he defines (as floating-point numbers) to the graphics package and in which he expresses his coordinates. A good graphics package allows a user to work in his personal world coordinates in all aspects of his work, except when referring to the screen or viewport directly, e.g. for placing text or for defining the viewport.

Wire-frame graphics: Three-dimensional objects defined by the edges of their surfaces are most frequently displayed on the screen by showing only these edges (all edges are shown). This results in a 'wire-frame' representation, which quickly gets confusing when many edges are involved. Only dynamic motion such as rotation, or very selective presentation will make such pictures useful. More ambitious presentations using surface notions (hidden-edge removal, surface shading, or at least, area fill) are possible technically, but not available in most graphics packages (see Fig. 4.4).

Write-through: See storage display.

4.5.2 Data access methods and databases

We have come across the problem of organizing i.e. structuring data in an earlier section (Section 4.1). In fact, the form of the organization of the large volumes of data necessary for analysis is a critical decision, of vital influence on the conception, writing, and later understanding of algorithms. Data organization has boundaries given by the availability of data access systems: As long as data are only needed internally in a program, it is the program language or the auxiliary package that is used, which impose their conventions. If data are to be communicated across different programs, maybe spread over different computer systems, the use of data access and transport packages becomes inevitable. If, in addition, frequent interactive access is needed to these data, e.g. for human updating operations, the use of a full commercial 'Data Base System' (DBS, sometimes DBMS for 'management system') may be indicated.

High-energy physicists have understood, over the last few years, that there is much to learn in the domain of data storage and access from experience in commercial data processing environments, and from studies of their theoretical foundation. Early reports like Jeffery (1982) or Hart, *et al.* (1983) underlined this connection. Recent large experiments have seriously gone through the process of evaluating various products, including some home-grown data access packages. These usually compare well in efficiency with commercially offered packages. They are tailor-made for high-energy physics problems, and in comparison, commercially available products show an apparent inefficiency and some lack of functionality for our applications (Mount 1987). The relevant aspect that has to be judged is the question of whether future demands on data access methods in high-energy physics will move ever closer to the commercial solutions which are being offered on the market (usually for organizing the personnel, stores, production data of a company). These systems, too, will evolve, of course. Experience shows that tailor-made systems have definite merits: no unused functionality, full understanding, maintenance by the originator. They also have their shortcomings: If the functionality can no longer be adapted to evolving needs, and a redesign is needed because the

4.5 Technical aids

problem was misestimated at the original design level, the apparent savings turn into a net loss as costs and efforts multiply by large factors. A reasonable compromise is often sought, using standard products in conjunction with home-grown extensions.

Let us now try to give at least some substance to the terms we have been using: What is a data access method, a database, what do commercial systems do, to which categories do they belong?

Data organization, first, is vital because our thinking concepts and our memory work with structures. We can only remember details *in the context* of some higher-level 'chunk', i.e. a larger unit of information. Depending on the context, more or less detail may be relevant. A wire may be in need of a detailed description in terms of position, or depend upon calibration constants for associated electronics. But this description is meaningless until defined as part of a chamber, which, in turn, is likely to be part of a higher unit like a barrel detector, and all of these have a meaning only in the context of a given experiment. Thus here we have defined a *hierarchical* 'universe', which is populated by detectors, and can be broken down into chambers, calorimeters, magnets, drift volumes, wires, scintillators, etc. The same experiment's universe contains, of course, components for construction or development, from cables to subroutines, physicists, publications and conference talks, and many other items that do not enter data analysis. Often, such data are also carried along in data structures, and take their own advantage of computerized data management.

We have thus defined a *data model*, in this example a hierarchical model, which is representable as a 'tree structure', as in Fig. 4.5. It consists of data entities and the relations between them. A data model more general than that of hierarchy is the *relational data model*, which has been pushed to some theoretical perfection (e.g. Schmidt and Brodie 1983). The simple hierarchy is, of course, a possible subset of a relational data model. In a relational model, all elements of the same type are grouped into a linear list usually called a 'table'. The relations between tables are expressed using 'keys', names which allow data elements to be linked to columns and tables in practically unrestricted patterns. Such links are not carried explicitly in the data structure, but are established by a search algorithm when the relation is needed. Figure 4.6 shows an example of tables and relations through keys in a relational data base. A further example is given in Subsection 5.4.3.

At the moment of interrogation, the structures in a relational data model are, of course, not more complicated than in a hierarchical model. The most

376 *Statistical data analysis*

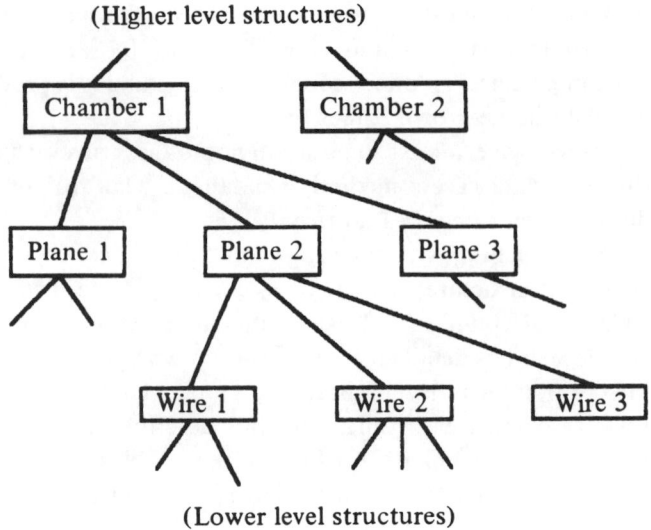

Fig. 4.5 Detector data taken as an example for a hierarchical data model.

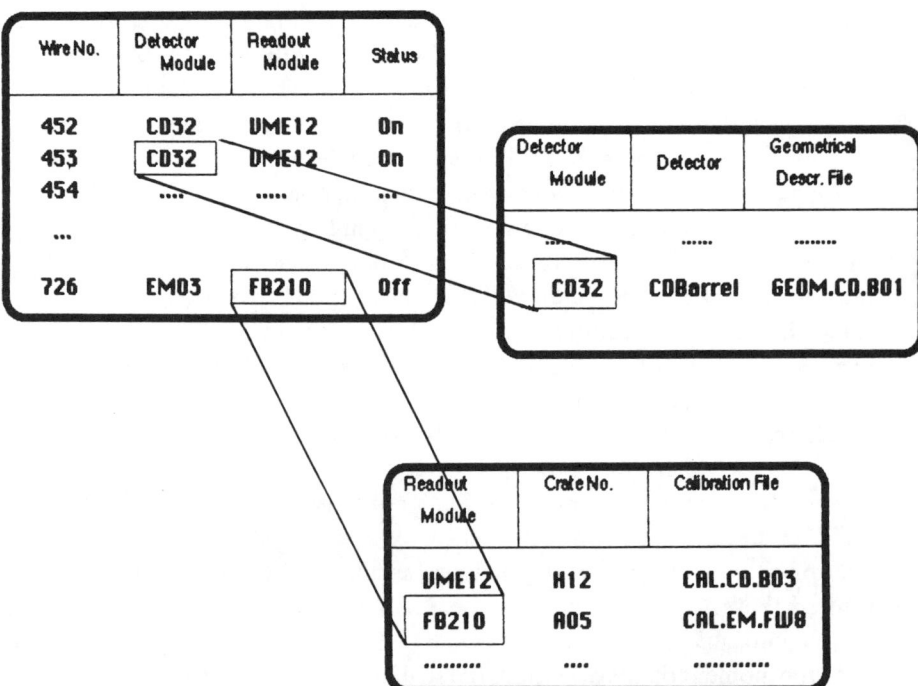

Fig. 4.6 Detector data taken as an example for a relational data model.

4.5 Technical aids

marked difference between the two models is given mostly by the flexibility inherent in the relational model, which allows the relations to be established *as needed or desired*. An added advantage in implementing the relational model consists in its automatic 'instanciation': elements that are referred to frequently and in different parts of a structure, are grouped and referred to by means of a key, and appear only once. The duplication, necessary for this in a simple hierarchy, is thus avoided in a relational model. Instanciation not only provides saving in data storage space, but primarily guarantees simplicity in the data updating mechanisms: only one 'instance' of every data item exists. A potential drawback of a relational model is the complexity of search mechanisms, which may result in response time problems.

These different ways of organizing data are, of course, important mainly for data with *long life-times* and which are accessed by many programs or people, i.e. for *multiple access* data. For these, one will then have carefully to choose the data *access method*, i.e. the implementation of the data model. In the simplest case, the access method may make use of the *file handling system* of the computer(s) on which the analysis is carried out. Partitioned data sets under IBM/MVS or the file naming hierarchy allowed by UNIX-based systems invite the use of this simple solution, and have been employed for this purpose in experiments. Its drawback lies in the lack of data portability; even when data can be ported, the file names cannot. It is also necessary in this implementation to specify which relations exist for each connected group of data. There are no keys nor explicit links, there is no automatic checking that the data satisfy any model which one may want to impose. Any directories will have to be hand-defined. The naming of data elements contained in such a collection (to avoid calling them a database) is, of course, not foreseen, everything is by convention and position in the file.

A somewhat more general way of accessing data is provided by the packages developed in the high-energy physics community. Originally these were introduced for managing and structuring data in memory only ('memory management'). Popular packages are BOS and derivatives, (Blobel 1979; Quarrie 1983) and ZEBRA (Brun and Zoll 1987). They are FORTRAN-based packages, and provide facilities for grouping data into blocks ('banks'), for expressing relations between banks, and for storing the data and their relations in random access files. For BOS, minimal structural support is provided, whereas ZEBRA data are hierarchically structured. At a high level (entire data structure), naming of banks is possible (UNIX-like in ZEBRA, linear in BOS). At a lower level, ZEBRA carries explicit links which provide the ultimate in efficiency, albeit at the expense of documentary value, ease of programming, and data checking possibilities. More recently the Aleph experiment at CERN has written a package (ADAMO) which implements, for a FORTRAN environment, a relational data model (Qian, et al. 1987) based on the 'entity–relationship model' of the excellent theoretical analysis of Chen (1976).

Finally, the most general level of a data access method is reached when moving to a rigorous data description which then deserves the name *database*. A database is a collection of data associated to a DBS. And a proper definition for a DBS might be *a convention supported by software tools concerning how to implement data structures, how to document them, and how to access the data from either a program or, interactively, from a terminal.*

DBSs are distinguished from the simple access methods by the *naming* of all, or at least the structurally relevant, data elements, by protection mechanisms against accidental or voluntary data destruction, and by a *query language* used for retrieving information. Flores (1970), Palmer (1975) and many others provide a good general introduction to the database problematic. A recent general overview with many references is found in Bestongeff (1986).

Most general and flexible in their structure are *relational DBSs*, which provide data in tables without *a priori* assumptions about their relations. Tables are made of columns, and both tables and columns are given mnemonic names. Relations between columns are then, in a data access session, established by these names ('keys'). The search mechanisms can quickly turn into a response time problem; they can be made more efficient by freezing a set of data relations.

High-energy physicists have found various applications for such general DBSs. State-of-the-art reports (and introductions in physicist language) are provided by Mount (1987) and Putzer (1987). As it stands, the DBSs available commercially today seem to be in need of substantial additional software to become efficient tools for high-energy physics experiments. Some of their advantages thus seem to be offset; however, both the products themselves and the experience in using them are evolving rapidly.

4.5.3 Ready-made tools for statistical analysis

Final data analysis from experimental data, abstracted, selected, and stored on tape or disk is the heart of any physics experiment. Physics intuition, a thorough understanding of the data inasmuch as the detector

4.5 Technical aids

and data processing have influenced them, and a sense for abstracted presentation to colleagues or a wider public have to be combined. The process is inherently an iterative one. Ideas are tried out, their results inspected, feedback is generated by discussing with widening audiences, and finally results are made to converge towards a more-or-less formal publication.

For many years, lack of readily available hardware has caused this activity to be inadequately (if at all) supported in a systematic way in high-energy physics laboratories. Even today, the majority of physicists does not have access to tools for nonexperts that would assist them in final-state physics analysis. In some laboratories, they have to resort to truly ancient methods of analyzing data, by combining graphs on paper with numbers copied or derived from numerical computer output.

Modern low-cost equipment like workstations or personal computers and the improved availability of interaction and graphics is about to effect a major change in this activity. Working *interactively* with data in graphical form has become a widespread computer application in industry. If in the field of high-energy physics the acceptance is slow, it is as usual a problem of software. Commercial analysis packages adequate for the research environment are not easy to locate, and physicist-developed general software takes some time to mature. In order for such an analysis package to be accepted over the established methods, the importance of its ease of learning for the casual user (which most physicists are and want to be), cannot be overestimated.

4.5.3.1 Home-made statistical packages

Several tools of this type have been developed by groups at CERN (Brun, *et al.* 1981), DESY (Bassler 1985) and SLAC (Burnett 1983). The experience gained in their development and use has resulted in a project to combine ideas from existing packages into a relatively complete and coherent new package. This is now known under the name Physics Analysis Workstation (PAW), and documented in Bock, *et al.* (1987b). PAW systematically assembles the functionality of its predecessors in a single system, attempting to solve the problem as far as possible by combining existing and largely portable modules. PAW software puts the modern workstation hardware to use for the high-energy physics community, but maintains the capability of working in a very similar way on nonintelligent terminals connected to a host's time-sharing system.

In the following we describe briefly the functionality and the main components of such a system. We use PAW as the most recent example of a home-grown package, it being understood that there is very substantial functional overlap with the older systems mentioned above.

Functionality

Access to data generated in batch jobs or in interactive sessions, possibly spread over several files, with a hierarchical naming convention for easy mnemonics and structure.

Manipulation and presentation possibilities for one- or two-dimensional histograms.

Projection and selection possibilities for events abstracted to few numbers (n-tuples).

Access to simple graphics operations from typed-in or batch-generated data.

Multiuser access to data files, and a documentation capability for data on files.

Interaction with the package allowing various degrees of expertise, from fully commented menu-driven operation for the novice to shorthand command typing for experts.

Largely self-documenting, with a full computer-maintained user guide available in addition.

Availability of command macros.

Capability to store, amend, and combine graphics images generated in a session, by editing; this implies the ability to store/retrieve graphical information in portable form.

Possibilities of extension beyond the foreseen commands, by easy access to a programming language level (FORTRAN-like).

General component overview

The command processor is an interface program for any application in need of an interactive input of commands. From the application, it is seen as a slave which supplies *the next command*, with associated parameters if desired. It takes care of user input and ensures a standard checking on command syntax and parameters. If constructed as a separate process, the command processor may run in a detached device like a PC.

The command processor has two main modes of interacting with the user, by *menu/mouse* (i.e. proposing commands to the user), or by *direct input* (i.e. the user knows the command names and the associated

4.5 Technical aids

parameters). The user can switch between these two modes at any moment, depending on his expertise.

The macro processor is an extension of the command processor. It allows the establishing of 'character objects', text containing commands to the command processor. Macros can be stored, edited, and fed into the command processor. They also may contain constructs not available in typed commands (like loops) or parameters.

The data manager provides capability for managing data objects, optimally structured hierarchically (in analogy to modern file management schemes) and in portable form. For PAW, a rather general data access system is used which is itself portable: this is ZEBRA (Brun and Zoll 1987). Particular use is made of ZEBRA's direct access input-output package (RZ), which provides a convenient programmer interface for devices allowing direct access (i.e. disks), and includes facilities for transporting files to different computers in an exchange format. ZEBRA RZ permits the build-up of an efficient data structure with hierarchical naming conventions. For the PAW application, it has been amended by a systematic object addressing scheme using structured names and allowing 'wildcarding', abbreviations expressed by special characters for the addressing of groups of objects.

The Language Interpreter allows the interpretation programme modules of the FORTRAN subroutine type, linking them into the program which is being already executed, if needed. This allows the utmost flexibility in programming selection algorithms for data (e.g. n-tuples), or the nonstandard input and/or transformation of data.

Histogramming and histogram manipulation is the core part of a data presentation package. In the case of PAW it is based on a subroutine package with many years of history, which is also familiar to many batch users. The package contains the components of HBOOK (histogram creation and manipulation) and HPLOT (histogram presentation). They are documented in the CERN program library (Brun, *et al.* 1981).

Simple graphics with vector variables is a necessary ingredient in any interactive data presentation package; a user cannot be restricted to manipulating experimental data in connection with HBOOK. The existing packages therefore all contain facilities to present arrays generated in ordinary FORTRAN programs graphically, or to generate such arrays directly in an interactive session from functions or on the keyboard.

The portable graphics interface serves to ease porting of the data presentation package to different computer systems, and provides a computer-independent storage possibility for graphical data. PAW relies on HIGZ, an intercept of the graphics package GKS, which adds no functionality other than the possibility of storing and porting graphical data. HIGZ has also been interfaced to other graphics packages (e.g. the Core package DI3000). HIGZ uses as data storage format standard ZEBRA banks and has, in addition, compression possibilities. In practice, this makes HIGZ output data greatly superior, in volume and portability, to the 'standard' metafiles, which are only comprehensible to a specific graphics package.

4.5.3.2 Commercial tools

A large choice of recent products, which run on personal computers or workstations, and which cater for the presentation of statistical data, is available. It might seem a very reasonable alternative for a physics experiment or department (wherever such decisions are taken), to base physics analysis on a product of this type, instead of relying on its own home-grown software or importing some other laboratory's product. To our knowledge, no major physics analysis project has resorted to this possibility. The packages developed inside the physics community, instead, enjoy continued popularity. True familiarity with existing commercial packages does not, therefore, exist. We have tried to understand the reasons for this somewhat surprising fact.

From one of the rare and quickly out-of-date overviews (Caporal and Hahn 1984) we have extracted several examples. Of these, we have looked more closely at SAS (Cody and Smith 1985) used widely in industry and also employed by administrative and computer centre groups in high-energy physics for extracting their daily statistics. SAS, too, has found its way into only very few physics applications. This can be explained as follows: On the one hand, the package is supported by a well-known institute and installed literally in thousands of sites. It has much to offer in terms of graphics and statistical analysis. SAS is, however, suffering somewhat from its age (it originated as a tool in batch environments), and is centred around its own batch-inspired control language. Although extensively documented, a build-up of special expertise is needed before its potential can be put to use ('high learning threshold'). For physicist applications, SAS further provides neither data portability between

4.5 Technical aids

computers, nor an obvious way of interfacing with the popular batch packages for physics analysis. SAS thus is sure to need software amendments and expert support in any laboratory where it is installed. This (coupled with the rental cost of SAS) seems to act as a major deterrent for potential customers in the physics community.

Some other general products (Dataman, DataViews, SPSS/PC) we also evaluated briefly. They seem visibly rather 'closed' products, quasi-impossible to interface to our environment and unlikely to be a good basis for upgrading to the functionality considered necessary for high-energy physics applications. The same problems, low functionality and difficult interfacing, seem to be characteristic of the often impressive analysis tools available on specific modern small workstations (including MacIntosh, Atari, IBM/PC or Amiga). Ease of extension by software, and documented, stable interfaces of sufficient generality (hence an 'open' system), seem features which are valued very highly in the research community, but which are not provided by most commercial products in this area.

5

Program development and software management

5.1 Characteristics of programs in large experiments

The big experiments in high-energy physics are designed and operated by a large number of physicists. The number of collaborators for experiments at the colliders is of the order of 50–400, the number of institutes ranges from 5 to 50. Approximately half of the physicists are involved in data analysis and are affected by data processing. It is the *large* number of people and institutions which creates a new type of problem. When a programmer wants to solve a small problem on his computer he or she writes down an algorithm in a mathematical form and transforms this algorithm into a programming language. Then he or she types the program into a computer, compiles and loads it and executes the program. It is characteristic for this kind of work that most of the effort goes into the formulation of an *algorithm*. In large collaboration much work is required for internal organization. A high fraction of work is of *nonalgorithmic* nature.

The number of computers in the entire collaboration is typically of the order of the number of institutes participating in the experiment. In the *real-time* environment of the experiment itself the number of mini- or microcomputers is of the order of 10 to about 100 or even 1000 not including computer farms with 10–100 computers acting on a single task. The data must be transported from one computer to another-raising all the problems of networking.

The maintained code volume of a collaboration is of the order of 100 000–500 000 lines. About two thirds of the code consists of executable code, one tenth is declarations and the rest are comments. Only a very small fraction of code is not written in FORTRAN. A large part of the analysis program must be maintained and updated in all laboratories. This immediately leads to a related question: How is the code to be managed? Must a collaboration be supported by a *code-management system*?

5.1 Programs in large experiments

The programs must run on computers from different manufacturers. The internal addressing schemes can produce different results on different computers. PDP11 or VAX computers address bytes and halfwords from *right to left*. IBM computers address from *left to right*. This can have direct consequences on the results if variables of different lengths are forced to use the same memory locations by an 'equivalence' statement.

In addition to the different addressing schemes computers have different *floating-point presentation* (IEEE-format-DEC-format, IBM-format) and different presentations for *alphanumerical data* (ASCII, EBCDIC). Manufacturers offer different compilers with specific extensions and different calls to system-dependent routines such as day and time. The handling of error conditions or input/output errors changes from one manufacturer to another.

This is just a brief selection of the problems encountered when generating program code for a large collaboration; the problems involved lead to the following consequences:

(1) The program code must be written in such a way that it is portable from one computer to another. One should not use computer manufacturer's specific extensions.
(2) All system-specific subroutine calls must be collected at one place to allow a simple adaption to another computer.
(3) Input and output should preferably be done in a computer-independent way. Data should be declared on the output medium so that they themselves explain what is integer, floating point, or character. An automatic procedure can then do the conversion for different data formats.
(4) Data in the user's application program should be screened off from the internal data structure as data are actually stored. This allows some flexibility if the structure changes in future.
(5) In order to exchange program code between institutes in a unique and simple way the collaboration should use a code-management system.
(6) To reduce the large number of global names for COMMON blocks and to store data in a 'structured' way a *memory-management system* can be very useful (see Subsection 4.5.2) (Blobel 1979; Brun, *et al.* 1985). A memory-management system can be avoided if the programming language used allows the description of data structures (e.g. PASCAL).
(7) A collaboration certainly needs a set of *coding rules*. This helps to avoid clashes of global names, makes the code readable, associates variable names with subdetectors and unifies the units of physical quantities.

(8) In order to exchange information between all the computers the collaboration must use a computer network.

The typical lifetime of an experiment is about 8–15 years. During this time the code and the experiment will not be stable and unaltered. There will be changes in the experimental set-up, in the detector development, in the understanding of the detector, in the evolution of computer hardware and software. The code must be adapted to future needs and maintained during the duration of the experiment. The turnover of programmers is high. Many people will participate in the experiment for only part of the time: Students stay for a few years and then change their job. The collaborations are so large and widespread that it is impossible to get a stable team of people together in one place.

5.2 Main steps in event off-line analysis

The main motivation of a physicist when confronted with data processing is to extract results as fast as possible using some kernel programs and some program packages for histogramming, statistics, and graphics presentation. The steps in an analysis chain are:

(1) Separate good events from background processes. Reconstruct tracks of particles or energy clusters from the raw detector data. To select events one applies some criteria such as e.g. the total visible energy above a certain threshold, tracks only from the main interaction point, the number of tracks above a given minimum.
(2) Correct data for different calibration constants.
(3) Improve the quality of the events by track and vertex fitting and by improving the fit parameters.
(4) Run programs to correct data for energy loss, radiation correction, multiple scattering, inefficiencies of detectors, or triggers etc.
(5) Estimate measurement errors.
(6) Compute the incoming particle flux or luminosity to compute cross sections.
(7) Run Monte Carlo programs to compute the acceptance of the apparatus.
(8) Extract physics results by applying cuts on momentum, transverse momentum, kind of particle, correlations.
(9) Compare these results with theoretical predictions by running Monte Carlo programs and treating the generated events in the same way as the data.

Detector specific programs are written by those people who are responsible for the different components of the detector such as inner wire chambers, the electromagnetic or hadronic calorimeter, muon detectors, detectors in the forward or the backward directions. These different activities must be coordinated to link the results together.

The programs for radiation corrections and Monte Carlo programs for the simulation of particle showers or the generation of events under the assumptions of some physical processes are available to the physics community and an interface to the specific experiment is required. These programs are documented and contain only a few bugs because many people have used them. In extracting the physics results physicists use program packages for histogramming, for geometrical presentation, and for statistics. These programs are well described and supported. The actual code used to extract the results is written by the physicist and is normally not documented. When the experiment has finished or the physicist starts working on another experiment the code is useless and is discarded.

5.3 Characteristics of real-time programs

Real-time programs are characterized by their strong coupling between experimental hardware, the on-line computers, and the man–machine interface to the physicist on shift. The kernel of these programs is written by a small team of experts who are intimately familiar with the architecture of the on-line computers. Programmers in this area must be experts in several fields besides their own profession. They must know how to design real-time systems, how to program in assembly and high-level languages, how to use and optimize the operating system and to adapt device drivers.

They are supported by a small number of programmers who are responsible for different parts of the detector. When writing code for an experiment the programmer has to know each bit of the connected hardware interface, when to disable or enable an interrupt etc. Many errors may occur in the connected hardware such as power failures of a crate or some other electronic breakdowns. The on-line program must be able to detect these errors and report them to the operator. But even if something goes wrong other parts of the program's functionality must be in good shape so that the operator can interact with the system. In other words, the on-line programs must be *robust*. But some external faults may result in a crash of the operating system if a bus cycle could not be finished, if the external DMA hardware is broken in such a way that data are not written into the corresponding buffer but into the system's area or if an external device generates interrupts at such a high rate that the operator cannot enter a dialogue with the computer.

It is therefore necessary to give the operator, besides software tools, some visual indicators, as well, showing him that the general address pattern of an input/output device is wrong or that the number of interrupts is high. On-line programs are of the size of 100 000–200 000 statements using 1–5 on-line computers and 10–100 microprocessors. These processors are connected to each other and form an embedded system.

In these complex systems it is difficult to detect errors. Much error detection and recovery must therefore be integrated into the hardware. One problem with on-line sofware is testing. It is easy to test a program if all the components are switched on and are in good shape. But a programming rule says that each statement of a program should be executed at least once. This, in fact, requires that one needs to be able to generate errors in the hardware. It is simple to switch off a crate or to pull out a cable or an electronic module. But how can parity errors or spurious interrupts or wrong register settings be generated?

On-line software requires good documentation for the physicist on shift who does not know much about either the internal operation of a program or the on-line computer if it differs from the computer that the physicist is normally used to. Also the user interface decouples the person on shift from the normal system commands to investigate the state of the computer.

Experience of large systems has shown that good error detection, error handling, and error reporting require, in addition, about 50% of the 'normal' code and that the code for a comfortable man–machine interface adds another 40%.

5.4   Program development management

In the previous section we described the problems which occur when large and complex programs are handled by a large number of people belonging to a large number of institutes. In this section we will describe possibilities which may help to solve some of these problems. How can one decompose the work? How can one describe large systems? What tools are available to support a project? We will here describe some techniques only briefly and refer the reader to the literature for further information (Page-Jones 1980; Ward and Mellor 1985; DeMarco 1978; Kennedy 1983; Auerbach 1984).

5.4 Program development management

5.4.1 Program productivity and group size

In the literature (e.g. Auerbach (1984)) it is agreed that the yearly productivity for normal straightforward batch programs is of the order of 5000 statements. Excellent programmers can reach 10 000 statements per year if they work under good conditions and if nobody disturbs them. For complicated programs such as compilers the productivity goes down to 2000 statements per year and for operating systems or input/output drivers one is down at 600 statements per year or 3 statements per day. The latter number must be used for real-time programs when complicated equipment is read out by special input/output drivers taking all possible errors into account. The programmer has to read documents about the hardware (e.g. FASTBUS manuals, bus specifications) or, when these are not available for home made devices, he or she has to spend much time with the engineer who produced the electronics.

We have seen that the productivity for programming is fairly low for a single programmer. How does the productivity change for a team of programmers? Each communication path reduces the productivity by something like 250 statements per year. In a team of five programmers each programmer has four colleagues which reduces the productivity to 4000 statements per year. Taking these numbers seriously a group of 20 programmers would not produce any code at all. A way out of this conflict is to decompose a problem and to have small teams working on each problem separately with limited interfaces to each other.

It should be pointed out that the numbers given above are of a 'theoretical' nature but they give the orders of magnitude when one wants to plan a big software project.

5.4.2 Phases of software life-cycle

The software life-cycle is the time from initialization of a software system until the end of its use. In traditional development this cycle can be divided into several phases:

(1) *Requirement analysis*: The requirement analysis should give answers to the following questions: What should the system do? How much manpower do we have or is required? What are the costs? Can existing software be used? How can progress be controlled and monitored? Which milestones can be defined?
(2) *Specification*: The specification describes the functions of a system for

users or physicists on shift. In this phase much communication between programmers and users takes place. Which type of computer will be used? Which storage medium will be used (tapes or disks for multiple access)?

(3) *Design*: In the design phase the whole project is decomposed into modules. The interfaces and data flows between modules will be described. In addition error handling, alarms, and message handling will be designed. At the end of the design phase a document will be available which can be given to the programming teams and to other institutes. In practice such a document has never been written for a high-energy physics experiment.

(4) *Coding*: The coding is done in a top-down way. This enables an easy module integration if new modules are available and better control of the progress. For some parts of the project, such as input/output routines, a bottom-up approach might be more suitable to give an early indication of how the hardware will behave in reality.

(5) *Module testing*: The modules are tested individually with all sorts of data. The behaviour of programs, their speeds, and response times are checked against the specifications.

(6) *Integration testing*: The complete system is tested and checked against the specification. In practice, data taking starts during this phase.

(7) *Maintenance*: During this time errors are corrected, updates due to new experimental requirements are implemented in software and hardware.

As given in the literature (e.g. Auerbach (1984)) the results of the studies of several large projects show that 30% of time is spent in analysis, 20% in coding and 50% in testing. One has to look for ways to improve the low productivity of programmers and to increase the efficiency. The main reason for the low productivity is the fact that programs are 'handwritten' and not produced automatically. The term efficiency has changed its meaning during the last few years. Code used to be efficient when the program was small and fast. Nowadays code is efficient if it is easy to maintain and if the programming effort is small. According to Auerbach (1984) structured analysis and structured design should help to improve this sort of efficiency even if the programs are 10% larger and 10% slower. A program developed by Structured Analysis Structured Design (SASD) needs 40% for the analysis, 25% for the design, and only 35% for programming, implementation, and testing. From these numbers one can

5.4 Program development management

see that the actual coding starts fairly late which may cause some conflicts with traditional programmers who are eager to code. A good design also helps to avoid effects such as the one which is called the 100 programmers rule: 100 programmers will produce 100 modules even if only 10 modules are required.

5.4.3 Some principles for developing complex systems

Many pieces of software and hardware are combined in large experiments. These parts interact in a nonsimple way with each other and form a *complex system* which is more than the sum of the parts.

To build a complex system partitioning is needed. This has the side effect that the different parts must be integrated at a later stage. In addition the teams spend more time in communication. This communication time is lost for work and decreases efficiency. When things are developed at separated locations it is likely that a lot of effort is spent developing similar things in parallel.

Following a paper from Manzo (Manzo 1987) we summarize some management rules which can be applied to the development of large systems.

(1) Large projects must be planned.
(2) Find and fix as many errors as early as possible. It is easier and less expensive to correct them at an early stage.
(3) Bring the elements of a system together as soon as possible even if only in an abstract or simulated form.
(4) Make testing an integral part of the development cycle. Design the parts in a way that makes them easy to test.
(5) No complex adaptive system will succeed in adapting in a reasonable amount of time unless the adaption can proceed subsystem by subsystem.
(6) Divide a system in such a way that the parts can be implemented, fixed, and changed with minimal effect on the rest of the system.
(7) Separate functionality to decrease the degree of coupling between modules.
(8) Define the interfaces between modules.
(9) Tell people developing a subsystem *what* they should do and not *how* they should do it.

5.5 Structured analysis, structured design (SASD)

During the first phases of a software project many people of different disciplines and skills must work together. This requires a description or a model of the whole system which can be understood by everybody. Data processing is a new discipline which has not yet established a unique vocabulary. Even experts have difficulty speaking to each other e.g.:

On IBM computers one 'attaches a task'.
On DEC computers one 'creates a process'.
On Norsk Data computers one 'RTs an RT-program'.

In all these cases the same thing happens but it is hard for one expert to understand his neighbour and it is impossible to understand anything if you work in another discipline. Structured analysis was introduced to overcome these problems and to describe the system in a simple unique way.

Structured analysis should lead to a model of the final system which makes it possible to decompose the work into small tasks and to allow easy maintenance. If the specifications change it is then possible to adapt the descriptions and programs to the new requirements without too much work. To make the description simple and easily understandable pictures are used instead of long tracts of English text. The programmer does not want to waste his time reading all 15 folders of a system description some of which contain some of the information already described in others and which normally do not contain the information the programmer really needs. The graphical presentation of a programming system starts in structured analysis with data-flow diagrams.

A good overview of the use of structured analysis in high-energy physics is given in Kellner (1987).

5.5.1 Data-Flow Diagrams (DFD)

A DFD is a model of the logic of a system without regard to its physical implementation. A typical data-flow diagram is shown in Fig. 5.1. Only a few symbols are used in data-flow diagrams to make things simple. People should be able to understand immediately what is going on when they look at a DFD. The symbols are (Fig. 5.2):

(1) A *square* represents a data source or a data sink. Data sources are detectors, operator input, voltage signals etc; data sinks are line printer

5.5 Structured analysis, structured design

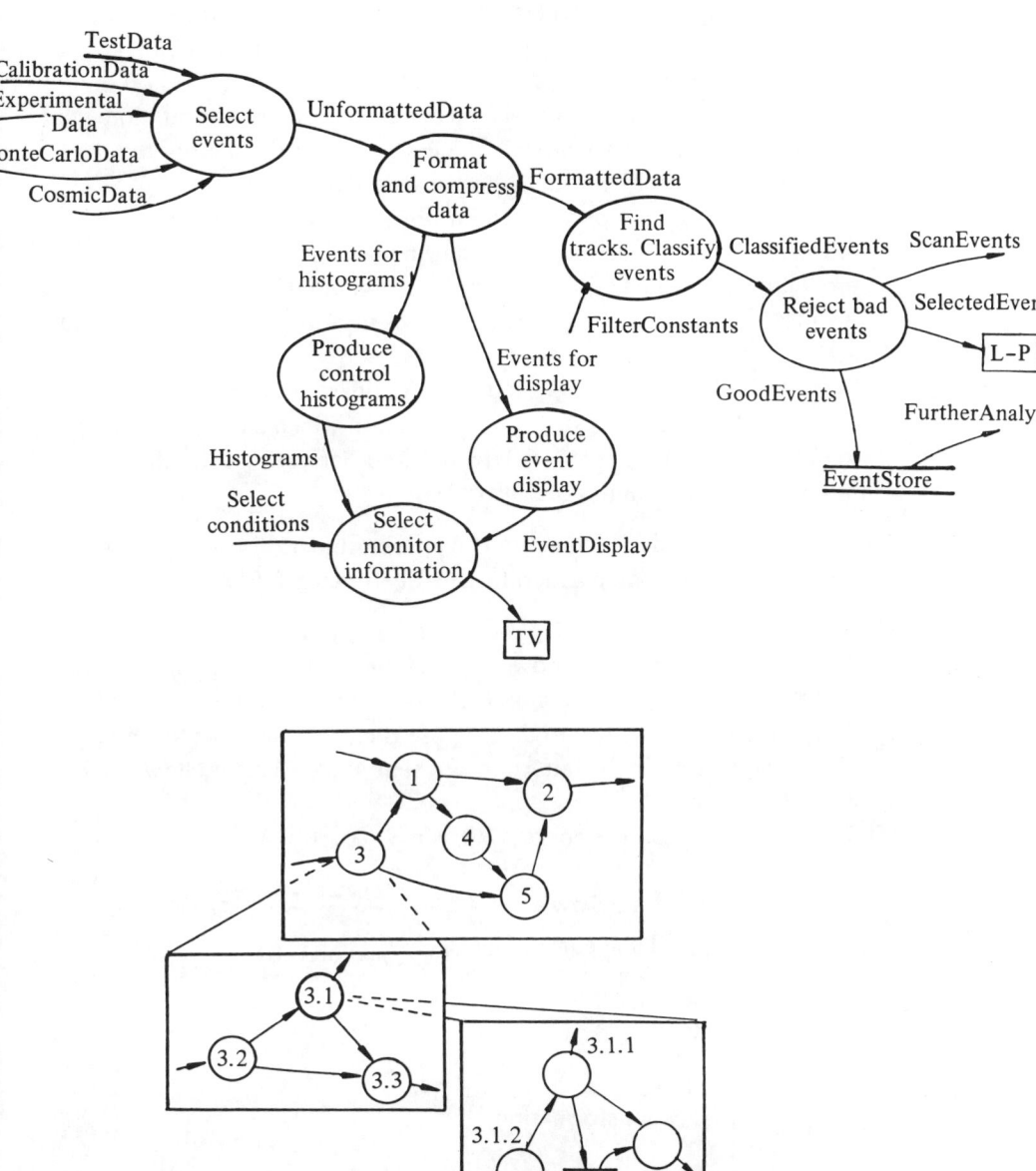

Fig. 5.1 DFD for a high-energy physics experiment. Data enter the system on the left hand side, are transformed by several processes and are sent for further analysis at the right hand side. Control information is not shown.

output, event displays on a TV, monitoring information for the operator or a publication of the results etc. A file is not a data sink. It is not the task of a system to produce files. A file is an interface.
(2) An *arrow* with a label is a data flow. Data flows are calibration data, trigger information, detector set-up conditions or run information etc. Each arrow must have a label which should clearly identify the data flow as CalibrationData or DetectorSetupConditions.
(3) A *circle* represents a process. Input data are transformed into output data. An abbreviation indicates what the process will do. All processes are numbered which allows the organization of DFDs in a hierarchical form. Processes do not generate control information such as READ data.
(4) A *horizontal line* represents a file. A file in this sense is not necessarily a part of a disk. A file is a storage medium into which data can be stored temporarily to be processed later by another process. In this sense buffers in main memory are also files.

DFDs are easy to understand not only by programmers but also by users or operators. These are some general rules for drawing DFDs:

(1) Do not draw more than 7 ± 2 processes per page (Miller 1956).
(2) Label all data flows. This shows whether you understand their purpose.
(3) Do not overload the diagrams with control flows and error paths.
(4) In large systems one has to decompose the processes. The lower-level diagrams must have the same number of input/output data flows as the parent processes.
(5) Files should be neither sources nor sinks.

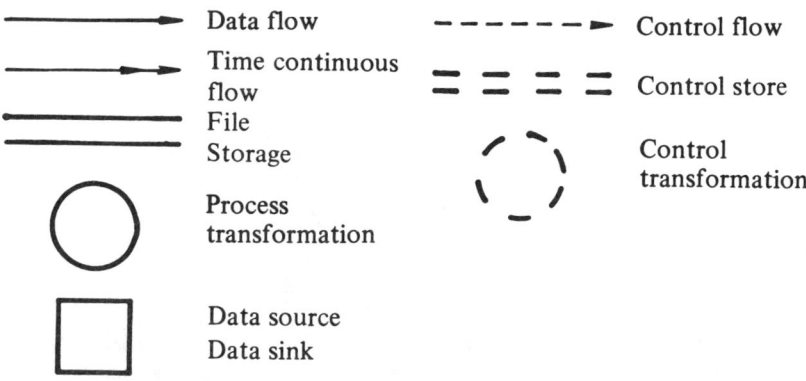

Fig. 5.2 Symbols used in DFDs.

5.5 Structured analysis, structured design

(6) If there are several data flows between the two processes combine them into a single data flow which will then be decomposed in the data dictionary.

A process with its incoming or outgoing data flows can be further decomposed at lower levels. This allows a programmer to 'zoom' into this process to 'see' more. Decomposition into several levels leads to 'layered' software. Errors occurring at a lower level are preferably handled and corrected there.

5.5.2 Structure charts

In the design phase of a project one has to transform the DFDs into a notation which represents a software design. From this design it should then be easy to start the actual coding. In structured design *structure charts* are used to model the software. Starting with a DFD one has to transform that diagram into structure charts. Firstly the processes must be replaced by modules. The modules are arranged in a hierarchical way. A main routine is responsible for all the functions of the system. The final structure chart shows the relationships between modules and data flows which will be transmitted by parameters. Before we describe in an example how to transform the DFD into a structure chart, we shall define the notation of these charts: An arrow between boxes indicates a normal transfer or subroutine call and a dashed arrow indicates an asynchronous transfer or task activation. Information exchange is indicated by small arrows. Arrows with an open circle transfer parameters from one module to another. Arrows with a filled circle transfer control information as end-of-file conditions or error messages. A number on a control arrow refers to an input/output table if there is too much information exchanged. If a module calls several subroutines a diamond symbol shows a decision made by a high-level module to call only one of the modules while a circle segment indicates that the upper module contains a loop to call the lower-level routines several times.

We will now transform the DFD in Fig. 5.1 into a structure chart. First one selects the input data flows or *afferent* data flows. The term afferent is taken from anatomy and describes a nerve sending information towards the brain. Second one collects all output data flows or *efferent* data flows. The processes between the afferent and efferent data flows perform the main *transformations* of the system. There are two possible ways to find a main

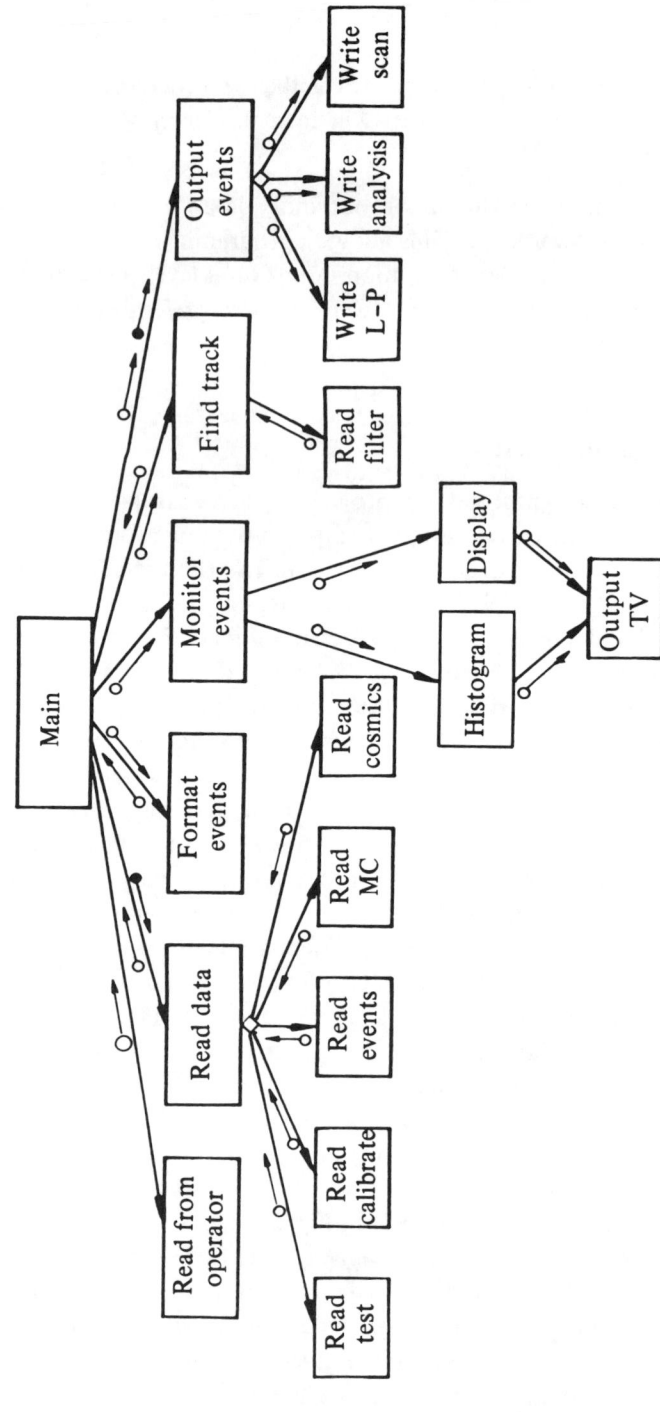

Fig. 5.3 Structure chart for a data-taking system. The input routines are on the left, the output routines are on the right hand side. Transformation routines are in the middle.

module: Either one of the main transfer processes becomes the main module or one introduces a new main module which controls the afferent, efferent, and transform modules. In structure charts afferent modules are drawn on the left and efferent modules on the right hand side of a paper. One possible transformation of the data flow from Fig. 5.1 is shown in Fig. 5.3. In addition to the DFDs structure charts also contain modules for READ or WRITE and control flows for error handling.

5.5.3 Entity Relationship Diagrams (ERD)

DFDs give a graphical representation of a model of the sytem. In a similar way ERDs are convenient for describing the data structures and their relations: A detector is built by a collaboration which consists of physicists belonging to different institutes. A detector can be subdivided into several subdetectors. A subdetector is a physical device: related to each subdetector are also electronic components, e.g. high voltage control, alarm circuits, gas control and software to analyse the data. The electronics can be grouped into racks, crates, slots inside a crate and channels for each slot. The objects mentioned above are *entities*: detectors, physicists, subdetectors, racks or electronic channels. The associations between two or more objects are called *relations*. A physicist *is employed* by an institute. The institute *is located* in a country. He or she *is responsible* for a component in a subdetector. The symbols used for ERDs are:

(1) A *rectangular box* is used to describe the objects or entities; one box per entity.
(2) A *diamond* is used to describe the relations. Either the relation is written in the diamond or a number inside the diamond points to a dictionary describing the relation.
(3) *Numbers* at the connections between entities describe the type of relation. $(1,1)$ is a one-to-one relation: An institute is guided by *one* director. $(1,*)$ describes a relation of several entities to one entity: several physicists are employed by one institute. $(*,*)$ is a relation between two groups of entities: several programmers work on several programs.

A typical entity relationship diagram is shown in Fig. 5.4. When data are described by an ERD they can be stored by a DataBase Management System (DBMS). A good data structure allows the data to be accessed by information retrieval in an easy way.

5.5.4 State-Transition Diagrams (STD)

In DFDs data enter a process, are transformed and leave the process. All incoming data are transformed in the same way, independent of previous data. The output only depends on the input. With this model it is not possible to describe control flows in a satisfactory way. Control flows do not contain 'data'. A control flow just occurs at a certain time. Control flows and control transactions are described by dashed arrows and circles, respectively (Fig. 5.2). The physicist at a control panel just presses one button. If he or she presses the button at the beginning of data taking this means: START experiment. If he or she presses the button again during data taking this signals: STOP experiment. The transformation must combine the occurrence 'push button' with its internal memory or status to select the correct output control flow. This internal presentation is described by STDs which contain the following symbols:

(1) A *rectangular box* describes the internal status of the system: experiment running or experiment stopped. In many state-transition models the status is represented by a circle but we will use a box to distinguish status description from data-flow processes.

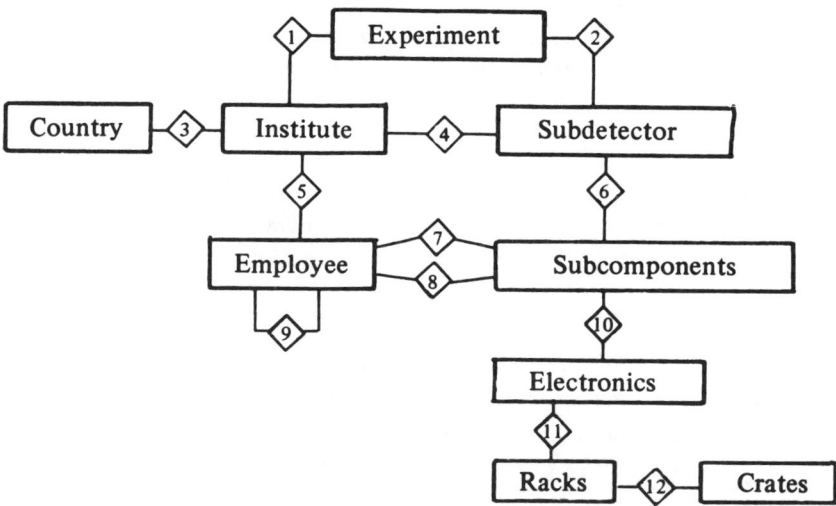

Fig. 5.4 An ERD. An experiment consists (2) of sub-detectors. Several institutes operate (1) an experiment. Institutes are located (3) in a country. Institutes are responsible (4) for a subdetector. A subdetector can be divided (6) into subcomponents. Employees belong (5) to an institute and build (7) subcomponents. One employee is responsible (8) for a subcomponent, another is the groupleader (9). Subcomponents are readout (10) by electronics housing (11) in racks containing (12) crates.

5.5 Structured analysis, structured design

(2) An *arrow* indicates a transition.
(3) *Information beside the transformation arrows* gives more information about the action (such as PUSH, PULL) and its effect on the system (enable data taking, remove red light).
(4) An *arrow with no source* points to the initial state of the system (initial state: power on).

A typical STD for a data-taking system is shown in Fig. 5.5.

5.5.5 Data Dictionary (DD)

The DD contains the definitions of data flows, processes, and files. It is the central reference system for all information. The nomenclature used

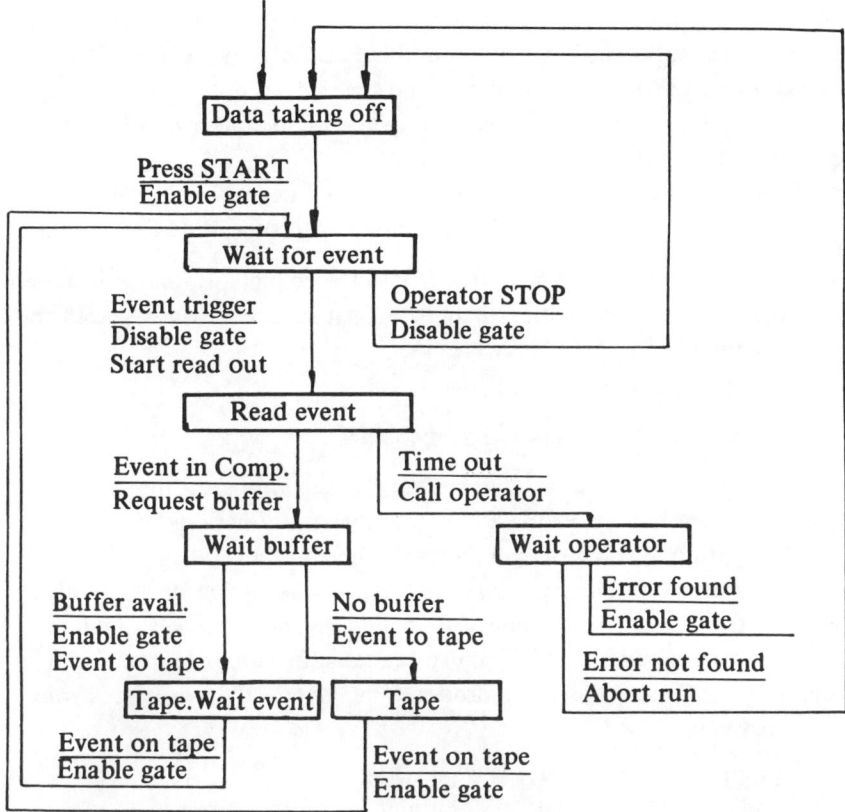

Fig. 5.5 An STD. The initial state is data taking off. When the operator presses START the data enable gate is opened and the system waits for events.

to write a data dictionary is similar to the nomenclature used to define program languages and to build compilers (Wirth 1977). How should one clearly define a decimal number such as 1 or 124 with up to 7 digits? 'A decimal number is defined as 1–7 iterations of decimal digits and optionally a decimal point'. Now we have to define a decimal digit: 'A decimal digit is defined as a number from 0 to 9'. The basic elements for a data dictionary are:

(1) '=' *is defined as*: BinaryDigit = [0, 1]. A binary digit is defined as 0 or 1.
(2) '+' *sequence*: FloatingPointNumber = IntegerNumber + Decimal Point (+ IntegerNumber). A floating-point number is defined by an integer number followed by a decimal point e.g. 123.
(3) '[ ]' *selection*: OctalDigit = [0, 1, 2, 3, 4, 5, 6, 7]. An octal digit is one element of the elements 0, ..., 7. Hence 8 and 9 are not octal digits.
(4) '{ }' *iteration*: OctalNumber = 6{OctalDigit}6. IntegerNumber = 1{DecimalDigit}7. An octal number is defined by six octal digits. In this sense 123 (not six octal digits) and 123997 (9 is not an octal digit) are not octal numbers. An integer number is defined as a number of 1–7 decimal digits.
(5) '( )' *option*: PositiveNumber = (PlusSign) + IntegerNumber. A positive number is defined as an integer number with or without a + sign.

A DD is preferably stored in a DBMS with the help of automated facilities which can provide alphabetical listings, cross referenced listings, search and query facilities and consistency checks.

5.5.6 Process-Description Language (PDL)

DFDs show how the system is decomposed, the DD describes the data flows and the PDL describes the functions of a process. The description must be short and exact. There are several approaches for describing a process such as narrative text, graphs, tables, decision tables, decision trees, and pseudocode. Pseudocode is a sort of very high-level programming language which is also readable by nonprogrammers. The key words can be chosen according to a given environment. Typical keywords could be:

IF ... THEN ... ELSE ... ENDIF	for selection
FOR EACH ... DO ... ENDDO	for loops, iterations
REPEAT ... UNTIL ...	for loops, iterations
GET, PUT, TYPE, PRINT	for input/output

UPDATE, STORE   for file access
WAIT, POST   for task synchronization

5.6 Quality measures for good design

We are sure that the reader has read some articles on how to write efficient, fast code (e.g. Metcalf (1985)). The quality measures for a good design are sometimes incompatible with the goal of writing fast executable code. A good design should minimize the efforts for program development, maintenance, and flexibility. The general code will be about 10% slower and larger than code written in such a way as to produce efficient code in a classical sense. In structured design it is better to pass several parameters to a module which transforms the data and sends the results back by parameters instead of using COMMON blocks. The programmer who must maintain the system can treat this module as a black box: known data go in, are digested and come back. The well-known FORTRAN practice of transferring information via COMMON blocks which are accessible by 100 subroutines makes the maintenance of such a system impossible. The programmer does not know which variable was modified by which subroutine under which condition. We summarize here the main guidelines needed to improve the design of a system:

(1) Modules should be handled like black boxes. The user must only know which data goes in and which comes back out. The coupling between modules should be minimal.
(2) Do not use stamp coupling. Two modules are 'stamp coupled' if they refer to the same data structure but not to the same data. Stamp coupling is often used in FORTRAN by COMMON blocks. If one routine changes the COMMON structure by a larger array all other routines will be affected. COMMON blocks must be modular.
(3) Minimize control flags. Passing down a control flag to tell the lower subroutine 'what to do' indicates a close coupling between modules. The passing back of an error flag is acceptable.
(4) Modules should not control too many or two few subroutines. About seven is a good number. Low-level decisions should be made by low-level routines.
(5) Constants should be defined on a project-wide basis. When a constant changes its value it should be changed in one place only.
(6) Avoid tramp data. Tramp data are pieces of information that are

transferred around the system. They are only needed by the first and last modules and are useless to all modules through which they pass.
(7) A module should perform a single, well-defined task (good cohesion). If one can write one sentence that accurately and completely describes what the module does cohesion is normally good.
(8) Avoid writing modules which do everything as 'do all input' or 'man–machine interface' (logical cohesion). These modules become bulky, are hard to maintain and contain in the end a mixture of different tasks which cannot be separated. Implementation details on the other hand must be separated and 'hidden' from the application program.
(9) Avoid modules which initialize everything such as 'open a device', reset some fields and initialize some variables (temporal cohesion). Related parts of a program should be close together e.g. open file, read file and close file. Do the initialization as late as possible.

At the end of the design the programs are coded and implemented in a 'top-down' approach. The routines of the top-level diagrams are written first. Routines which are not yet available are replaced by dummy or stub routines. These routines generate some data or make simple transformations on the data. Later these routines are replaced by the final versions. This technique is called *step-wise refinement*. Preferably the output routines are written and implemented first because they are needed in the early stages of testing the system. To optimize the programs we remind the reader of the 90–10 rule which means 90% of computer time is spent in 10% of the code. Certain system service routines allow a spectrum of computer time spent per statement to be produced. Optimization is only done around the hot spots of the spectrum – if at all.

5.7 Support tools for program development

We have seen that productivity is low when writing programs. One reason for this is the fact that programs are hand-made. We have shown that there are methods to describe large systems which avoid the programmer having to read a monolithic block of documentation. Commercial tools are available to support the programmer not only in coding and debugging but also in the production of documentation for flow diagrams and DDs: tools are available which produce DFDs, structure charts, and DDs (SASD 1986).

These programs run on a minicomputer or a personal workstation with graphics capabilities. With an editor one can enter or modify DFDs. If one

5.7 Support tools for program development

goes from one process to the next level down all data flows are visible at the edge of the screen and need to be routed to the subprocesses. Formal checks and checks against the DD are performed with verification programs.

We have already mentioned that other tools are available to generate DDs and to print cross references or perform queries. These programs are normally connected to a database system. Other tools transform the DFDs into structure charts. The structure charts can then be split into smaller components. The process description is automatically associated with the structure chart module. For all these products plot facilities for various hardcopy devices are available.

In the coding phase several tools are available from the manufacturers such as editors, compilers and debuggers. In addition there are some editors on the market which are syntax directed and which can understand the language. These special editors can be used to change a specific class of statements or can extract ordered lists of variables. The user can provide extra tokens which are then expanded to standard headers for subroutines or statements (e.g. OPEN) which contain many parameters.

With static analysers the final code can be checked (e.g. Grote 1984). Certain types of errors are found by code checkers e.g. variables which are used but never set or set but never used, mismatches in calling parameters, unreachable sections of code, inconsistent length of COMMON blocks. Part of these tasks is performed by good compilers. Code checkers can also be used by the customer to check whether the conventions for portability between institutes or name conventions for detector components are consistent. In addition documentation can be provided for calling trees and matrices for COMMON block variables set or used.

Debuggers are widely used to monitor, control, and modify executing and memory contents of programs. These debuggers work interactively, one can control the programs by single steps or breakpoints, one can refer to variables by their names as well as by their addresses. Some debuggers show the source code as it is executed.

Path coverage tools show which paths of a program have been executed and which have or not. These tools can also be used for performance tests.

5.7.1 Final remark on software management

Some of the numbers and management rules cited in this section cannot perhaps be directly transferred to high-energy physics experiments. They can only be used as a guide when large and complex systems are designed. But one thing is obvious: tools are necessary.

6

Some final remarks

Man is a tool-using animal. Without tools he is nothing, with tools he is capable of everything. This sentence was written by Carlyle, more than a 100 years ago. It perfectly describes the situation of the researcher specializing in high-energy physics research: without modern technologies from superconductivity to control engineering no modern accelerators could be built and operated; without novel materials and gases coupled with fast and sensitive electronics, there would be no detector to match the accelerator; without the timely advent of digital methods and their culminating expression, computers and the software surrounding them, accelerators and detectors would have produced data that could not be recorded and processed. A large number of different technologies had to develop in parallel over the last two decades, to produce that glimpse into the secrets of nature with which particle accelerators, those giant microscopes of high-energy physics have provided us.

Tools are, of course, conceived by the human mind and implemented by humans. Many of the tools we have referred to have been shaped with some, often relevant, contributions from members of our physics research community. Most have a more general application, which makes them interesting to fellow-researchers in other fields also. Some have been invented for very different purposes, and the specific application to our problems is far from obvious.

We would like to think that our book is appreciated for bringing closer together the tools of data analysis, in general, and their high-energy physics applications. Any of the tools is described more exhaustively in the specialized literature about electronics, pattern recognition, statistics, or software engineering, to quote but a few headings. Our intention was to provide a dramatic precis of this literature, extracting those parts which we judged most relevant, and presenting them in physicists' language and using physicists' examples.

Some final remarks

Data analysis is the last link in the chain that starts with accelerating particles. Despite the high cost of computing equipment, analysis does not require financial resources at nearly the same level as the building of accelerators or detectors. But it certainly needs as much ingenuity and hence human resources to extract information from complex detectors, as it takes to conceive and build these detectors.

If our book succeeds in avoiding some duplication that might not be due to novel ideas, but simply caused by the difficulty of accessing information, and if it helps to avoid the use of nonoptimal analysis methods, we have not spent our effort in vain.

References

Aarnio, P. A. (1986): *FLUKA86 User's Guide* CERN, Geneva, CERN-TIS-RP/168 Divisional Report

Abramowicz, H., et al. (1981): The Response and Resolution of an Iron-Scintillator Calorimeter for Hadronic and Electromagnetic Showers between 10 GeV and 140 GeV, *Nuclear Instruments and Methods* **180**, 429

Abramowitz, M. and Stegun, I. A. (1970): *Handbook of Mathematical Functions*. Dover Publications Inc., New York

Abshire, G., et al. (1979): Measurement of Electron and Pion Cascades in a Lead-Acrylic Scintillator Shower Detector, *Nuclear Instruments and Methods* **164**, 67

Adiels, L., et al. (1986): A π^0 and η Spectrometer of Lead Glass and BGO for Momenta up to 1 GeV/c, *Nuclear Instruments and Methods in Physics Research* **A244**, 380

Advanced Micro Devices (1978): *The Am2901 Family*. Sunnyvale

Ahlen, S. P. (1980): Theoretical and Experimental Aspects of the Energy Loss of Relativistic Heavily Ionizing Particles. *Review of Modern Physics* **52**, No 1, 121–73

Akesson, T., et al. (1984): Properties of Jets in High-E_T Events Produced in pp Collisions at $\sqrt{s} = 63$ GeV, *Zeitschrift für Physik* **C25**, 13

Akesson, T., et al. (1985): Properties of a Fine-sampling Uranium-Copper Scintillator Hadron Calorimeter, *Nuclear Instruments and Methods in Physics Research* **A241**, 17

Akesson, T., et al. (1986): Three-Jet Production in pp Collisions at $\sqrt{s} = 63$ GeV, and a Determination of α_s, *Zeitschrift für Physik* **C32**, 317

Albajar, C., et al. (1987): Events with Large Missing Transverse Energy at the CERN collider, I. W $\to \tau\nu$ Decay, *Physics Letters B* **185**, 233

Allen, A. O. (1978): *Probability, Statistics and Queuing Theory*. Academic Press, London

Allison, W. and Wright, P. (1987): The Physics of Charged Particle Identification: dE/dx, Cherenkov and Transition Radiation, *Experimental Techniques in High Energy Physics*, Addison-Wesley Publishing Company Inc. Menlo Park, California (Editor Ferbel, T.), 371–417

Amaldi, U. (1981): Fluctuations in Calorimeter Measurements, *Physica Scripta* **23**, 409.

Anders, B., et al. (1986): *Performance of a Uranium-Scintillator Calorimeter* DESY 86-105.

Andrews, H. C. (1972): *Introduction to Mathematical Techniques in Pattern Recognition.* Wiley Interscience, New York

Anthonioz, J., Joosten, J., Letheren, M., and van Praag, A. (1981): User Aspects of Mice, *Topical Conference on the Application of Microprocessors to High Energy Physics Experiments.* CERN, Geneva, CERN 81-07

Appel, J. A., et al. (1985): Measurement of the $\sqrt{s}$ Dependence of Jet Production at the CERN $\bar{p}p$ Collider, CERN, Geneva, Internal note *CERN-EP/85-111*

Appel, J. A., et al. (1986): Measurement of $W^\pm$ and Z^0 Properties at the CERN $\bar{p}p$ Collider, *Zeitschrift für Physik* **C30**, 1

Apsimon, R. J., et al. (1986): A Ring-Imaging Cherenkov Detector for the CERN Omega Spectrometer – Design and Recent Performance, *Nuclear Instruments and Methods in Physics Research* **A248**, 76

Aquilar-Benitez, M., et al. (1983): The European Hybrid Spectrometer – A Facility to Study Multihadron Events Produced in High Energy Interactions, *Nuclear Instruments and Methods* **205**, 79–97

Armstrong, T. W., et al. (1972): *Oak Ridge Nat. Lab. Report ORNL-TM-3667*

Armstrong, T. W., et al. (1984): A Comparison of High Energy Fission Models, *Nuclear Instruments and Methods in Physics Research* **222**, 540

Arnison, G., et al. (1983a): Experimental Observation of Isolated Large Transverse Energy Electrons with Associated Missing Energy at $\sqrt{s} = 540$ GeV, *Physics Letters* **122B**, 214

Arnison, G., et al. (1983b): Hadronic Jet Production at the CERN Proton-Antiproton Collider, *Physics Letters* **132B**, 214

Arnison, G., et al. (1986): Analysis of the Fragmentation Properties of Quark and Gluon Jets at the CERN SPS $\bar{p}p$ Collider, *Nuclear Physics* **B276**, 253

Arnold, R., et al. (1986): Photosensitive Gas Detectors for the Ring-Imaging Technique andthe Delphi Barrel RICH Prototype, *Nuclear Instruments and Methods in Physics Research* **A252**, 188

Aubert, J. J. and Broll, C. (1974): Track Parametrization in the Inhomogeneous Field of the CERN Split Field Magnet, *Nuclear Instruments and Methods* **120**, 137–41

Auerbach Publishers Inc. (1984): *Data Processing Management*, Boston, Mass.

Avilez, C. (1984): A Data Driven Parallel Pipelined Hardware Reconstruction Processor. *Proceedings of the Symposium on Recent Development in Computing, Processor and Software Research for High Energy Physics.* Universidad Autonoma de Mexico, Guanajuato, Mexico

Axelrod, T. S., Dubois, P. F., and Eltgroth, P. G. (1983): A Simulator for MIMD Performance Prediction – Application to the S-1 MKIIa Multiprocessor. *Proceedings of the 12th International Conference on Parallel Processing*, 350

Bakken, J. A., et al. (1985): Study of a BGO Calorimeter Using Electron and Hadron Beams from 1 GeV to 50 GeV, *Nuclear Instruments and Methods in Physics Research* **A228**, 294

Banner, M., et al. (1982): Observation of Very Large Transverse Momentum Jets at the CERN p̄p Collider, *Physics Letters* **118B**, 203

Barbe, D. F., et al. (1980): *Charge-Coupled Devices*. Springer, Berlin

Barrelet, E., et al. (1982): A two-dimensional Single Photoelectron Drift Detector for Cherenkov Ring Imaging, *Nuclear Instruments and Methods in Physics Research* **200**, 219

Bartel, W., et al. (1986): Experimental Studies on Multijet Production in e^+e^- Annihilation at PETRA Energies, *Zeitschrift für Physik* **C33**, 23

Bassler, E. (1985): *GEP – Graphics Editor Program, User Manual*. Internal Report DESY R02-81/02, revised June 1985

Bassompierre, G., et al. (1986): *JETSET: Physics at LEAR with an Internal Gas Jet Target and an Advanced General Purpose Detector*, PSCC Proposal P97, CERN/PSCC 86-23, 25 March 1986

Becker, J. J., et al. (1984): *A New Approach to Track Finding and Fitting in Vector Drift Chambers*. SLAC-PUB-3442

Beer, A., et al. (1984): The Central Calorimeter of the UA2 Experiment at the CERN p̄p Collider, *Nuclear Instruments and Methods in Physics Research* **A224**, 360

Behrend, H. J. (1981): The Fast Track-Finder for the CELLO Experiment at DESY, *Computer Physics Communications*, **22**, 365–74

Ben-Ari, M. (1982): *Principles of Concurrent Programming* Prentice Hall International, Englewood Cliffs, New Jersey

Berkelman, K. (1981): *Track Finding and Reconstruction in the Cleo Drift Chamber*, Cornell University, Ithaca NY Report CBX-81-6

Bernstein, D. (1980): SNOOP module CAMAC Interface to the 168/E Microprocessor. *IEEE Transactions on Nuclear Science* **NS-27**, 587

Bestongeff, H. (1986): *Applications of Databases and Knowledge-based Systems to Scientific Data*. CODATA Bulletin 64 (1986) 18

Bichsel, H. (1970): Straggeling of Heavy Charged Particles: Comparison of Born Hydrogenic-Wave-Function Approximation with Free-Electron Approximation, *Physical Review B*, **1**, No 7, 2854

Billoir, P. (1984): Track Fitting with Multiple Scattering: A New Method, *Nuclear Instruments and Methods in Physics Research* **225**, 352–66

Billoir, P. (1986): *Propagation of Transverse Errors for Charged Tracks*, CERN, Geneva, CERN-DELPHI 86-66, PROG 52

Billoir, P. (1987a): *Precise Tracking in a Quasi-Homogeneous Magnetic Field*, CERN, Geneva, CERN-DELPHI 87-6, PROG 65

Billoir, P. (1987b): *Error Propagation in the Helix Track Model*, CERN, Geneva, CERN-DELPHI 87-4, PROG 63

Billoir, P., Frühwirth, R., and Regler, M. (1985): Track Element Merging Strategy and Vertex Fitting in Complex Modular Detectors, *Nuclear Instruments and Methods in Physics Research* **A241**, 115–31

Birsa R., et al. (1977): Reconstruction of the Momentum of a Particle Moving in an Axially Symmetric Magnetic Field., *Nuclear Instruments and Methods* **146**, 357–65

Blobel, V. (1979): *BOS – Bank Organisation System Manual.* DESY Internal Report F14-79/02

Blobel, V. (1984a): Function Minimization. In: *Formula and Methods in Experimental Data Evaluation,* Vol. 3, L 1-49 European Physical Society, Geneva

Blobel, V. (1984b): Least Squares Methods. In: *formulae and Methods in Experimental Data Evaluation,* Vol. 3, L 1-31 European Physical Society Geneva

Blucher, E., et al. (1986): Tests of Cesium Iodide Crystals for an Electromagnetic Calorimeter, *Nuclear Instruments and Methods in Physics Research* **A249**, 201

Bock, R. K., Hansl-Kozanecka, T., and Shah, T. P. (1981): Parameterization of the Longitudinal Development of Hadronic Showers in Sampling Calorimeters, *Nuclear Instruments and Methods* **186**, 533

Bock, R. K., et al. (1984a): Lorentz Force: Trajectory of a Charged Particle; Units. In: *Formulae and Methods in Experimental Data Evaluation,* Vol. 1. European Physical Society, Geneva

Bock, R. K., et al. (1984b): Errors in Track Reconstruction. In: *Formulae and Methods in Experimental Data Evaluation,* Vol. 1. European Physical Society, Geneva

Bock, R. K., et al. (1987a): *HIGZ: High-level Interface to Graphics and Zebra,* CERN, Geneva, CERN Computer Program Library Q120, Long Writeup

Bock, R. K., et al. (1987b): PAW-Towards a Physics Analysis Workstation, *Computer Physics Communication* **45**, 181

Booth, K. S. (1979): *Computer Graphics.* IEEE Computer Society, Long Beach (Compcon 79)

Botner, O., et al. (1987): Preliminary Analysis of the Performance of a RICH Counter for Low-p_T Electron Identification at the CERN $p\bar{p}$ Collider, *Nuclear Instruments and Methods in Physics Research* **A257**, 580

Bouclier, R., et al. (1974): Proportional Chambers for a 5000 Wire Detector. *Nuclear Instruments and Methods* **115**, 235-44

Brafman, H., et al. (1983a): *A Fast General Purpose IBM Hardware Emulator.* Weizmann Institute, Dept. of Nucl. Physics., Internal Report, January 1983, Rehovot, Israel

Brafman, H., et al. (1983b): A Fast General Purpose IBM Hardware Emulator. *Proceedings of the Three Day In-Depth Review on the Impact of Specialized Processors in Elementary Particle Physics,* Padova, Italy, March, 1983

Brammer, K. and Siffling, G. (1975): *Kalman-Bucy-Filter.* R. Oldenburg Verlag, München-Wien

Brandt, S. (1970): *Statistical and Computational Methods in Data Analysis.* North Holland, Amsterdam

Brandt, S. (1984): Elements of Probability and Statistics. In: *Formulae and Methods in Experimental Data Evaluation,* Vol. 3. European Physical Society, Geneva

Brandt, S. and Dahmen, H. D. (1979): Axes and Scalar Measures of Two-Jet and Three-Jet Events, *Zeitschrift für Physik* **C1**, 61

Braunschweig, W., et al. (1986): Inclusive π^0 Production by e^+e^- Annihilation at 34.6 GeV Center of Mass Energy, *Zeitschrift für Physik* **C33**, 13

Braunschweig, W., et al. (1988): *Results from a Test of a Pb-Cu-Liquid Argon Calorimeter*. Nuclear Instruments and Methods in Physics Research **A265**, 419

Breiman, N., et al. (1984): *Classification and Regression Trees*. Wadsworth, Belmont California

Brun, R., et al. (1981): *User Guides for HBOOK and HPLOT*. CERN, Geneva, CERN Computer Program Library, Long Writeups Y250/Y251. Also unpublished notes on the packages HTV, ZCEDEX and GUI

Brun, R., et al. (1985): *The GEANT3 Electromagnetic Shower Program and a Comparison with the EGS3 Code*. CERN, Geneva CERN Internal Note DD/85/1, March 1985

Brun, R., et al. (1986): *GEANT3 User Guide*. CERN, Geneva, CERN Internal Note DD/EE/84-1, May 1986

Brun, R., et al. (revised 1979): *MUDIFI: Multidimensional Fit Program*. CERN, Geneva, CERN DD/US/69

Brun, R. and Zoll, J. (1987): *ZEBRA User Guide*. CERN, Geneva, CERN Program Library, Long Writeup Q100

Brun, R., Hansroul, M. and Kubler J. (1980): *LINTRA – A Principal Component Analysis Program*. CERN, Geneva, CERN/DD/US/70

Brückmann, H., Behrens, U., and Anders, B. (1987): *Hadron Sampling Calorimetry, a Puzzle of Physics*, DESY 86–155, to be published in *Nuclear Instruments and Methods in Physics Research*. Also: H. Brückmann, et al., (1987): *On the Theoretical Understanding and Calculation of Sampling Calorimeters*. DESY 87-64 and in *Lepton Nucleon Interactions at High Energies, 23–27 Feb*, 1988, Seville, Spain. Eds. F. Barreiro and J. L. Sanchez-Gomes, World Scientific, Singapore, 1988

Buchanan, C. D. (1982): *Proportional Mode Calorimeters of the TPC Facility, Proceedings of the Gas Calorimeter Workshop*. Fermilab, October 1982

Bugge, L. (1986): On the Determination of Shower Central Positions from Lateral Samplings, *Nuclear Instruments and Methods in Physics Research* **A242**, 228

Bugge, L. and Myrheim, J. (1981): Tracking and Track Fitting, *Nuclear Instruments and Methods* **179**, 365–81

Burnett, T. (1983); *IDA Interactive Data Analysis*, SLAC Mark III Memo 1/83–6

Burrows P. N. (1987): Multicluster Analysis of Hadronic Data at C. M. Energies between 12.0 and 46.8 GeV, *Proceedings of the 1987 International Symposium on Lepton and Photon Interactions at High Energies*, North Holland, Amsterdam,

Bursky, D. (1985): Digital GaAs ICs, *Technology Report, Electronic Design*, December 1985

Busi, C., et al. (1983): *Proposal to the CERN SPSC*, CERN Geneva, CERN/SPSC/83-24 (P186)

Calvi, M. (1987): *HPC Pattern Recognition*, Internal Note, Milano, May 1987
Caporal, P. M. and Hahn, G. J. (1984): General Software for Statistical Graphics – A Survey, *Statistical Software Newsletter* **10** Nr. 1, p. 3., published by the Gesellschaft für Strahlen- und Umweltforschung, München
Cashmore, R., et al. (1985): Monte Carlo Studies Towards the Design of Iron/Scintillator and Uranium/Scintillator Calorimeters, *Nuclear Instruments and Methods in Physics Research* **A242**, 42
Cassel, D. G. and Kowalski H. (1981): Pattern Recognition in Layered Track Chambers Using a Tree Algorithm, *Nuclear Instruments and Methods* **185**, 235–51
Charpak, G. (1978): Multiwire and Drift Proportional Chambers, *Physics Today* **31**, 23–30
Charpak, G. and Sauli, F. (1987): High-Resolution Electronic Particle Detectors *Experimental Techniques in High Energy Physics*, Addison-Wesley Publishing Company Inc. Menlo Park, California Ed. T. Ferbel, p. 189–255
Chen, P. P.-S. (1976): The Entity-Relationship Model—Toward a Unified View of Data, *ACM Trans. on Database Systems* **1**, 9
Cochet, C. et al. (1986): The Central Electromagnetic Calorimeter of UA1, *Nuclear Instruments and Methods—Physics Research* **A243**, 45
Cody, R. P. and Smith, J. K. (1985): *Applied Statistics and the SAS Programming Language.* North Holland, Amsterdam
Coffman, D. M. (1987): *Properties of Semileptonic Decays of Charmed D Mesons*, Thesis, CALTECH Internal note CALT-68-1415
Conetti, S. (1984): A Review of Triggers and Special Computing Hardware for the Fermilab Fixed-Target Program. *Proceedings of the Symposium on Recent Development in Computing, Processor and Software Research for High Energy Physics.* Universidad Autonoma de Mexico, Guanajuato, Mexico
Corden, M. et al. (1985): Central Hadronic Calorimeter of UA1, *Nuclear Instruments and Methods in Physics Research* **A238**, 273
Corporaal, H. (1979): Estimating the Error in the Inverse Momentum for Particle Tracks in High Energy Physics Using Wind's Quintic Spline Method, *Nuclear Instruments and Methods* **158**, 127–8
Cox, M. G. (1982): *Practical Spline Approximation.* National Physics Laboratory Report DITC 1/82, Teddington, Middlesex TW11 OLW, UK
Dahl-Jensen, E. (1979): *Track Finding in the R807 Detector.* CERN, Geneva, CERN/R807/8 internal note
Das, S. R. (1973): On a New Approach for Finding all the Modified Cut-sets in an Incompatibility Graph, *IEEE Transactions on Computers* **C-22**, no. 2, 187–93
DeMarco, T. (1978): *Structured Analysis and System Specification.* Yourdon Press, New York
Dewdney, A. K. (1985): Computer-Recreations, *Scientific American*, April 1985
Diddens, A. N. et al. (1980): A Detector for Neutral-Current Interactions of High Energy Neutrinos, *Nuclear Instruments and Methods* **178**, 27
Dietl, H. et al. (1985): Performance of a BGO Calorimeter with Photodiode Readout and with Photomultiplier Readout at Energies up to 10 GeV, *Nuclear Instruments and Methods in Physics Research* **A235**, 464

Dieudonné, J. (1979): *Eléments d'analyse*, Vol. 1. Gauthiers-Villars, Paris
Dobinson, R. W. (1982): *Bus Basics*. CERN Computer School, Geneva CERN 83-03
Dorenbosch, J. et al. (1987): Calibration of the CHARM Fine-grained Calorimeter, *Nuclear Instruments and Methods in Physics Research* **A253**, 203
Drescher, A. et al. (1986): Calibration and Monitoring of the Argus Shower Counters, *Nuclear Instruments and Methods in Physics Research* **A249**, 277; The Time-Projection Ring Imaging (RICH) Counter – New Experimental Results, *IEEE Transactions on Nuclear Science* **NS-31**, 949
Dreyfus, H. L. (1979): *What Computers Can't Do: The Limits of Artificial Intelligence*. Harper and Row, New York
Drijard, D., Ekelöf, T., and Grote, H. (1980): On the Reduction in Space Resolution of Track Detectors caused by Correlations in the Coordinate Quantization, Proceedings of the Second International Wire Chamber Conference, *Nuclear Instruments and Methods* **176**, 389–95
Eadie, W. T., Drijard, D., James, F. E., Roos, M., and Sadoulet, B. (1971): *Statistical Methods in Experimental Physics*. North-Holland, Amsterdam
Eek, L. O. et al. (1984); The Time-Projection Ring Imaging Chamber Counter – New Experimental Results, *IEEE Transactions on Nuclear Science* **NS-31**, 349
Efron, B. (1982): *The Jackknife, the Bootstrap, and Other Resampling Plans*, SIAM, Bristol, 1982
Eichinger, H. (1980): Global Methods of Pattern Recognition, Proceedings of the Second International Wire Chamber Conference, *Nuclear Instruments and Methods* **176**, 417–24
Eichinger, H. and Regler, M. (1981): *Review of Track Fitting Methods in Counter Experiments*. CERN, Geneva, CERN 81-06
Eisenhandler, E., et al. (1984): *Electron Shower Profiles in the Gondolas*, Internal Note UA1/TN 84-64 (August 1984)
Ellsworth, R. W. et al. (1982): A Study of Albedo from a Hadronic Calorimeter for Energies $\sim$ 100–2000 GeV, *Nuclear Instruments and Methods in Physics Research* **203**, 167
Enderle, G., et al. (eds.) (1986): *Advances in Computer Graphics, Eurographics Seminars*. Springer, Berlin
Enderle, G., Kansy, K. and Pfaff, G. (1984): *Computer Graphics Programming: GKS – The Graphics Standard*, Springer, Berlin
EUR 4100 (1972): *CAMAC, A Modular Instrumentation System for Data Handling*
EUR 4600 (1972): *CAMAC, Organization of Multi-Crate Systems*
Eyges, L. (1948): Multiple Scattering with Energy Loss, *Physical Review* **74**, 1534–5
Fabjan, C. W. (1985): Calorimetry in High Energy Physics, *Concepts and Techniques in High Energy Physics* III, p. 281 T. Ferbel ed., Plenum Press, New York 1985.
Also in: Internal note CERN-EP/85-54, CERN, Geneva

Fabjan, C. W. and Ludlam, T. (1982): Calorimetry in High Energy Physics, *Annual Review of Nuclear and Particle Science* **32**, 335–89

Fabjan, C. W., et al. (1977): Iron Liquid-Argon and Uranium Calorimeters for Hadron Energy Measurement, *Nuclear Instruments and Methods* **141**, 61

Farrell, E. J. (1987): Visual Interpretation of Complex Data, *IBM System Journal* **26/2**, 174

FASTBUS (1983): *A Modular High Speed Data Acquisition System for High Energy Physics and Other Applications*. Esone Committee, Esone/FB/01

FASTBUS (1985): *IEEE Standard FASTBUS Modular High-Speed Data Acquisition and Control System*. The Institute of Electrical and Electronic Engineers, Inc, ISBN 0-471-84472-1

Ferbel, Th. (1987): *Experimental Techniques in High Energy Physics*, Frontiers in Physics, Addison Wesley, Menlo Park, California

Fernow, R.C. (1986): *Introduction to Experimental Particle Physics*, Cambridge University Press, Cambridge

Fesefeldt, H. (1985): *The Simulation of Hadronic Showers, Physics and Applications*, RWTH Aachen Report PITHA 85/02

Fidecaro, C. et al. (1980): Measurement of the Polarization Parameter in pp Elastic Scattering at 150 GeV/c, *Nuclear Physics* **B173**, 513–45

Flores, I. (1970): *Data Structures and Management*, Prentice Hall, Englewood Cliffs, New Jersey

Foley, J. D. and Van Dam, A. (1982): *Fundamentals of Interactive Computer Graphics*. Addison-Wesley, Menlo Park, California

Ford, W. T. et al. (1982): The MAC Calorimeters, *Proceedings of the International Conference on Instrumentation for Colliding Beams*. Stanford, SLAC-250 June 1982

Forden, G. E. and Saxon, D. H. (1985): *Improving Vertex Position Determination by Using a Kinematic Fit*. RAL-85-037, Rutherford

Freeman, H. (1980): *Interactive Computer Graphics*, IEEE Computer Society, Long Beach (Compcon 80)

Freytag, D. R. and Walker, J. T. (1985): Performance Report for the Stanford/SLAC Microstore Analog Memory Unit, *IEEE Transactions on Nuclear Science* **NS-32**, 622

Friedman, J. H. (1977): A Recursive Partitioning Decision Rule for Nonparameteric Classification, *IEEE Transactions on Computers* **C-26**, 404

Friedman, J. H. and Tukey, J. W. (1974): A Projection Pursuit Algorithm for Exploratory Data Analysis. *IEEE Transactions on Computers* **C-23**, 881

Friedman, J. H., McDonald, J. A. and Stuetzle, W. (1987): An Introduction to Real-Time Graphical Techniques for Analyzing Multivariate Data, *Computer Physics Communication* **45**, 149

Frodesen, A. G., Skjeggestad, O., and Tøfte, H. (1979): *Probability and Statistics in Particle Physics*. Universitetsforlaget, Bergen, Oslo and Tromsø

Fröhlich, A., Grote, H., Onions, C. and Ranjard F. (1976): *MARC-Track Finding in the Split Field Magnet*, CERN, Geneva, CERN/DD/76/5

Frühwirth, R. (1986): Estimation of Variances in a Linear Model Applied for

Measurements of Trajectories, *Nuclear Instruments and Methods in Physics Research* **A243**, 173–80

Frühwirth, R. (1987): Application of Kalman Filtering to Track and Vertex Fitting, *Nuclear Instruments and Methods in Physics Research* **A262**, 444–50

Frühwirth, R. (1988): Application of filter Methods for the Reconstruction of Tools and Vertices in Events in Experimental High Energy Physics. Thesis, University of Technology, Vienna

Fucci, A. and Storr, K. M. (1983): Using 3081/E Emulators in On-Line and Off-Line Environments. *Proceedings of the Three Day In-Depth Review on the Impact of Specialized Processors in Elementary Particle Physics*. Padova, Italy, March, 1983

Gabriel, T. A. and Bishop, B. L. (1978): Calculated Hadronic Transmission Through Iron Absorbers, *Nuclear Instruments and Methods* **155**, 81

Galaktinov, Y. et al. (1986): The Performance of a Uranium Gas Sampling Calorimeter, *Nuclear Instruments and Methods in Physics Research* **A251**, 258

Galitz, W. O. (1985): *Handbook of Screen Format Design*. North Holland, Amsterdam

Gelb, A. Ed. (1975): *Applied Optimal Estimation*. MIT Press, Cambridge, Mass.

Georgiopoulos, C. H., Goldman, J. H., Levinthal, D. and Hodous, M. F. (1986): A Non-numerical Method for Track Finding in Experimental High Energy Physics Using Vector Computers, *Nuclear Instruments and Methods* **A249**, 451–4

Gluckstern, R. L. (1963): Uncertainties in Track Momentum and Direction due to Multiple Scattering and Measurement Errors, *Nuclear Instruments and Methods* **24**, 381–9

Grant, A. (1975): A Monte Carlo Calculation of High Energy Hadronic Cascades in Matter, *Nuclear Instruments and Methods* **131**, 167

Grassmann, H. and Moser, H. G. (1985): Shower Shape Analysis in Longitudinally Sampled Electromagnetic Calorimeters, *Nuclear Instruments and Methods in Physics Research* **A237**, 486

Greenhalgh, J. F. (1984): A Trigger Processor for a Fermilab Di-Muon Experiment, *Proceedings of the Symposium on Recent Development in Computing, Processor and Software Research for High Energy Physics*, Universidad Autonoma de Mexico, Guanajuato, Mexico

Greville, T.N.E. (1969): *Theory and Applications of Spline Functions*. Academic Press, New York

Grote, H. (1981): *Data Analysis for Electronic Experiments*. CERN 81-03 pp. 136–81

Grote, H. (1984): *'FLOP' User's Guide and Reference Manual*, CERN Data Handling Division, DD/US113

Grote, H. and Zanella, P. (1980): Applied Software for Wire Chambers, Proceedings of the Second International Wire Chamber Conference, *Nuclear Instruments and Methods* **176**, 29–37

Grote, H., Hansroul, M., Lassalle, J. C. and Zanella, P. (1973): Identification of

Digitized Particle Trajectories. *Proceedings of the International Computing Symposium Davos*, North Holland Amsterdam, pp. 413–21

Hart, J. C., et al. (1983): Databases and Bookkeeping for High-energy Physics Experiments, *ECFA Working Group on Data Processing Standards*, Report ECFA/83/78

Hauptmann, J. (1979): *Calorimeter Analysis.* Internal Note HEE-121, UCLA High Energy Group, July 1979.

Hayakawa, S. (1969): *Cosmic Ray Physics.* J. Wiley and Sons, New York

Highland, V. (1975): Some Practical Remarks on Multiple Scattering, *Nuclear Instruments and Methods* **129**, 479–99

Holder, M. et al. (1978): Performance of a Magnetized Total Absorption Calorimeter between 15 GeV and 140 GeV, *Nuclear Instruments and Methods* **151**, 69

Hopgood, F. R. A. and Duce, D.A. (1986): *Graphics Standards – The Current State.* RAL Internal Report RAL 86–081

Hopgood, F. R. A., Hubbold, R. J. and Duce, D. A. (1987): *Advances in Computer Graphics* II, Springer, Berlin

Hopgood, F. R. A. et al. (1983): *Introduction to the Graphics Kernel System (GKS).* Academic Press, London

Hungerbuehler, V. (1981): UA2 Trigger and Data Acquisition, *Topical Conference on the Application of Microprocessors to High Energy Physics Experiments.* CERN, Geneva, CERN 81–07

Iwahori, J. et al. (1986): Performance of a BGO Calorimeter below 1.6 GeV, *Nuclear Instruments and Methods in Physics Research* **A248**, 309

Iwata, S. (1980): *Calorimeters for High Energy Experiments at Accelerators*, Nagoya University Report DPNU-13-80

Jackson, J. D. (1962): *Classical Electrodynamics.* John Wiley and Sons, Inc., New York

James, F. (1972): Function Minimization, *Proceedings of the 1972 Cern Computing and Data Processing School*, CERN, Geneva, CERN 72-21

James, F. (1981): Determining the Statistical Significance of Experimental Results, *Proceedings of the 1980 CERN School of Computing*, Geneva, CERN 81-03.

James, F. (1983): Fitting Tracks in Wire Chambers Using the Chebyshev Norm Instead of Least Squares *Nuclear Instruments and Methods* **211**, 145–52

James, F. and Roos, M. (1986): *MINUIT–'Function Minimization and Error Analysis'.* CERN Computer Program Library, Geneva

Jaroslawsky, S. (1977): Processor for the Proportional Chamber of the TASSO Experiment at DESY designed by S. Jaroslawsky, Imperial College, London

Jeffery, K. G. (1982): Some Database Applications in High Energy Physics, *Workshop on Software in High Energy Physics, Proceedings*, CERN, Geneva CERN 82-12.

Johnsen, K. (1987a): Linear $e^+ e^-$ Colliders. *Proceedings of the Workshop on Physics and Future Accelerators*, CERN, Geneva CERN 87-07

Johnsen, K. (1987b): *Introduction to Accelerator Physics.* Elementary Particles: XCII Corso; Soc. Italiana di Fisica, Bologna, Italy

Jonker, M. *et al.* (1982): The Response and Resolution of a Fine-Grain Marble Calorimeter for Hadronic and Electromagnetic Showers, *Nuclear Instruments and Methods in Physics Research* **200**, 183

Kaplan, D. M. (1984): A Parallel, Pipelined Event Processor for Fermilab Experiment 605, *Proceedings of the Symposium on Recent Development in Computing, Processor and Software Research for High Energy Physics*, Universidad Autonoma de Mexico, Guanajuato, Mexico

Kellner, G. (1987): Development of Software for ALEPH using Structured Techniques. Proceedings of the International Conference on Computing in High Energy Physics, Asilomar, 2–6 February 1987, *Computer Physics Communications*, **45**, 229–43

Kendall, M. G. and Stuart, A. (1967): *The Advanced Theory of Statistics*, Volume II (Interference and Relationship, second edition). Charles Griffin and Company Limited, London

Kennedy, A. (1983): *Structured Software Methods*, Software Course, Department for Computing. Imperial College, London

Kleinknecht, K. (1986): *Detectors for Particle Radiation*. Cambridge University Press, Cambridge

Kostarakis, P. *et al.* (1981): A Fast Processor for Di-Muon Trigger, *Topical Conference on the Application of Microprocessors to High Energy Physics Experiments*. CERN, Geneva, CERN 81-07

Kowalski, H., Moehring, H.-J. and Tymieniecka, T. (1987): *High Speed Monte Carlo with Neutron Component NEUKA*. Invited talk at the Argonne Workshop on Detector Simulation for SSC DESY 87–170

Kunszt, Z. (1987): Large Cross Section Processes, *Proceedings of the Workshop on Physics and Future Accelerators*, CERN, Geneva CERN 87-07

Kunz, P. (1976): The LASS Hardware Processor, *Nuclear Instruments and Methods* **135**, 435

Kunz, P. (1981): Use of Emulating Processors in High Energy Physics, *Physica Scripta*. **23**, 492

Kunz, P. *et al.* (1983): The 3081/E Processor, *Proceedings of the Three Day In-Depth Review on the Impact of Specialized Processors in Elementary Particle Physics*, Padova, Italy, March, 1983

Lala, P. K. (1985): *Fault Tolerance and Fault Testable Hardware Design*, Prentice Hall, London

Lankford, A. J. (1984a): A Review of Trigger and Online Processors at SLAC, *Proceedings of the Symposium on Recent Development in Computing, Processor and Software Research for High Energy Physics*. Universidad Autonoma de Mexico, Guanajuato, Mexico

Lankford, A. J. (1984b): The ASP Energy Trigger, A Review of Trigger and Online Processors at SLAC, *Proceedings of the Symposium on Recent Development in Computing, Processor and Software Research for High Energy Physics*. Universidad Autonoma de Mexico, Guanajuato, Mexico

Lassalle, J. C., Carena, F., and Pensotti, S. (1980): TRIDENT: A Track and Vertex Identification Program for the CERN Omega Particle Detector System. Proceedings of the Second International Wire Chamber Conference, *Nuclear Instruments and Methods* **176**, Nos 1, 2, 371–9

Laurikainen, P. (1971a): On the Geometrical Fit of Bubble Chamber Tracks by Treating Multiple Scattering as Measurement Error, *Commentatione Physico-Mathematical* **41**, 131–48

Laurikainen, P. (1971b): *Multiple Scattering and Track Reconstruction*, Report Series in Physics 35, University of Helsinki

Laurikainen, P., Moorhead, W. G. and Matt, W. (1972): Least Squares Fit of Bubble Chamber Tracks Taking into Account Multiple Scattering, *Nuclear Instruments and Methods* **98**, 349–59

LeCroy (1985): *Catalog for fast electronics*. LeCroy Research System Corporation, New York

Leo, W. R. (1987): *Techniques for Nuclear and Particle Physics Experiments (A How-to Approach)*, Springer Verlag, Berlin

Leroy, C., Sirois, Y., and Wigmans, R. (1986): An Experimental Study of the Contribution of Nuclear Fission to the Signal of Uranium Hadron Calorimeters, *Nuclear Instruments and Methods in Physics Research* **A252**, 4

Levit, L. B. and Vincelli, M. L. (1985): A Modular Implementation of Sophisticated Triggers, *Nuclear Instruments and Methods* **A235**, 396–406

Litchfield, P. (1984): Partial Wave Analysis. In: *Formulae and Methods in Experimental Data Evaluation*, Vol. 2. European Physical Society, Geneva

Longo, E. and Luminari, L. (1985): Fast Electromagnetic Shower Simulation, *Nuclear Instruments and Methods in Physics Research* **A239**, 506

Longo, E. and Sestili, I. (1975): Monte Carlo Calculation of Photon-initiated Electromagnetic Showers in Lead Glass, *Nuclear Instruments and Methods* **128**, 283

Louis, F. and Verkerk, F. (1975): *REPMAG – Fit of Three-Dimensional Expansion in Terms of Chebycheff Polynomials to a Magnetic Field Volume* CERN Computer Program Library, Geneva Entry POOL-W 1023

Lütjens, G. (1981): How can Fast Programmable Devices Enhance the Quality of Particle Experiments. *Topical Conference on the Application of Microprocessors to High Energy Physics Experiments*. CERN, Geneva, CERN 81-07

Manzo, J. (1987): On Managing Large Scale Projects, Proceedings of the International Conference on Computing in High Energy Physics. Asilomar, 2.–6. February 1987. *Computer Physics Communications*, **45**, 215–28

Maples, C. (1984): Experience with Scientific Applications on the MIDAS Multiprocessor System, *Proceedings of the Symposium on Recent Development in Computing, Processor and Software Research for High Energy Physics*, Universidad Autonoma de Mexico, Guanajuato, Mexico

Margenau, H. and Murphey, G. M. (1964): *The Mathematics for Physics and Chemistry*. Van Nostrand, Toronto

Marx, J. N. and Nygren, D. R. (1978): The Time Projection Chamber, *Physics Today* **31**, 46–53

McCarthy, R., et al. (1986): Identification of Large-Transverse-Momentum Hadrons Using a Ring-Imaging Cherenkov Counter, *Nuclear Instruments and Methods in Physics Research* **A248**, 69

Mecking, B. (1982): On the Accuracy of Track Reconstruction with Inhomogeneous Magnetic Detectors. *Nuclear Instruments and Methods* **203**, 299–305

Mess, K. H., Metcalf, M. and Orr R. S. (1980): Track Finding in a Fine-Grained Calorimeter, *Nuclear Instruments and Methods* **176**, 349–54

Messel, H. and Crawford, D. F. (1970): *Electron-Photon Shower Distribution Function*. Pergamon Press, Oxford

Metcalf, M. (1974): *Analysis of the SFM Field*. CERN, Geneva, CERN DD/OM/AP-10

Metcalf, M. (1985): *Effective FORTRAN 77*. Clarendon Press, Oxford

Metcalf, M. (1986): *Computers in High Energy Physics*. Advances in Computers, Vol. 25, Academic Press Inc. New York, Ed. M. Yovitis, 277–334

Metcalf, M. and Regler, M. (1973): Solution of the Equation of Motion of a Charged Particle in a Stationary Magnetic Field. *Journal of Computational Physics* **11**, No. 2, 240–249

Metcalf, M., Regler, M., and Broll, C. (1973): *A Split Field Magnet Geometry Fit Program: NICOLE*. CERN, Geneva, CERN 73-2

Miller, G. A. (1956): The Magical Number Seven, Plus or Minus Two: Some Limits on Our Capacity for Processing Information. *Psychological Review*, **63**, No. 2

Mitaroff, W. (1986): In *Report on Global Track and Vertex Fitting in the DELPHI Detector*. CERN, Geneva, CERN-Delphi 86-99, PROG-61

Mitaroff, W. (1987): *Status of Utility Library for Helix Tracking and Error Propagation*. CERN, Geneva, CERN-DELPHI 87-51, PROG-85

Monolithic Memories (1985): *PAL Programmable Array Logic Handbook*

Moorhead, W. G. (1960): *A Program for the Geometrical Reconstruction of Curved Tracks in a Bubble Chamber*. CERN, Geneva, CERN 60-33

Morse, P. M. (1958): *Queues, Inventories and Maintenance*. John Wiley and Sons Inc., New York

Motorola (1977): *Bit Slice Processor 10800*

Mount, R. P. (1987): Database Systems for High Energy Physics Experiments, *Computers Physics Communication* **45**, 299

Mourou, G. A., Bloom, D. M. and Lee, C. H. (1986): Picosecond Electronics and Optoelectronics. *Proceedings of the Topical Meeting, Lake Tahoe, Nevada, 1985*, Springer, Berlin

MULTIBUS II (1984): *Multibus II Bus Architecture Specification Handbook*, Intel Corp

Muraki, Y. et al. (1985): Radial and Longitudinal Behaviour of Hadronic Cascade Showers Induced by 300 GeV Protons in Lead and Iron Absorbers, *Nuclear Instruments and Methods in Physics Research* **A236**, 47

Murzin, V. S. (1967): *Progress in Elementary Particle and Cosmic-Ray Physics*, Vol. IX p. 247, North-Holland, Amsterdam, Ed. J. G. Wilson

Myers, D. R. (1983): *A Background to Computer Graphics*, CERN, Geneva, CERN 83-07

Myers, D. R. (1988): Interactive Computer Graphics and PHIGS, *Computer Physics Communication* **50**, 143–57

Myrheim, J. and Bugge, L. (1979): A Fast Runge Kutta Method for Fitting Tracks in a Magnetic Field, *Nuclear Instruments and Methods* **160**, 43–8

Nagy, E. et al. (1978): Measurement of Elastic Proton-Proton Scattering at Large Momentum Transfer at the CERN Intersecting Storage Rings, *Nuclear Physics* **B150**, 221–65

Nash, T. et al. (1986): The Fermilab Advanced Computer Program Multi-Microprocesor Project, *Computing in High Energy Physics, Proceedings of the Conference held in Amsterdam, The Netherlands, 25–28 June, 1985*, North Holland, Amsterdam

Naumann, Th. and Schiller, H. (1984): Multidimensional Data Analysis, in: *Formulae and Methods in Experimental Data Evaluation*, Vol. 3, European Physical Society, Geneva.

Nelson, W. R., Hirayama, H. and Rogers, D. W. O. (1985): *The EGS4 Code System*, Internal Report SLAC-265.

Notz, D. (1981): Microprocessors at DESY, *Topical Conference on the Application of Microprocessors to High Energy Physics Experiments*. CERN, Geneva, CERN 81-07

Notz, D. (1982): *The Input/Output Software for the 370/E Emulator*. DESY, Internal Report F1-82/01

Notz, D. (1984): A Review of Triggers and Special Computing Hardware at DESY, *Proceedings of the Symposium on Recent Development in Computing, Processor and Software Research for High Energy Physics*. Universidad Autonoma de Mexico, Guanajuato, Mexico

Notz, D. (1985a): The 370/E Emulator at DESY, *Nuclear Instruments and Methods* **A235**, 380–2

Notz, D. (1985b). *A Data Processing System Based on the 370/E Emulator*. DESY 85-46.

Notz, D. and Rehlich, K. (1980): *A Microprogrammable Computer for the Fisher CAMAC System Crate*. Internal Report DESY F1-80/01

OCCAM, (1984): *OCCAM Programming Manual*. INMOS Limited, Prentice Hall International, Englewood Cliffs, New Jersey

Olsson, J., Steffen, P., Goddard, M. C., Peace, G. F., and Nozaki, T. (1980): Pattern Recognition Programs for the JADE Jet-chamber, *Nuclear Instruments and Methods* **176**, 403–7

Page-Jones, M. (1980): *The Practical Guide to Structured Systems Design*. Yourdon Press, New York

Palmer, I. R. (1975): *Database Systems: A Practical Reference*, QED Information Sciences, Wellesley, Mass.

Particle Data Group (1986): Review of Particle Properties, *Physics Letters* **170B**

Pernicka, M., Regler, M., and Sychkov, S. (1978): Drift Time Relations in Rotated MWPCs and Their Advantage in Practice. Proceedings of the First International Wire Chamber Conference, *Nuclear Instruments and Methods* **156**, Nos. 1, 2, 147–57

Pimiä, M. (1985): *Track Finding in the UA1 Central Detector at the CERN $\bar{p}p$ Collider*. University of Helsinki HU-D45

Platner, E. (1976): **Paper presented at** the 1976 Nuclear Science Symposium and Scintillation and Semiconductor Counter Symposium, October 20–22, 1976, New Orleans, Louisiana, *IEEE Transactions on Nuclear Sciences* NS-24(1) Feb. 1977

Putzer, A. (1987): *Database Systems in High Energy Physics Experiments*, CERN School for Computing 1987, CERN 83-03.

Qian, Z., et al. (1987): Use of the ADAMO Data Management System within ALEPH, *Computer Physics Communication* **45**, 283

Quarrie D. (1983): *YBOS User Guide*, CDF Note 156, 25.10.1983

Regler M. (1968): *EQUMOT–'Equations of Motion of a Particle in a Static Magnetic Field'*. CERN Computer Program Library

Regler, M. (1977): Vielfachstreuung in der Ausgleichsrechnung, *Acta Physica Austriaca* **49**, 37–45 (English translation in *Formulae and Methods in Experimental Data Evaluation*, Vol. 2, G I–II. European Physical Society, Geneva, 1984)

Regler, M. (1981): Influence of Computation Algorithms on Experimental Design, Proceedings of the 4th Europhysics Conference on Computational Physics, *Computer Physics Communications* **22**, 167–75

Regler, M. and Frühwirth, R. (1989): *Reconstruction of Charged-Particle Trajectories*, Plenum Publishing Corporation

Rehlich, K. (1980): Processor for the Vertex detector of the TASSO experiment at DESY designed by K. Rehlich, DESY, Hamburg

Reingold, M., Nievergelt, J., and Deo N. (1977): *Combinatorial Algorithms, Theory and Practice*. Prentice Hall Inc., New Jersey

Richman, J. (1986): PhD Thesis, CALTECH Internal note CALT-68-1231

Rossi, B. (1965): *High Energy Particles*. Prentice Hall Inc., New York

Rossi, B. and Greisen, K. (1941): Cosmic Ray Theory; §23. The Distribution Function, *Review of Modern Physics* **13**, 265–8

SASD (1986): SASD tools are available from: *Structured Analysis Tools, User's Manual*, Tektronix part no. 070-5478-00, Product Group 61, April 1985: *Structured Design Tools, Tektronix User's Manual*, Tektronix part no. 070-5912-00, Product Group 61, April 1986: *Teamwork/SA and /SD* from CADRE Technologies Inc: *PCSA*, available from StructSoft: *ProMod*, available from GEI: *Excelerator*, available from Index Technology Corporation: *Analyse/Design Toolkit*, Available from Yourdon, New York: *ARGUS II*, developed by Boeing Computer Service in internal use

Sauli, F. (1978): Limiting Accuracies in Multiwire Proportional and Drift Chambers. Proceedings of the First International Wire Chamber Conference, *Nuclear Instruments and Methods* **156**, Nos 1, 2, 147–57

Sauli, F. (1987): Principles of Operation of Multiwire Proportional and Drift Chambers. *Experimental Techniques in High Energy Physics*, Addison-Wesley Publishing Company Inc., Menlo Park, California, Ed. T. Ferbel, pp. 79–188

Schildt, P., Stuckenberg, H.-J. and Wermes, N. (1980): An On-line Track

Following Microprocessor for the Petra Experiment Tasso, *Nuclear Instruments and Methods* **178**, 571
Schmidt, J. W. and Brodie, M. L. (1983): *Relational Database Systems*. Springer, Berlin
Schorr, B. (1974): *Introduction to Reliability Theory*. CERN, Geneva, CERN 74-16
Schorr, B. (1976): *Introduction to Cluster Analysis*. Lecture given in the CERN Academic Training Programme 1975/76, CERN internal report DD/76/3
Schroeder, V. (1981): The Charged Particle Trigger of the Cello Detector, *Topical Conference on the Application of Microprocessors to High Energy Physics Experiments*. CERN, Geneva, CERN 81-07
Schulz, H. D. (1984): The Argus Trigger Processor "Little Track Finder", *Proceedings of the Symposium on Recent Development in Computing, Processor and Software Research for High Energy Physics*. Universidad Autonoma de Mexico, Guanajuato, Mexico
Schulz, H. D. and Stuckenberg, H.-J. (1981): A Trigger Processor for Argus, *Topical Conference on the Application of Microprocessors to High Energy Physics Experiments*. CERN, Geneva, CERN 81-07
Scott, W. M. (1963): The Theory of Small-Angle Multiple Scattering of Fast Charged Particles, *Review of Modern Physics*, **35**, No. 2, 231-313
Scott, J. E. (1982): *Introduction to Interactive Graphics*. John Wiley and Sons, New York
Séguinot, J. and Ypsilantis, T. (1977): Photo-Ionization and Cherenkov Ring Imaging, *Nuclear Instruments and Methods* **142**, 377
Shuey, D. et al. (1986): PHIGS: A Standard Dynamic Interactive Graphics Interface, *IEEE Computer Graphics and Applications* **6/8**, 50
SLD (1984): *SLD Design Report*, SLAC-273, Stanford Linear Accelerator Center, Stanford, May 1984.
Sonderegger, P. (1987): Fibre Calorimeters: Dense, Fast, Radiation Resistant, *Nuclear Instruments and Methods in Physics Research* **A257**, 523
Sternheimer, R. M. (1952): The Density Effect for the Ionization Loss in Various Materials. *Physical Review*. **88**, 851-9
Stuckenberg, H.-J. (1968): *Nukleare Elektronik I*. DESY F56-1
Stuckenberg, H.-J. (1981): Two Levels Triggering in Storage Ring Experiments, *Topical Conference on the Application of Microprocessors to High Energy Physics Experiments*. CERN, Geneva, CERN 81-07
Suffert, M. (1985): New Materials for a Crystal Barrel, *Third LEAR Workshop, Tignes 1985* Editions Frontières, Gif-Sur-Yvette
Synertek (1976): The C10115 is produced by Synertec. 2050 Coronado Drive, Santa Clara, California 95051
Ten Hagen, P. J. W. (ed.) (1986): *Eurographics Tutorials, Eurographics Seminars*, Springer, Berlin
Toki, W. et al. (1984): The Barrel Shower Counter for the Mark III Detector at Spear, *Nuclear Instruments and Methods in Physics Research* **219**, 479

Tysarczyk, G., Mättig, P. and Lohrmann, E. (1985): *Separation of π^0 and Single γs at High Energies.* DESY, Tasso Note 351

VMEbus specification Manual (1985): *The Parallel Sub System Bus of the IEC 821 Bus.* Revision C, November 1986

Walenta, A. H., Heinze, J. and Schürlein, B. (1971): The Multiwire Drift Chamber, A New Type of Proportional Wire Chamber, *Nuclear Instruments and Methods* **92**, 373–80

Waloschek, P. (1984): *Fast Trigger Techniques.* DESY 80-114

Ward, P. T. and Mellor, S. J. (1985): *Structured Development for Real-Time Systems.* Yourdon Press, New York

Weilhammer, P. (1986): *Experience with Si Detectors in NA32.* CERN, Geneva, CERN-EP/86-54

Weinberg, G. M. (1972): *The Psychology of Computer Programming.* Computer Science Series, van Nostrand Reinhold Company, New York

Wigmans, R. (1986): *On the Energy Resolution of Uranium and Other Hadron Calorimeters.* CERN, Geneva, CERN-EP/86-141

Wind, H. (1972): Function Parametrisation. *Proceedings of the 1972 CERN Computing and Data Processing* CERN, Geneva CERN 72-21, 53–106

Wind, H. (1974): Momentum Analysis by Using a Quintic Spline Model for the Track, *Nuclear Instruments and Methods* **115**, 431–4

Wind, H. (1978): An Improvement to Iterative Tracking for Momentum Determination, *Nuclear Instruments and Methods* **153**, 195–7

Wind, H. (1979): The Use of a Non-diagonal Weight Matrix for Momentum Determination with Magnetic Spectrometers, *Nuclear Instruments and Methods* **161**, 327–9

Wind, H. (1984): Interpolation and Function Representation. In: *Formula and Methods in Experimental Data Evaluation*, Vol. 3, M. European Physical Society, Geneva.

Wirth, N. (1977): *Compilerbau.* Teubner, Stuttgart

Workshop on Compensated Calorimetry, Transparencies, Internal Report CALT-68-1305, Sept. 1985.

Wu, S. L. (1984): e^+e^- Physics at Petra – The First Five Years, *Physics Reports* **107**, 95

Young, T. Y. and Calvert, T. W. (1974): *Classification, Estimation, and Pattern Recognition.* Elsevier Publishing Company, Amsterdam

Ypsilantis, T. (1981): Cherenkov Ring Imaging, *Physica Scripta* **23**, 371

Ypsilantis, T. (1987): Future of Ring Imaging Cherenkov Detectors, Presented at the Conference on Position-Sensitive Detectors London, Sept. 1987 Proceedings to be published in *Nuclear Instruments and Methods in Physics Research*

Zacharvo, V. (1982): *Parallelism and Array Processing*, CERN School of Computing, Geneva, CERN 83-03

Zahn, C. T. (1973): Using the Minimum Spanning Tree to Recognize Dotted and Dashed Curves. *Proceedings of the International Computing Symposium* Davos, North Holland, Amsterdam (1974) pp. 381–7

Zeuner, W. (1984): *Der Anteil der Photonen an der Energiebilanz hadronischer Endzustände aus der e^+e^- Vernichtung bei 34 GeV*, Diplomarbeit, Universität Hamburg

Zupančič, Č, (1986): A Simplified Algorithm for Momentum Estimation in Magnetic Spectrometers, *Nuclear Instruments and Methods in Physics Research* **A248**, 461–70

Index

ability to discriminate 266
accelerator 1, 2, 7–9, 11, 13
acceptance 22, 61, 386
acceptance region 259, 340
accidentals 22
addressing
   broadcast 119, 120, 126
   geographical 119, 125, 126
   logical 119, 126
advanced computer project (ACP) 114, 115
afferent data flow 395, 397
albedo in calorimeters 198
alternative hypothesis 259
ambiguity 262, 267
Amdahl 104
AMD–2901 99, 100
amplifier 88
analog-memory unit (AMU) 96, 97
analog-to-digital converter (ADC) 18, 19, 21, 54, 58, 95, 96, 98, 115, 119
ansatz 246
arbitration 115, 124, 127
ARGUS experiment 69, 70
arithmetic and logic unit (ALU) 99–101
ASCII 385
ASP experiment 74, 75
association of tracks to a vertex 319
associative memory 66
asymptotic
   normality 256
   property 252
asynchronous transfer 117, 125, 126, 128
attenunation 95, 96
autoincrement 120
avalanche 131
Avogadro number 10, 292
axially symmetric magnetic field 280
azimuth angle 272

background 7, 15, 28, 52, 60, 84, 85, 258, 340, 386

background contamination 260, 340
backscattering in calorimeters 198
beam line 281
beam tube 273
beam transport system 9
bending force 298
bending magnet 273
Bethe–Bloch formula 296
Bhabha scattering 17, 48, 74
bias 244, 247, 250, 262, 276, 284
biassed summation 289
binning 348
bit map 357
bit-slice processor 19, 23, 74, 99
block transfer 120, 121, 126, 128
Boolean algebra 55
bootstrap 350
bottom-up programming 390
boundary condition 286
boxes
   of the field model 278
   number of 289
   non-overlapping 289
boxwise field representation 291
breakpoint 263, 287, 294, 308
bremsstrahlung 48, 191, 269, 291
broadcast addressing 119, 120, 126
bubble chamber 18, 272
bunch 11, 61
bus master 115–17
bus standards 100
   CAMAC 99, 104, 115, 117, 119, 121, 123, 125, 127
   FASTBUS 104, 115, 117, 119, 121, 123–7, 389
   G 64, 117
   MULTIBUS II 115, 117, 127
   VME 104, 114, 115, 117, 121, 123, 126, 127
   VMS 128
   VSB 128

Index

bus transfer
  asynchronous 117, 125, 126, 128
  synchronous 117, 125, 126

calorimeter 3, 5, 20, 51, 76, 130, 150, 387
  albedo 198
  backscattering in 198
  calibration 198, 208
  cell structure 189, 194
  clusters in 190
  compensation in 197
  containment of 196
  electron/hadron factor in 197
  energy flow in 190
  granularity 194
  hermetic 189
  homogeneous 194
  invisible energy in 193
  physical processes in 190
  sampling 194
  sandwich 194
  simulation 205
CAMAC 99, 104, 115, 117, 119, 121, 123, 125, 127
canonical discrimination 225
carry look-ahead generator 99
Cartesian coordinates 321
cascade 188
cathode 130
  pad 135
  plane 131, 169
  strip 131
CDF experiment 115
cell structure in calorimeters 189, 194
CELLO experiment 60, 62
censored mean 254
centre-of-mass energy 2, 10
central detector 272
CERN 2, 10, 84
CESR 10
charge-coupled device (CCD) 84, 97, 98
charge division 52, 82, 131, 179
charged hadron 9, 15, 51
charm quark 52
Chebycheff inequality 256
Chebycheff polynomial 283, 289
Cherenkov counter 4, 8, 18, 49, 50, 231
Cherenkov radiation 3
Cherenkov threshold 231
chi square 246, 259
  cut 260
  distortion of the $\sim$ distribution 284
  distribution 256, 257, 259, 260, 284
  empirical $\sim$ distribution 256
  effective pseudo $\sim$ distribution 254
  evaluation 323

filtered 325
increment 265
probability density function 257
smoothed 325
test 259
total 265
chronotron 56
circle 156, 180
circular accelerator 8
class
  object 138, 143
classical electron radius 294
classification space 138
CLIC 8
clipping of electronic signals 20
closeness 353
cluster 140, 164, 180
cluster size 253
clustering 77, 78
clusters in calorimeters 190
coaxial cylinder 272
coaxial cylindrical detector 312
code-management system 384, 385
coding of software 389, 390
cohesion 402
  logical 402
  temporal 402
coil 288
coincidence 22, 54
collider 2, 4, 9, 12, 51, 52, 76, 385
colour graphics 359
combinatorial 179
combinatorial effects 326
comparison of coefficients 280
compatibility graph 153
compensation in calorimeters 197
Compton wave length 293
concurrent programming 111
confidence coefficient 260
confidence interval 255
confidence level 260
conservation of expectation value 257
consistency 251
constant field approximation 280
constraint 258
  additional 258
  vertex 319
constraint equation 147
  kinematical 327
constraint surface 138, 244, 259
containment of calorimeters 196
contamination 260, 340
content addressable memory (CAM) 23, 66, 69, 71, 74, 86, 93, 94, 101
contention 108, 109, 115
continuity condition 286

continuous medium 263
convergence property 254, 318
coordinate space 249
CORNELL 10
correctness in programming 113
correlation
  positive 256
covariance matrix 137, 138, 144, 246, 249, 323
  infinite 324
  inversion of the 268
  non-diagonal 317
CPU time 323
Cramer–Rao inequality 252
crate controller 119, 125
critical region 259
critical value 259
cross section 10, 23, 46, 53, 69, 386
cumulative distribution function 258
curvature 285
  change in the 286
  sign of the 269
cut parameter 340
cylindrical coordinates 321
cylindrical set-up 312

daisy-chained bus line 123–5
data 395, 397, 399, 400, 402, 403
data abstraction 335
data access method 374
data model 375
data selection 339
data structure 375
data-flow diagram 392, 394, 397, 398, 402
database system 374, 378, 397, 400, 403
de Morgan's law 1, 55
dE/dx 254, 297, 315
dead time 7, 19–22, 26, 28, 30–2, 34, 35, 57, 69, 88, 113, 299
debugger 403
DEC floating point format 385, 392
decision function 138, 166, 341
decision quality diagram 341
deflection 269, 273
deflection angle 292
defocussing 282
degrees of freedom 167, 254, 326
  number of 257, 259
Dekker's algorithm 112, 113
delay line 22, 53, 95, 96, 98
delta ray 253, 300
density of atoms 295
derivative(s) 275, 289
  of the field model 278
  of the LSM ansatz 317
  parallel integration of 276, 277

design of software 390
DESY 3, 10, 12, 57, 60, 115
detector(s) 129, 246
  high precision vertex 312
  layout 246
  position-sensitive 269, 298
  resolution 243, 249
  surface 290
deterministic approximation 275
deterministic function 244
deterministic process 298
deterministic step function 250
deterministic track model 275
DFD 392, 394, 397, 398, 402
di-muon events 52, 79, 81
diagonal 248, 283
differential cross-section 292
differential equation 276
differentiator 88, 89
diffractive cross section 14
digital-signal processor (DSP) 19, 23, 102
  TMS-32020 102
digital-to-analog converter (DAC) 79
dip angle 272, 280
dipole field 272
dipole magnet 8, 9
discontinuity 278
discrete mass hypothesis 328
discriminator 18, 20, 86, 89
  nonupdating 87, 88
  updating 87, 88
dispersion matrix 150
distribution
  Gaussian 257
  Poisson 12, 14, 29
  random 244
DMA 102, 119, 387
dominant variable 298
DORIS 10
drift chamber 18, 23, 54, 56, 57, 59–61, 63–5, 69, 71, 74, 121, 132, 299, 300
drift direction 287
drift space 133, 135
drift time 132
DSP 19, 23, 102
duty cycle 12
dynamic system 266

EBCDIC 385
ECL 68, 86, 126
effects of matter 291
efferent data flow 395, 397
efficiency
  detection 131, 170
  reconstruction 137, 165, 166, 172
  relative 253

Index

efficient estimator 253
elastic cross section 14
elastic scattering 259
electromagnetic interaction 291
electromagnetic shower 49, 78, 190
electron mass 48, 49, 294
electron/hadron factor in calorimeters 197
elementary charge 270
emitter coupled logic circuit (ECL) 68, 86, 126
emulator 86, 103, 104
emulsion detector 18
energy-momentum conservation 254
energy flow in calorimeters 190
energy loss 275, 286, 291, 295
   per unit length 295
entity relationship diagram 397, 398
equation of motion 244, 269
   explicit solution 272
   solution 246
equation (s)
   differential 276
   kinematical constraint 326
   Maxwell's 269
   measurement 323
   system of linear 317
equidistant 311
error(s)
   experimental 244, 298
   Gaussian 258
   measurement 251
   of the first kind 260
   of the second kind 260
   on (the momentum) P 311
   on the theta direction 311
   on the relative momentum 313
   quasi random 299
   systematic 254
   type I 260
   type II 260
error classification 349
error matrix 250
error propagation 246, 253–5, 286, 321
error tuning 263
estimate 243
   ad hoc 317
   final 315
   linear 246
estimation
   of track parameters 266
   interval 246, 249, 255
   parameter 276
   point 249
   recursive 266
estimation theory 248
estimator 248, 249
   asymptotically a minimum variance bound 252
   efficient 253
   explicit 249, 250
   implicit 249
   linear 251, 252
   minimum variance bound 252
   optimal 253
   robust 253
   unbiased 251, 252, 299
evaluation of material effects 279
event rate 259
event topology 323, 326
expansion point 246
expectation value 243, 244, 298
experiment
   fixed target 273
   storage ring 272
experimental design 246
explicit estimator 249
explicit solution of the equation of motion 272
extrapolation 161, 175

F particle 52, 84
fairness in programming 113
FASTBUS 104, 115, 117, 119, 121, 123–127, 389
feature extraction 138, 146, 164, 175
feature space 141, 143
Fermilab 10, 67, 69, 114
Fermi plateau 297
field 97
   analytical property of the 279
   global $\sim$ model 290
   homogeneous magnetic 271, 276
   inhomogeneous magnetic 275
   magnetic 289
   model 277, 290
   quadrupole 281
   solenoidal 272
   static magnetic 269
   transverse $\sim$ gradient 277
field component 289
field derivative 277
field gradient 277, 282, 287
field programmable logic array (FPLA) 61, 63, 64, 92, 93
field representation 278, 288–90
filter equations 324
final event analysis 348
fit(ting)
   geometrical 316
   global track 246, 287
   goodness of 258, 259
   kinematical 319, 327

fit(ting) *(Contd.)*
  progressive 265
  track 243, 246
  vertex 246, 319
fitted parameter 251
fitted value 244
fixed target 2, 3, 270
fixed target experiment 273
flash ADC 19, 98
flat top 12
flip-flop 18, 69, 79
FORTRAN 86, 104, 105, 115, 385, 401
forward region 270
Fourier transform 102
FPLA 61, 63, 64, 92, 93
free parameters 271
frequency of occurrence 252
Fujitsu 104
functional units 105
function optimization 316

G 64, 117
GaAs technology 90, 105
gamma beam 8, 9
Gauss–Markov theorem 251, 253
Gaussian distribution 246, 250, 253, 257, 294
Gaussian error 258, 259
genuine detector noise 253
geographical addressing 119, 125, 126
ghost(s) 253
ghost track 134
global fitting 246, 287
global method 174, 179
gluon 1, 48
goodness of fit 258, 259
Gram–Schmidt transformation 284
granularity of calorimeters 194
graph 151, 153
graphical communication 354
graphics glossary 362
grid 288
grid search algorithm 288

hadron 3, 4
hadroic shower 192
handshake 117, 126
Harvard architecture 102
heavy atom 296
helix 272
helix axis 272
helix tracking 276
HERA 3, 9, 10, 13
hermetic calorimeter 189
Hessian matrix 318

Higgs particle 47, 48
high-level graphics 359
histogram 162, 180, 348
hole 167
homogeneous calorimeter 194
hyperplane 139, 244
hyper-surface
  five dimensional 275
hypothesis 248
  alternative 259
  null 259
  pattern recognition 249

IBM floating point format 104, 385, 392
IEEE floating point format 385
impact 248
  on a reference surface 271
impact parameter 295
impact point 299
importance sampling method 37
independent measurement 247
induction 288
induced pulse 131
inefficiency 57–9, 61, 63, 93, 386
inelastic cross section 14
inflection point 318
information 252, 315
  loss of 276
  non-geometrical 315
  prior 323
inhomogeneous matter 280
initialization 174, 175
insignificant term 289
instanciation 377
instrumented iron volume 289
integrated circuit
  AMD-2901 99
  Motorola-10800 99
  RAM C-10115 57, 58
integration constant 270
integration parameter 275, 286
interactive graphics 356
interpolation 161, 177, 263, 288
  optimal 266
interpolation constraint 287
interpolation formula 289
interpolation polynomial 289
interrupt 122
invariant mass 50, 52, 55, 78–81
invisible energy in calorimeters 193
ionization
  average ~ potential 295, 296
ionization chamber 130
ISR 2
iterative algorithm 323
iterative procedure 317

Index

Jacobian matrix 322
jet 4
jets of particles 190, 302
Josephson junctions 105

Kalman filter 262, 266
 inverse 326
 with variable dimension 323
Kaon 3, 49, 50
kinematical constraint 327
kinematical fit 319, 327
kinks 323
knot
 additional 287
Kronecker symbol 247

laboratory frame 270
Lagrangian multiplier 327, 328
Λ particle 52, 84
Laplace equations 280, 289
latch 18, 56, 115
latency time 124
LEAR 12
least square(s)
 joint $\sim$ ansatz 265, 315
 property of the $\sim$ method 252
least squares method (LSM) 246, 250, 254
least variance 251
left–right ambiguity 133, 135, 178, 179
LEP 10, 13
level of significance 260
lever arm 273
LHC 13
life time measurement 326
light atom 296
likelihood
 maximum $\sim$ method (MLM) 249, 349
likelihood ratio 249
linear accelerator 2, 12
linear expansion 246, 250, 275
linear feature extraction 138, 146, 164, 175, 225
linear model 246, 259
linear system 266
Lithium hydride 9
local method 175
logical addressing 119, 126
look-at-me (LAM) 123
look-up table 56, 59, 60, 67–9, 71–5, 82, 84, 90
Lorentz factor 49
Lorentz force 269
Lorentz invariant form 9
losses of tracks 262, 340
low energy 259
luminosity 10, 13, 14, 386

lumped delay line 56, 96

machine precision 278
magnet yoke 288
magnetic field 289
many-processor system 106
MARK III experiment 56
mass assignment 316, 326
master point 179
material effects 275
matrix
 block structure of a 323
 diagonal 248, 283
 dispersion 150
 error 251
 positive definite 256
 rank of a 247, 317
 singular 321
 symmetric 256
matrix inversion 247, 319
matrix operations 248
maximum likelihood method (MLM) 249, 349
Maxwell's equations 269
mean ionization 296
mean time between failure (MTBF) 40–2, 44, 45
measurement(s)
 deviating 253
 unbiased 298
 virtual 321
measurement equation 323
measurement error 298
measurement space 244, 275
measurement vector 136, 244, 249
mechanical precision 288
memory
 analog (AMU) 96, 97
 associative 66
 content addressable (CAM) 23, 66, 74, 86, 93, 94
 logic unit (MLU) 90, 91
 PROM 63, 69, 92, 93, 99
 RAM 60, 61, 63, 86, 90, 102
memory management 377, 385
merging of track segments 267
message passing 111
metric 143, 353
micro instruction 74, 101, 102
microcode 101
microstrip detector 84
MIMD 106
minimization 247
minimization technique 316
minimum
 of a function 316

430   Index

minimum condition 317
minimum ionization 296
minimum number of coefficients 289
minimum variance 243, 244, 246
minimum variance-bound estimator 252
minimum spanning tree 150, 151, 172, 184, 221
mirror symmetry 280
missing energy 51
missing momentum 51
model
    linear 246, 259
Moliere's formula 294, 295
momentum and energy conservation 319
Monte Carlo (MC) 45, 105, 111, 289, 386
most relevant coefficients 283
Motorola–10800 99
MULTIBUS II 115, 117, 127
multidimensional analysis 352
multidimensional test 341
multiple scattering 138, 176, 262, 275, 291
multiplicity 47, 50, 64, 72
multistep integration 278
multistring model 47
multivibrator 88
multiwire proportional chamber (MWPC) 18, 49, 60, 61, 65, 131, 250, 302
muon 4, 49
mutual exclusion 111

n-tuple 349, 380
NaI 48
NA 32 experiment 84, 85
neutral particles 2, 4, 8, 15, 51
neutrino 48, 51
neutron 3, 49
Newton method 317
NIM 86
NMOS 104
noise 275
noise signal 253
non-central parameter 326
non-geometrical information 315
non-updating discriminator 87, 88
NORD 100, 104
normal distribution 254
normal pressure 297
normalization of the chi-square distribution 254
nuclear instrument module standard (NIM) 86
number
    of atoms 10
    of coefficients 283
    of operations 322

numerical differentiation 276, 277
numerical integration 276, 290
numerical table of the magnetic field 289
numerical techniques 155

object class 138, 143
object classification 138
OCCAM 103
off-diagonal weight matrix 247
offset of a parameter 326
open collector 116
optimal estimator 253
optimal prediction 266
optimization 311
order of the polynomial 289
orthogonal functions 283
orthogonal polynomials
    low order 289
orthogonalisation 284
oscillation frequency 296
outlier(s) 253, 263, 266, 290, 347
    detection of 254
    multi $\sim$ problem 325
overdetermination 259
overinstrumentation 262, 313

page fault 288
pair production 9, 48, 79
parabola 155, 175
parallel processing 105, 110
parallel-to-series interconnection 39
parameter(s) 243, 257
    adjustable 257
    choice of track 246
    fitted 251
    free 271
    impact 295
    non-central 326
    track 244, 272, 275, 322
    true 249
parameter estimation 276
parameterization 138, 162, 175, 283, 290
    global 283
particle flux 10
particle identification 297, 315
particle trajectory 243
PASCAL 385
path length 270
pattern recognition 129
pattern recognition hypothesis 249
pattern space 136, 139, 143, 146, 149
PDP 11 104 115
pedestal 82
PEP 10, 13
PETRA 10
phase space 137, 163

Index

photocathode 18, 53
photoelectric effect 18
photomultiplier 18, 53, 79, 85, 95
photon 3, 48, 49, 51, 52
pion 48–50
pipeline 95, 96, 98, 101, 105, 106
pipeline register 99, 100
pixel 357, 369
Planck's constant 296
point estimation 249
point of interaction 319
point removal 166
point-like particle 292
Poisson distribution 12, 14, 29
polar axis 272
polling loop 122
polynomial 155
portability 288
positive correlation 256
positron 9
power of test 326, 341
power series expansion 280
precision 312
predicted vector 324
prediction 263
  optimal 266
predictor-corrector method 278
pretrigger 69
primary electron 300
principal component analysis 149, 225
prior information 323
probability density 28, 250
  conditional ∼ function 249, 298
probability density function 257
  joint 249
process noise 325
process-description language (PDL) 400
processing elements 106
processor
  data driven 46
  program driven 46
  vector 106, 108, 110
productivity of programming 389, 390
program maintenance 338
programmable read only memory
  (PROM) 63, 69, 92, 93, 99
programmed array logic (PAL) 63, 92, 93
projectile 291
projection 168
proportional chamber 18, 49, 60, 61, 65, 131
Proton 2, 3, 49, 50
prototype 143, 147
pseudocode 400
pseudorapidity 227
pseudostochastic process 289

pull quantity 248, 256, 258, 286, 287
pull-up resistor 116
pulse former 18, 20, 86, 88, 89

QED events 17, 61, 65
quadratic average 254
quadratic behaviour 316
quadrupole magnet 9, 12
  defocussing 282
  focussing 282
quark 13, 52
queue length 32–4
queuing 26, 36
queuing simulation 36
queuing theory 26

radiation length 294
radiofrequency system 11
RAM C–10115 57, 58
random access memory (RAM) 60, 61, 63, 86, 90, 102
random distribution 244
random measurement vector 244
random quantity 243
READ–MODIFY–WRITE 113, 120
real path 266
real-time processing 248
realization 247, 250
  individual 298
recoil particle 277, 295, 316
recursive estimation 266
recursive formula 276
recursive method 319
recursive partitioning 219
recursive track fitting 262
reduced 102
reduced residual 256, 257
reduced test quantity 254
reference cylinder 273
reference plane 247
reference surface 244, 271, 273, 321
reference track 268, 315
regular 268
rejection region 259, 340
relation 397
relational data model 375
relative efficiency 253
relative error on the momentum 312
relativistic 2, 3
relativistic Lorentz factor 269
relativistic rise 297
reliability 41, 43–5
reliability theory 39
representative 259
representative sample 262
requirement analysis 389

Index

residual vector 265
resolution 243
  double pulse 20, 84, 87
  double track 84
  energy 4, 8
  spatial 3, 18, 53, 84, 85
  two particle 299
resolution function 298, 299
resolution time 20, 22
response 20, 99, 121, 126, 390
rest mass 269
ring imaging Cherenkov counter (RICH) 50, 231
rise time of scintillator 18
road 177
robust estimator 253
robustness 111, 113, 253, 254, 347
rotation invariance 283
Runge–Kutta method
  of Nystrom 276
  of Simpson 277
Rutherford formula 292

sample-and-hold ADC 96
sandwich calorimeters 194
SASD 390, 392
scattering
  contributions of continuous 314
  elastic 259
  elastic nuclear 291
  multiple 138, 176, 262, 275, 291
  of the electrons 291
scattering experiment 9
Schmitt trigger 88, 89
scintillation counter 1, 49, 50, 64, 69, 88, 90
scintillator 3, 53
screening 297
second derivative 318
segment cable 126
segment crate 126
segment interconnect 126
selected event topology 258
semiconductor detector 18
sense of rotation 272
sequencer 99
series expansion of helix 279
series-to-parallel interconnection 39
SERPUKHOV 10
short-lived particles 8, 84
shower 188
  electromagnetic 190
  hadronic 192
shower parameters 200
shower simulation 205
significance of test 341

significance test 259
SIMD 106
simulation 279
  event 105, 386
  exponential arrivals 37
  parallel processing 109
  particle shower 387
  queue 28, 36
  shower 205
similarity 144, 172, 184, 353
SLAC 8, 10, 12, 56, 74, 90
SLC 10, 13
smoothed chi square 325
smoothed vertex position 325
smoothing 266, 267, 287, 325
smoothing algorithm 315
software engineering 338
Sommerfeld's fine structure constant 294
space
  classification 138
  drift 133, 135
  feature 141, 143
  pattern 136, 139, 143, 146, 149
  phase 137, 163
space charge 131, 176
space point 136, 168
sparse data scan 122, 126
spatial vertex evaluation 313
SPEAR 10
specification of software 389
spectator quark 3
spectrometer 3, 4, 46, 47, 52, 65, 79, 80, 85
  downstream $\sim$ arm 273
  two lever arm 311
  with coaxial cylindrical detectors 312
spill 12, 45
spline 155, 157–60
  cubic 159, 285, 286
  doubly integrated cubic 287
  natural 159
  quintic $\sim$ model 285
spline approximation 284
spline interpolation 287
SPS 10
stability of the derivative value 278
standard deviation 247, 298
starting element 267
starting vector 250
state vector 323
state-transition diagram (STD) 398
statistic 259
statistical analysis 378
statistical property 246
step-by-step tracking 276
step length 277
step-wise refinement 402

Index

stochastic emission of synchrotron radiation 298
stochastic perturbation 275
storage ring experiment 272
stretch function 256
structure chart 395, 403
structured analysis 390, 392
structured data 338, 375
structured design 390, 392
subset of measurements 263
sufficient statistic 252
symbols 243
symmetry plane 283, 288
synchronous transfer 117, 125, 126
synchrotron radiation 2, 272
  stochastic emission of 298
system equation 323
systematic errors 254

target 291
target constant 10
TASSO experiment 57, 59–61, 64, 71, 82
Taylor expansion 278
template matching 145, 183
test
  power of the 326
  reduced $\sim$ quantity 254
  significance 259
test sample 284
test statistic 259, 341
thin lense 282
threshold 50, 51, 77, 81, 85, 88, 89, 389
time-of-flight (TOF) 19, 48, 61, 69, 87, 99, 315, 316
time projection chamber (TPC) 135, 168, 313
time-to-digital converter (TDC) 4, 19, 58 69, 115, 119
TMS–32020 102
token 122
top-down programming 390, 402
total cross section 55, 69
track 164
  choice of $\sim$ parameters 246
  five-dimensional $\sim$ model 286
  incompatible 153, 166, 171
  reference 268, 315
  undisturbed 249, 275
track association strategy 323
track candidate 165
track detector 272
track element 178
track element merging 313
track finding 164, 265
track fit(ting) 243, 246
  global 246, 287

recursive 262
track following 175
track model 246, 249, 256, 263, 269, 275
track multiplicity 259
track overlap 171
track parameters 244, 272, 275, 322
track quality 232
track road 177
track search 243
track segment 154, 161, 172, 179, 184, 313
track string hypothesis 243
tracking algorithm 291
tracking detector 269, 299
training sample 137, 144, 147, 163, 283, 284, 340
transistor-transistor coupled logic circuit (TTL) 86
transition diagram 399
transition radiation 3, 49
transputer 102, 103
transverse energy 51, 61, 71, 76, 78, 386
transverse momentum 273, 291, 386
tree 151
tree structure 375
trigger 4, 17
  data driven 46, 56
  fixed flow 46
  logical 46, 53, 55
  multilevel 22
  program driven 46
  variable flow 46, 53
trigger arithmetic 46
trigger coplanarity 48
trigger energy 74–9
trigger interaction point 52, 82–5
trigger invariant mass 52, 79–81
trigger missing energy 51
trigger momentum 47
trigger multiplicity 47
trigger track 56–74
trigger type of particle 48–50
triggering 315
trigonometric function 283
TRISTAN 10
tristate 116
true coincidence 22
true value 244
truncation 254
tuning of the detector alignment 267
two-dimensional grid 289
two-particle resolution 299
two-peak separation 299
two-photon cross section 17

UA2 experiment 76–8
unbiased 250

unbiased estimate 246
unbiased estimator 251, 252, 299
unbiased measurement 298
unbinned data 349
uncorrelated 247
UNIBUS 115
uniformly distributed 258
units 270
UNK (II) 10, 13
updating discriminator 87, 88

variance 246, 251, 298
  minimum 243
  of the residual vector 256
  per scattering process 292
  per unit length 293
variation 278, 328
VAX 104, 385
vector potential 280, 281
vector processor 146
Vees 323
velocity of light 270
vertex 23, 46, 52, 82–5, 129, 386
  high precision $\sim$ detector 312
  secondary 323

vertex constraint 319
vertex detector 15, 59
vertex fit(ting) 246, 319
vertex position 321
vertex reconstruction 313
virtual memory 288
VME 104, 114, 115, 117, 121, 123, 126, 127
VMS 128
von Neumann processor 66, 73
VSB 128

W gauge boson 76
Ware's model 108
weight 150
weight matrix 246, 250, 251, 298, 322
window discriminators 90
wire chamber 130

zero constraint fit 258
zero hypothesis 249
zero trajectory 276, 277
Z^0 gauge boson 15, 76, 79
ZEUS experiment 115